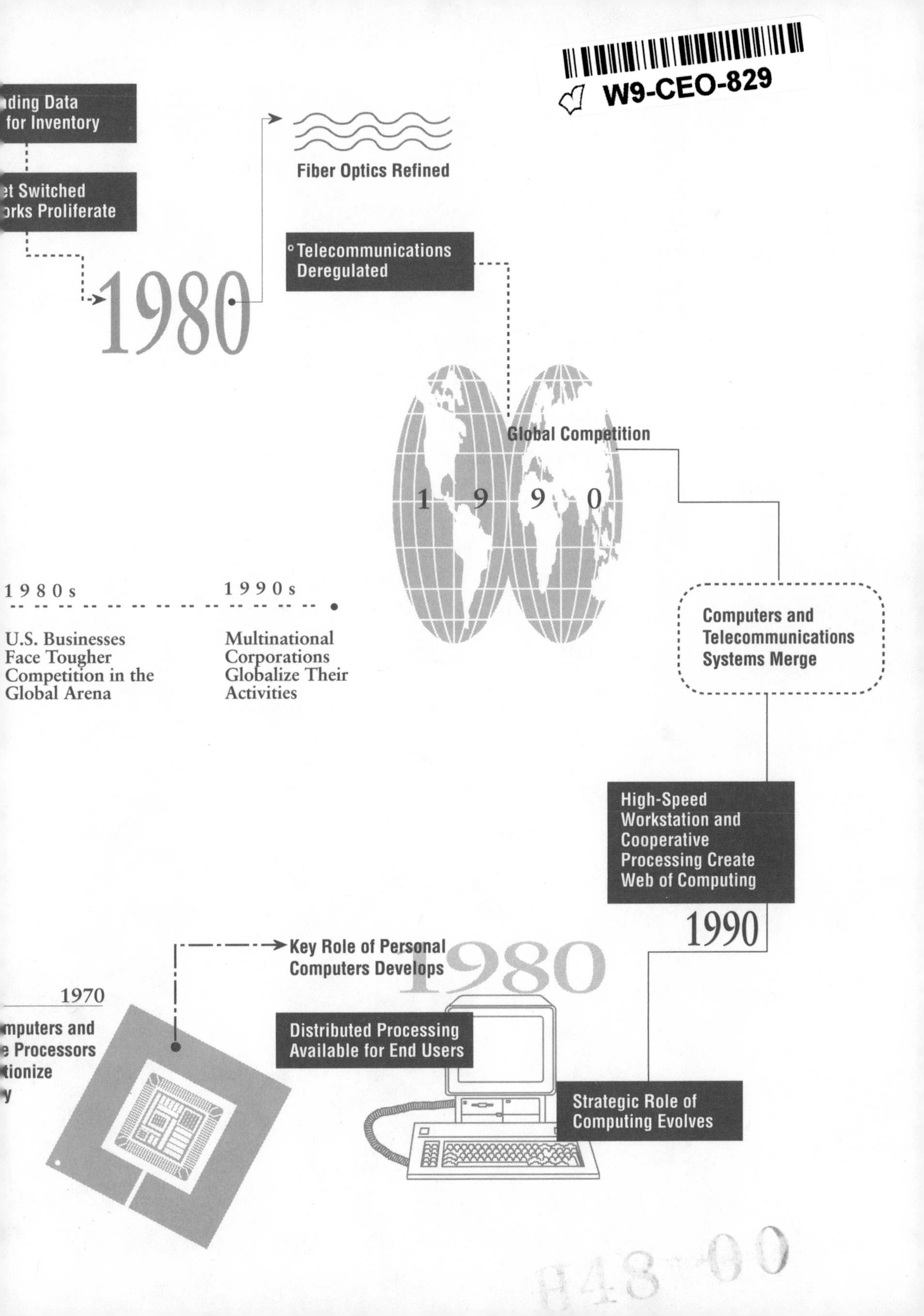
ding Data
for Inventory
et Switched
orks Proliferate
1980
Fiber Optics Refined
Telecommunications
Deregulated
Global Competition
1 9 9 0
1 9 8 0 s
U.S. Businesses
Face Tougher
Competition in the
Global Arena
1 9 9 0 s
Multinational
Corporations
Globalize Their
Activities
Computers and
Telecommunications
Systems Merge
High-Speed
Workstation and
Cooperative
Processing Create
Web of Computing
1990
Key Role of Personal
Computers Develops
1980
1970
mputers and
e Processors
tionize
y
Distributed Processing
Available for End Users
Strategic Role of
Computing Evolves

Telecommunications and Business Strategy

Telecommunications and Business Strategy

Edward M. Roche
University of Arizona

The Dryden Press
Chicago ▪ Fort Worth ▪ San Francisco ▪ Philadelphia
Montreal ▪ Toronto ▪ London ▪ Sydney ▪ Tokyo

Acquisitions Editor: DeVilla Williams
Project Editor: Karen Shaw
Art and Design Director: Jeanne Calabrese
Production Manager: Bob Lange
Director of Editing, Design, and Production: Jane Perkins

Interior Designer: Rebecca Lemna
Cover Designer: Mercedes Santos
Copy Editor and Indexer: David Talley
Text Type: 10/14 Sabon

Library of Congress Cataloging-in-Publication Data
Roche, Edward Mozley.
Telecommunications and business strategy / Edward M. Roche.
p. cm.
Includes index.
ISBN 0-03-032914-0
1. Telecommunication. 2. Strategic planning.
3. Telecommunication—Case studies. 4. Strategic planning—Case studies. I. Title.
HE7631.R63 1990
658.4'5—dc20 90-41377

Printed in the United States of America
123-040-987654321

Address orders:
The Dryden Press
Orlando, FL 32887

Address editorial correspondence:
The Dryden Press
908 N. Elm St.
Hinsdale, IL 60521

The Dryden Press
Holt, Rinehart and Winston
Saunders College Publishing

To

Hiroko, Katherine Fumi, Waylan, Jimmy Joe, Alexandra, Jim, Hugh, Kim, James, Trudy, Papa, Granny, Grandmother and Grandfather, and Stephen Justin

Also to Ken, Jane, DeVilla, and my friends at Booz, Allen: Marty and Rolf

The Dryden Press Series in Information Systems

Arthur Andersen & Co., Boynton and Shank
Foundations of Business Systems: Projects and Cases

Arthur Andersen & Co., Flaatten, McCubbrey, O'Riordan, and Burgess
Foundations of Business Systems

Bradley
Case Studies in Business Data Bases

Bradley
Introduction to Data Base Management

Bradley
Introduction to Data Base Management in Business
Second Edition

Burstein and Martin
Computer Information Systems

Goldstein Software, Inc.
Joe Spreadsheet Statistical

Goldstein Software, Inc.
Joe Spreadsheet Statistical, Macintosh Version

Gray, King, Watson, and McLean
Management of Information Systems

Harrington
Making Database Management Work

Harrington
Microsoft® Works: A Window to Computing

Harrington
Relational Database Management for Microcomputers: Design and Implementation

Laudon and Laudon
Business Information Systems: A Problem-Solving Approach

Laudon, Laudon, and Weill
The Integrated Solution

Liebowitz
The Dynamics of Decision Support Systems and Expert Systems

Martin and Burstein
Computer Systems Fundamentals

Parker
Computers and Their Applications
Second Edition

Parker
Computers and Their Applications
Second Edition
With Productivity Software Tools

Parker
Productivity Software Guide
Third Edition

Parker
Understanding Computers and Information Processing:
Today and Tomorrow
Third Edition

Price
Microcomputer Applications

Price
Microcomputer Applications Workbook

Roche
Telecommunications and Business Strategy

Preface

It now seems indisputable that telecommunications will continue to be one of the driving forces behind the growing integration of the world economy.

Financial Times, July 19, 1989, p. II.

This book is based on the premise that today's management strategists must work with telecommunications to successfully position their businesses for the 1990s. To continue to rely on worn-out financial gimmicks and quick-fix cost reductions, many of which merely alienate and destroy workers in order to pump up cash flow, is a recipe for disaster. This disaster is showing itself more clearly year by year as American companies are crushed by foreign competition or taken over by financiers. Instead it is necessary for industries to concentrate on innovation, trust of employees, and rebuilding of infrastructure. Telecommunications is a large part of that infrastructure.

No large firm can operate today without a complex web of information and telecommunications technology serving as a nervous system. Structuring and controlling information and communications are virtually equivalent to controlling operations themselves, even at the most detailed level. Managerial control cannot be extended without telecommunications. Of course, without telecommunications movement of information and data would be impossible. In a sense, then, the study of the business and strategy of telecommunications is in reality a study of the firm and its operational structure.

In order to learn true management, therefore, today's undergraduates, MBA students, and professionals must master the fundamentals of telecommunications and business strategy. This book has been written for these undergraduate and MBA students, along with practicing professionals, who will be working with strategic business issues in their careers. One of the great problems with management of technology today is that few people who earn MBA degrees are engineers, and yet the issues they must address require clear understanding of highly technical subjects. Accordingly, this book is written at a nontechnical level to convey technical concepts important for the general manager to understand. It is also a good introduction to the role of information systems in business.

This book is distinguished from its competitors in that it does not get bogged down in page after page of arcane telecommunications concepts that never really come to the attention of today's managers. Many people take great pride in knowing these details, but in reality the technicalities do not help with business strategy. It is more important that managers know how local area networks are used for business transactions than that they understand the details of the protocols used to move data through circuits. It is more important that today's managers know how to use information and telecommunications systems to link suppliers, distributors, and customers together than that they know the details of network control algorithms and subtle variations in vendor solutions.

This attitude determines what appears in the book. Instead of a handful of business cases, which other books include as afterthoughts, this book concentrates on business problems, telecommunications solutions, and assessments of benefits. It uses case after case to make its points. Instead of presenting telecommunications concepts in a piecemeal fashion, it presents telecommunications business solutions holistically at the level where management needs to do its most intense thinking.

At the same time, this book is addressed to the telecommunications professional who has already mastered the highly complex technical details of telecommunications and must now learn how to communicate effectively with nontechnical management and how to think strategically about what telecommunications really does for business. In fact, much better communication is needed between telecommunications professionals and top strategic management. Neither group has all of the skills needed to build the complex business systems that will be necessary for corporate survival in the future.

The Productivity Crisis

Although few people may realize it, in the United States today approximately 50 percent of all new capital investment is going into information technology. This includes computer systems, private branch exchanges, telephone systems, and, of course, all types of telecommunications equipment.

As the amount of investment in these information processing and telecommunications technologies has grown and become much more visible, questions have been raised about whether scarce investment funds are being spent well. Some claim that it is impossible to identify any significant increase in productivity as a result of this massive investment. In fact, detailed analyses by such people as Martin Bailey at the Brookings Institution and Stephen Roach at the investment banking firm of Morgan Stanley have given strong evidence that investments in data processing and telecommunications technology do not pay off, as indicated in the accompanying graph.

These researchers are unable to find a direct relationship between this kind of investment and increased productivity and competitiveness in business. This research has raised very serious questions about further investment, and this message has reached the highest levels of today's corporations. The result is that information systems managers, who typically also control telecommunications for the firm, have often been placed on the defensive. They have had to explain many times to top management why they spent so much of the corporate budget, and what is required for this investment.

With such difficult questions being asked, the information systems manager—who increasingly is an MBA or business major—must be able both to define and to clarify the rationale for expenditures *and* relate these factors to the competitive business strategy of the corporation. It is no longer enough to be a technical whiz kid, giving mystical reasons for these expenditures in such terms as "critical" or "basic for

Exhibit P.1

From 1979 to 1986 there is no relationship between use of information technology and total factor productivity growth.

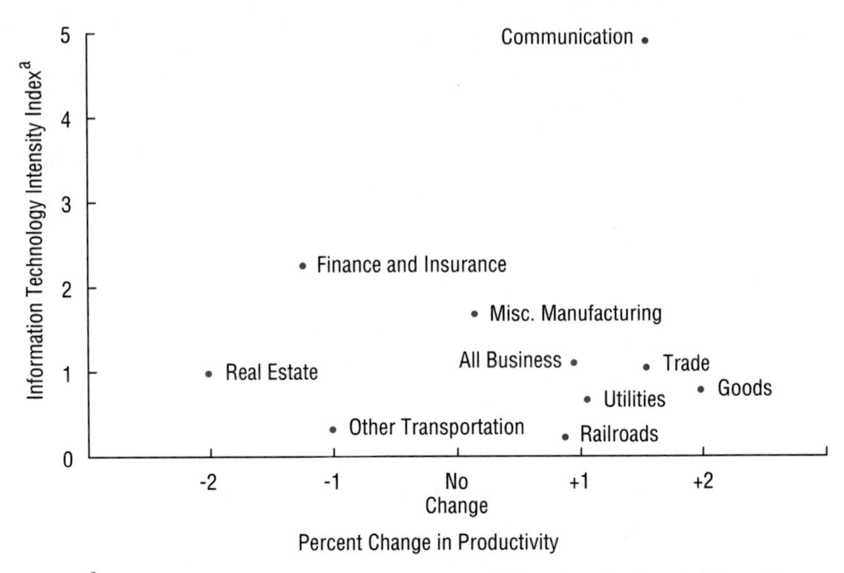

[a] Note: Roach used an index based on measurement of information technology capital as a share of each industry's overal capital stock.

---

the business." Technical people, although they are essential to the day-to-day operation of the information technology infrastructure of the business, nevertheless frequently fail to communicate adequately with top management. They might be able to answer detailed questions about the performance and future growth options for the information technology infrastructure (which, of course, includes telecommunications), but they frequently fail to understand and communicate how this technical magic actually promotes the success of the business and its strategy.

This cultural gap between different types of people—techies and managers—is a serious failure. Ultimately, it can result in difficulty relating business strategy to the technology needed to carry it out. Why does this gap occur? Because if either the technical side or the business

Exhibit P.2

Telecommunications planning must balance the influence of information systems and line management and use both hard and soft measurements.

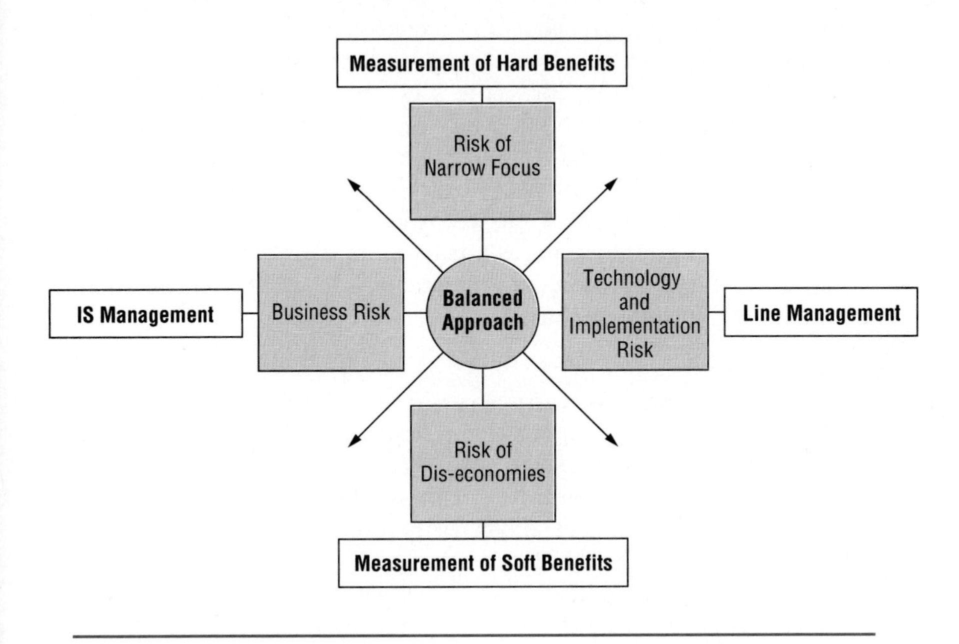

side, typically line management, dominates investment decision making for information technology and telecommunications systems, a wrong decision can be made, and a disaster can occur.

This danger derives from the concepts of *technology risk* and *business risk*. Both of these factors must be taken into consideration when planning a technology strategy that must mesh with business strategy. The accompanying exhibit illustrates the factors that affect decision strategy.

- *Technology Risk*
  Any new development in information and telecommunications technologies may fail because the technical plans do not work properly. When considering low-budget or small-scale projects, this factor can generally be ignored. However, when the firm is involved in developing a strategy for a major change in its

information and telecommunications infrastructure, the scope of a project can be very large, with very large associated costs. These can be multi-year projects. Under these circumstances, any possible failure of the project down the line can have serious consequences for the firm. First, the cost can escalate out of control. Second, the firm can lose the opportunity to gain competitive advantage against its competitors because it must wait until the problem is solved. This almost inevitably means a substantial delay. Many proposed systems make sense from the business side but founder on the hard rocks of technology. When this happens, the systems never get developed as quickly or as cheaply as first intended, and they ultimately deliver considerably less than their original stated objectives. In some cases, a failed project can kill or maim the entire company.

- *Business Risk*
  These large projects carry a business risk as well. If they are intended to create a specific competitive advantage for the firm, then their success must be measured against that criterion, although this is sometimes difficult to do. In some cases, these systems fail to achieve their business objectives. They may be technologically elegant, but regardless of how clever or efficient they are, or how well they work, if they fail to accomplish their business objectives, they are a failure to the firm. Any technologically complex undertaking has, therefore, a significant degree of business risk.

  Some have argued, for example, that the move into home banking was a failure for many banks. These banks developed complex information technology and telecommunications systems through which customers could use personal computers or simple data terminals to access their accounts on the bank's mainframe computers. A great deal of effort and expense went into developing the technologies for these systems. The systems worked fine. The customers who used the systems were satisfied. But the problem was too few customers, and, as a result, many home banking systems failed. They were technological successes but business failures. This is the essence of business risk of new systems.

- *Balancing Risks*
  The successful business and technological strategy must balance between these two types of risk. If either line management or technical engineers dominate planning and decision making, risk of business failure increases substantially.

> There are a variety of techniques by which to balance technical and business influences. These techniques involve developing measures of both feasibility and risk from both the technical and business perspectives. The objectives of the business are analyzed and then divided into more specific objectives. These objectives in turn are answered by technology. A technology plan is created by playing various technology strategies off against one another until the best solution is found *that still satisfies the business strategy.*

The skills needed to perform this type of analysis, or to oversee and manage this type of process, must be absorbed and mastered by the business professional. Given today's changing world economy, where each company must compete globally rather than locally, it is no longer enough for business professionals to think one-dimensionally. They must understand both technical issues and business issues. MBAs or business majors who work with technology strategies for their businesses must use both parts of their brains, the left side and the right side—the technical side and the business side.

That effort guides the intention of this book. It covers both technical and business matters. At the end of the book, several business cases integrate the information and concepts it presents into total problem-solving environments typical of real business situations.

This book is intended to teach the potential manager how to use telecommunications to achieve competitive advantage in business. As many examples as possible are given to demonstrate how other companies have found successful telecommunications-based strategies. Also, many analytical methodologies are presented, such as how to choose the cheapest linkages to give the business strategist a good box of tools for planning. In addition, much information is provided about the historical development of our telecommunications systems and the regulatory constraints and practices that have grown up in parallel to this development.

But the real goal of the book is to jog the mind of the student or executive, to teach him or her how to think about telecommunications. For example, when is telecommunications a strategic asset, or when is it just another back-office overhead item? How can one tell the difference? Depending on the overall importance of the telecommunications system in a company, how do the methodologies for analysis change? How do the techniques of cost accounting change? How does the role of management change? Finally, how does one match telecommunications system design against the goals of the organization and the generally hostile business environment in which it is forced to operate?

Organization of the Book

The book is divided into three parts:

- Part I—Technical Foundations
- Part II—Strategic Business Applications
- Part III—Minicases with Critiques

The first chapters examine the technical foundations of telecommunications. Differences between fiber optics and microwave transmission, packet switching, leased lines, Private Branch Exchanges (PBXs), satellites, network control centers, and other technologies are examined. These chapters also discuss the range of operations from wide area networks (WANs) to local area networks (LANs) and the options they provide for network planning and integration of computer networks. We also discuss newer types of telecommunications technologies, such as ISDN, which can carry voice, data, and images over the same lines at the same time!

A few later chapters focus on the types of decisions managers must make regarding telecommunications. They discuss special topics in telecommunications, concentrating on typical problem-solving strategies. By this time in the course, the student should have a good overview of the basics of telecommunications but would typically be weak in methodologies for problem solving. This section is intended to give practical examples of typical problems a manager might have to solve.

The next group of chapters is designed to give the student a broader historical and regulatory perspective on the development of global telecommunications. It paints the picture of the relationship between changing technologies, the development of international organizations (such as the International Telecommunications Union) that shape the global regime, and the strategies of business to cope with the complexity that results. As with all knowledge and practice, understanding of the historical context is very useful for addressing the fundamental issues before concentrating on the specific, technical details of a solution.

The Strategic Business Applications portion of the book is built around the idea of using telecommunications to achieve competitive advantage. First, a few basic concepts are presented, such as backward and forward linkages between the firm, its suppliers, and its customers. Next, the idea of the value-added chain is presented, then the discussion turns to several ways in which firms can use telecommunications to conduct their business, and possibly achieve competitive advantage in the marketplace.

The following chapters go through many examples of utilization of telecommunications. Readers can see for themselves how telecommunications strategy fits into the broader business strategy.

The final chapter presents a high-level executive overview of the field as a whole. This overview covers problem-solving techniques, diagnostic methodologies, general management of the telecommunications function, and an overview of the dilemmas managers will probably face in the 1990s.

At the end of the book, a series of "minicases" offers various management challenges and solutions. The solutions are presented as management team reports. After each minicase, a critique of the answer given is provided so the students can see the weaknesses and strengths of the plan presented by the telecommunications team.

Much of the information presented in the case studies is based on interviews and student research under the author's direction. It is based on information that is in the public domain or on insights and inferences pieced together from research on the firms being studied. The primary intent is pedagogical, and the best efforts have been made to maintain accuracy.

This, then, is the plan of the book. The intention is clear: the student must learn to understand both the business and the technological dimensions of business activities and be equipped with the methodological tools to handle them.

As the author surveys the state of American business, he sees a once-great giant humbled and decaying, wracked by foreign competition. Ultimately, the productivity, wealth, and comfort of our society depends on the knowledge and skills of our people. It is everyone's duty to help make a better and richer world through work and through careful economic management of all available resources. Consider a verse from the old Chinese classic, *The Great Learning:*

> Things being investigated, knowledge became complete. Their knowledge being complete, their thoughts were sincere. Their thoughts being sincere, their hearts were then rectified. Their hearts being rectified, their persons were cultivated. Their persons being cultivated, their families were regulated. Their families being regulated, their States were rightly governed. Their States being rightly governed, the whole kingdom was made tranquil and happy.

Information technology and telecommunications systems are critical now and will remain critical for at least another half century. Therefore, to the extent that this book helps improve the knowledge of future managers, its purpose will have been accomplished.

The author urges students: please study hard, not only this book but all books and knowledge with which you come into contact. Only through knowledge can you improve the world. You are studying not only for yourself but for your society and for future generations. In addition, you might make note of the fact that telecommunications experts with business training also attract high salaries!

This text comes with a complete instructional support package. The *Instructor's Manual* provides guidelines for conducting a consulting simulation as a class assignment at the end of the course. The author has tested this simulation many times and found it to be highly beneficial to the learning process. Please contact your local Dryden Press sales representative for further information.

Acknowledgments

This text would not have been possible without the input of many colleagues, reviewers, and students. I would like to express my gratitude to Dr. Hitoshi Watanabe and Dr. Nobuhiko Shimasaki of the Nippon Electric Corporation in Tokyo for their insights concerning the future of telecommunications as seen from the perspective of one of the world's largest providers. In addition, Jane and Ken Laudon, New

York University, provided much practical advice. I would also like to thank my graduate students at New York University for their work under my direction on many of the case studies.

My appreciation also goes to the following reviewers, who shaped the manuscript's final form and sequence: David Van Over, University of Georgia; Edward Stohr, New York University, Stern Graduate School of Business; Sue Wyrick, Kirkwood Community College; G. Torkzadeh, University of Toledo; Hasan Pirkul, The Ohio State University; and Sergio Davalos, Pima Community College. I should also mention the Microsoft Corporation, which donated its Word 4.0 program, including SuperPaint, for preparation of the drafts.

Finally, I owe many thanks to The Dryden Press, especially my remarkable acquisitions editor, DeVilla Williams, and her team, including Aimée Gosse, Jim Massey, Karen Shaw, Tom Hoffa, Becky Lemna, Jane Perkins, Sandy Lopez, Bill Schoof, Karen Schroeder, and Bob Lange. The marketing efforts at The Dryden Press were exceptional, and I owe special thanks to the marketing manager, Patti Arneson. I would also like to thank Susan Pierce and Kelly Gozdziak for their efforts.

Edward M. Roche
October 1990

About the Author

Edward M. Roche is an Assistant Professor of Management Information Systems at the University of Arizona. He earned an M.A. in International Relations from the Johns Hopkins University School of Advanced International Studies in Washington. He earned an M.Phil. in Political Economy and a Ph.D. in Political Science from Columbia University in New York City. His doctoral work focused on transborder data flows and international computer communication systems. He has done research on international computing issues in China, Japan, Korea, Brazil, and the USSR.

Prior to entering academia, Dr. Roche worked as a consultant for The Diebold Group Inc. and Booz, Allen and Hamilton, Inc., both in New York. His current research centers on the use of information and telecommunications technologies in multinational corporations. He lives with his wife, Hiroko, and daughter, Katie, in Manhattan.

Brief Contents

Contents

Telecommunications and Business Strategy

Chapter *1*

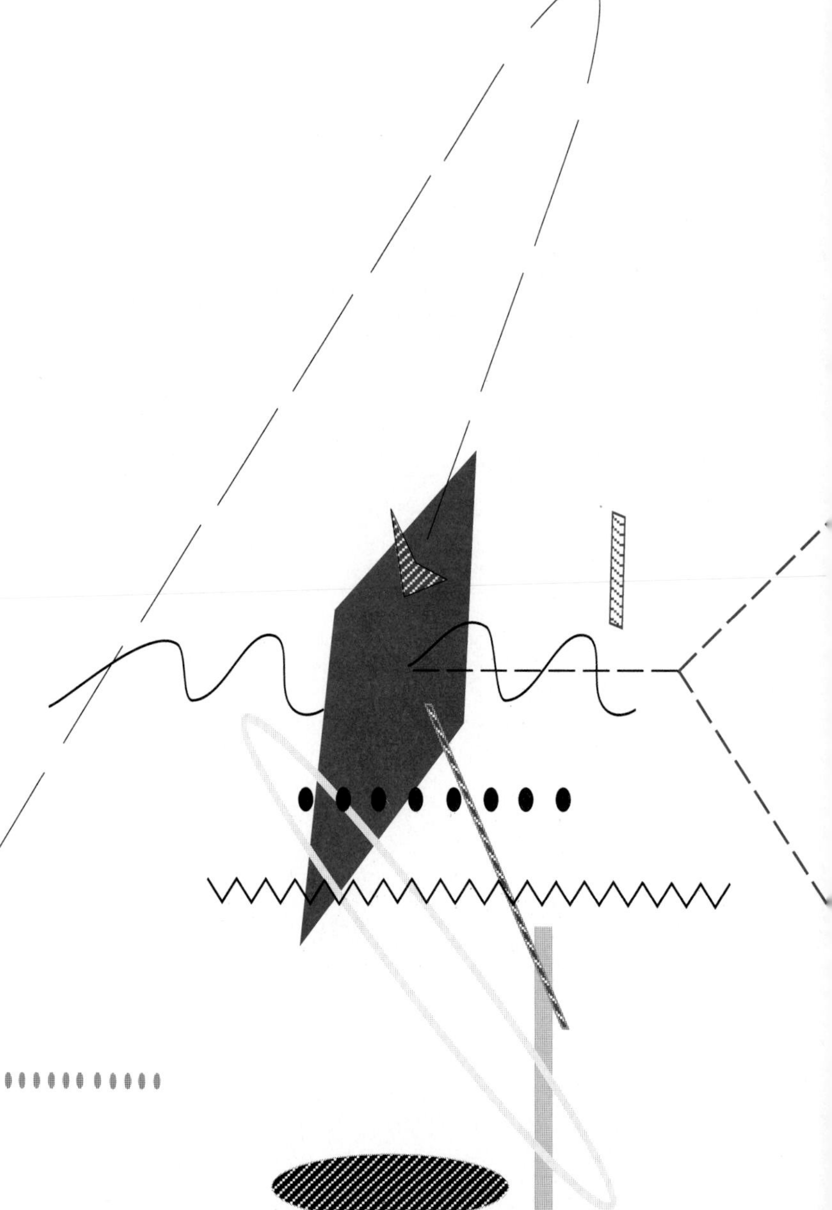

The Dawn of International Telecommunications

To understand the present, we must look to the past. Telecommunications is not a new technology. It is far older than the computer, and it has always been critical in business. In fact, from the very beginning with the electromechanical telegraph, telecommunications has provided the necessary information and data for business decisions.

But telecommunications has not remained static. It has changed and evolved through several generations of technology: from primitive wires carrying one signal at a time to fiber optics carrying the equivalent of the Oxford English Dictionary in less than a second, from simple news about the outcome of European wars to millions of electronic funds transfers per second. Telecommunications continues to power the world's business systems.

Chapter VII of the United Nations Charter concerns the actions the Security Council may take in response to acts of aggression. Article 41 states, in part, that these actions "may include complete or partial interruption of postal, telegraphic, radio and other means of communication." This clause properly emphasizes the strategic importance of communications in international relations, but to fully understand the structure and meaning of present-day communications and how they have come to achieve so much significance in international politics and business strategy, it is necessary to go back to the beginning of international telecommunications in the 15th century.

Those interested in the development of global communications and how humankind as a whole has reaped the fruits of these technological advances need to review a whole series of questions in light of historical analysis. How did the current organization of communication channels in the world come about? What institutional forms guided the development of the international media? What major technological changes prompted these developments? How have these benefits been distributed throughout the world? Who is served and who is harmed by the current structure? How is the political power of the nation state reflected in the current structure and what part did it play in guiding the development of the structure? Finally, what does this all mean for conducting international business and commerce in the closing decades of the 20th century?

History of Communications

Communication has always played a vital role in the consolidation of states. Before the development of modern states, various communications channels nurtured cultural relationships that held together the diverse subgroups of the world empires. China, Rome, the Levant, the Inca, the ancient states in India—all of these empires were held together by centralized administrative communication structures. These channels of communication disseminated the very identity of the state or empire. In the medieval era, the "international" exchange of information from purely intellectual pursuits was quite widespread.

Communication fostered growth in the scope and power of the Church bureaucracy, which attempted to monopolize communication through the notorious *Index liborium prohibitorum.* Papal censorship was outflanked by the invention of the moveable-type printing press in approximately 1450. Control over books had been possible only so long as the expense of production by hand copying kept their numbers small. This was not to be the last time technology transcended and made obsolete a current political arrangement. At that time, much of "business" was concentrated in the hands of international guilds and merchant associations, under the domination of large families.

The rise of printing was the motor that drove the Enlightenment and the scientific revolutions of the 16th and 17th centuries. Combined with expansive exploration overseas and other forms of scientific endeavor, European civilization ascended by tactics amounting to a type of barbarous economic rape of technically less advanced peoples, who were first culturally decimated then financially enslaved in capitalism. Scholarly exchanges were published and letters were carried by couriers back and forth over the continent leading to the founding and prosperity of national scientific academies. Scientists in Britain could know the work of scientists in Germany, France, and Italy. The data from overseas explorations came pouring into these centralized scientific bureaucracies.

Many of the major developments in communications began in the latter part of the 19th century, though several national systems had already progressed past the mere exchange of letters and invoices. In France, for example, as early as 1792 people were aware of the **Chappe system** of telegraphy. This system was essentially mechanical, with a series of telescopes trained on fixed points in the distance where semaphore flags were waved to and fro to spell out various messages. This system was serving the whole of France by the mid 19th century.

Communication in the United States, however, was quickly moving toward the future. In 1843 Congress approved monies for construction of a telegraph line between Washington, D.C., and Baltimore. The first message sent along this line on May 24, 1844 used the telegraphic coding system of Samuel F. B. Morse: "What hath God wrought!" In retrospect, this was an entirely appropriate exclamation; only a few far-sighted individuals could divine the tremendous significance of electronic communications. By 1860 most American cities were linked by telegraph lines.

The Morse system was put in use back in Europe in 1858 when 10 European countries paid Mr. Morse 400,000 francs as compensation. Electronic communication technology quickly made the Chappe system obsolete. Efforts at organization of communications progressed rapidly.

The first treaty designed to connect the telegraph systems of two states was signed in 1849 by Prussia and Austria. At this time there were essentially two independent systems, which had developed in parallel, centered in Western Europe and Central Europe. France, Belgium, Switzerland, Spain, Portugal, Sardinia, Hanover, the Vatican and the two Sicilies had one system in the West. Prussia, Austria, Bavaria, and Saxony had the other in the East. There were soon conventions in operation: the 1857 Austro-German Telegraph Union and a Brussels convention that mixed the two other systems. By 1860 Russia was hooked up. The International Telegraph Union was created in Paris in 1865. The French had envisioned a plan to map the world's telegraph systems which were then in the process of expanding rapidly to Moldo-Valachia, Asiatic Russia, Malta, Corfu, Asiatic Turkey, India, Persia, and Japan.

In 1866, the message "All right" was related across the Atlantic from Valentia, Ireland, to Heart's Content, Newfoundland using the first permanently successful trans-atlantic cable. Europe and America had been linked by submarine cable in 1858, but the cable was damaged in deep water after it had been in use only three months.

The 1875 St. Petersburg Telegraph Conference brought into being a communications regimen that lasted until 1932, almost 60 years. In 1903, Teddy Roosevelt sent the first around-the-world telegraph message as the London Conference was meeting to rename the hodgepodge system of conventions. Upon renaming, they became the International Telegraph *and Telephone* Union.

International Business

From its very inception, the system of world communications was inextricably tied with commerce. There are two basic reasons for this. First, the command and control of vast trading systems spread throughout the world necessitated an effective means of communication.

Second, resources could be located, identified, and gradually brought into the world trading system. Exotic plants, minerals, and other resources were discovered in this era of informational expansion.

During the initial phases of European imperialist ventures in the 15th and 16th centuries, the risks were great and no effective and systematic means of global communication existed. Once distant resources had been located and firms came to rely on a steady supply, the system of global communications, albeit primitive, became crucial in the success of trading relations. Political news of diplomatic importance; reports from wars; historical, cultural, and scientific data—all these types of information composed the substance of this early stage of international communication. The economic advantages to be gained from an effective international system of resource control powered the explosive growth of the system.

Furthermore, the information media were by no means neutral in their social effects on non-European peoples. Not merely the means for transmission of data, information systems were a dynamic element that promoted and even directly caused vast social changes. Like fertile ivy, the crawling growth of European civilization spread about the world; communications made possible effective material and political coordination.

These political-economic realities were the foundations of what may be called *l'ancien regime* in communications, a regimen that saw many effective efforts in international cooperation. The conferences noted above achieved startling efficiency in managing the problems of those times. Many of the fundamentals of the negotiating process were worked out for future consultations. Among those fundamentals were the rates, the languages accepted for messages, the crossing of borders, voting, administrative structures, and many more details. As the European system expanded to include almost the entire globe, so did its peculiar regimen of international communications.

By the First World War, the system was truly global, encompassing such geographically diverse members as Australia, the Belgian Congo, Bosnia-Herzegovina, Eritrea, French Indochina, Madagascar, New Caledonia, Siam, South Africa, and Uruguay. Immediately before that war, the 1906 Berlin Radiotelegraph Conference adopted regulations governing telegraphy by means of radio. These regulations were almost identical to the pre-existing wire regulations, but they were not destined to last long.

In 1903, the International Radiotelegraph Union had divided up the frequency spectrum for the first time, designating a special set of channels for maritime traffic and creating the famous S.O.S. distress signal. By 1912 all ships had to carry radiotelegraph equipment, regular weather reports were arranged, and frequency allocations for shore-based radiotelegraph stations were finalized. Still, these regulations did not sufficiently safeguard against disaster. In 1912 the *Titanic* sank 95 miles south of the Grand Banks of Newfoundland. Investigation revealed that a nearby oceanliner could have aided the sinking vessel had its radio operator been on duty. After the *Titanic* disaster, the 1913 London International Convention for Safety of Life at Sea required that every ship maintain a 24-hour radio watch. Merely being equipped with radiotelegraph, as the *Titanic* had been, was not enough.

The postwar peace conferences also touched upon communications, foreshadowing Article 41 of the present United Nations Charter. The Treaty of Versailles imposed important restrictions on Germany's use of the facilities of international communications.

During the interwar period, there was substantial consultation on the technical aspects of improving and regulating all types of communication services. For these purposes, the 1924 Paris meeting of the ***Comité Consultatif International des Communications Téléphoniques à Grande Distance*** and the 1925 ***Comité Consultatif International des Communications Télégraphiques*** played significant parts. Prior to this, on a separate track, the international regulation of the telephone was developed further. In 1896 at Budapest a time limit of three minutes had been placed on international calls, and in 1903 at London an even more comprehensive set of regulations had been drawn up. This frantic series of conferences indicates that technology was outpacing the abilities of governments to respond, as had been the case in the 15th century.

On Christmas Eve 1906, the world's first radio broadcast took place, leading in 1920 to voice transmissions through radio-telegraph instruments. By 1925 interference caused by stations broadcasting on overlapping frequencies had become a widespread problem. The 1925 London formation of the ***Union Internationale de Radiodiffusion*** marked the beginning of systematic allocation of bands to broadcasting services.

Meanwhile, American amateur **shortwave** broadcasts were heard in Europe in 1920, foreshadowing the adoption of even more complex regulations. By the time of the 1927 **Washington International Radiotelegraphic Conference**, eavesdropping on radio signals was becoming a problem, further indicating the shape of things to come. More important was the call for unification of both the radiotelegraph and the telegraph conventions. In 1932 at Madrid the **International Telegraph Conference** and the **International Radiotelegraph Conference** met simultaneously. This meeting saw the birth of the **International Telecommunication Union (ITU)**, a unification of the Radiotelegraph and Telegraph Conventions that encompassed telephone, telegraph, and radio under one organization. Allocation of frequencies occupied the 1933 Lucerne Conference, the 1938 Cairo Conference, and the 1939 Montreux Conference. As is well-known, frequency allocation has never ceased to be a major problem.

The Second World War brought the activities of the International Telecommunication Union to a standstill with heavy damage to international communications facilities. Also, for national security reasons during the war a great many restrictions were placed on communications of all types. Paradoxically, at the same time communication facilities were being systematically destroyed, much technical progress was made, especially in radio. Besides being used in military operations and for propaganda, radio became essential for civil defense information. Multiplexers, facsimile telegraphy, and the telex were all created during the global upheaval.

The 1945 **Bermuda Telecommunications Conference** led to a special meeting of the Big Five—the United States, the United Kingdom, France, China, and the Soviet Union—at the 1946 **Moscow Preparatory Conference.** Once again, frequency allocation became a major bone of contention. On the other hand, the arrangements for telegraph and telephone were set up without much conflict.

With the 1947 **International Telecommunications Conference** in Atlantic City, for which the Moscow Conference had prepared, one can see the beginning of almost 50 years of U.S. dominance in the field of telecommunications. The relative lack of war damage to the U.S. economic base and the many technological advances developed here gave the United States the most power to shape the future communications order at the Atlantic City conference. This was most evident in the fight over the participants in the conference. The United States had invited many countries that were not members of the union, bringing the very legality of the conference into question. This dispute was settled in favor of a more comprehensive viewpoint compatible with that of the United States.

Prior to this time, the ITU had been dominated by the European nations, which were often condescending and arrogant towards the United States. (They still are in many cases.) Once the ITU was placed under the **United Nations**, it became a much more globally oriented organization. It must be remembered that the ITU had never been a part of the **League of Nations**. Now, as a specialized agency, the ITU is bound to political coordination with the United Nations as provided in Article 41.

After the Atlantic City conference, international communications once again seemed to be under political control. This political control was to be shaken by the development of the communications **satellite**. The new communications structure brought about by satellite technology was a challenge to the political control of the ITU because the patterns of ownership were so radically different. Only a new political organization, **INTELSAT**, was capable of temporarily containing the onrush of technological progress, although no complete political solution exists at the moment.

As shown by the impact of printing on the 16th century, the impact of telegraphy on the 19th century, and the impact of radio and satellites on the 20th century, in communications, we can no more predict the future than the soul who wrote "What hath God wrought!"

The Emergence of Switching

It is traditional to say that telecommunications began in 1844 when the telegraph was used to transmit information across wires from one location to another. Samuel F. B. Morse's Morse Code served as the backbone of telegraphy for a century or more. Morse Code represented letters and numbers by binary signals, allowing the telegraph to become the first technology for movement of information through wires using electric signals. It was the beginning of telecommunications.

Approximately 30 years later in 1876, Alexander Graham Bell demonstrated that sound could be transmitted over the telegraph wires. He received the patent on the telephone. Because of the early development of telegraphy, there was an installed base of wires, although not enough to cover the world or even the United States. The invention of the telephone spurred a great advance to this development, but it took approximately 10 more years of development, particularly the laying of cables and the insulation of the transmission wires, before voice communications started to improve greatly in quality. For long distance transmission, the copper wire became the standard.

At about the same time (in 1879) A. B. Strowger invented a step-by-step switch that automatically switched calls, thus eliminating the need for operators in many locations. The switch was based on the on/off clicking associated with the older rotary telephones.

Technological developments such as the remote teleprinter, experimentation with radio telephony, and transmission of the pictures that would later become television developed rapidly. The first around-the-world telephone conversation took place in 1935. The world was shrinking rapidly, as can be seen by looking at the world's circuits illustrated in Exhibit 1.1.

By the 1940s, the crossbar electromechanical switch was rising in popularity. Also, during the 1940s and 1950s, radio telephone and microwave transmission technology entered the market. In 1954, color television transmission took place, carried over radio relay systems. By 1960, the computerized switch gradually began to replace older switches throughout the telephone company, allowing the installation of touchtone telephones. In the 1970s, the first modems, transmitting at approximately 300 baud, came onto the market and began to achieve acceptance.

In the mid 1970s, the electronic private branch exchange (PBX) was introduced, and the introduction of digital switches continued throughout the backbone of the telephone system. During the early 1980s, AT&T added digital services such as terminal interconnection and electronic mail. By the mid 1980s, use of high-capacity circuits such as the T-1 was increasing rapidly, particularly in private networks. Larger networks were being linked together with multiple T-1 circuits for even larger capacities and bandwidths. Also, various local area networks

Exhibit 1.1

International Telecommunications Circuits

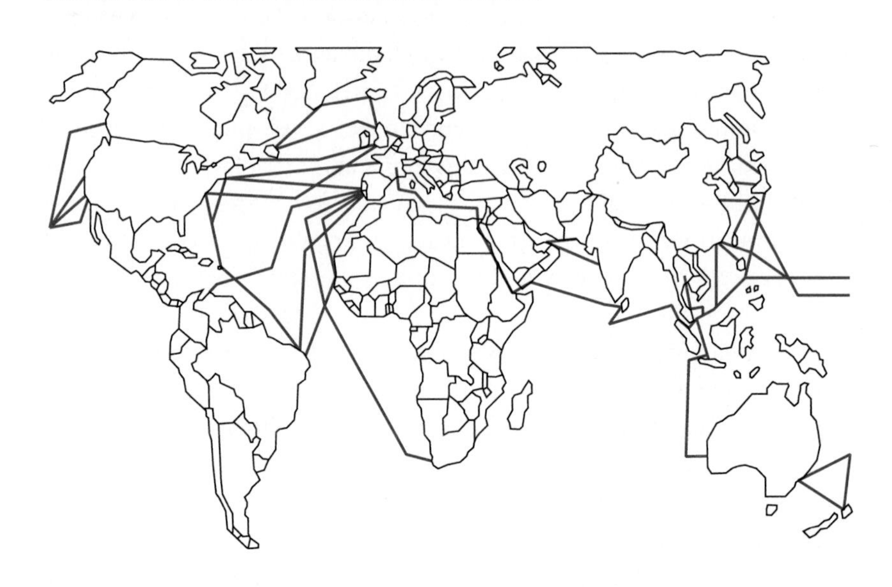

were being introduced, and they were expected to achieve large-scale adoption during the decade.

Telecommunications and Business

In order to understand the effect of modern computer communication systems, one has to imagine the time when telephones were not very common in business. Before the telephone, the postal system or personal face-to-face contact gave people the only options for conducting business. In class-based societies such as Great Britain, the system of private clubs and business clubs facilitated face-to-face meetings among members. This widened the gap between the owner-manager group and others since there was little if any mechanism of communication between people of different groups.

The telegraph greatly influenced the business operations of the world. Systems for transferring money, sending messages, confirming orders, etc., were all based on the telegraph. The telegraph and the telecommunications system it represented spread throughout the world hand in hand. Business telecommunications became the essential ingredient in operation of the worldwide cartels and businesses that existed in the imperialist nations of the world. Businesses in the British, Dutch, and French empires all used telecommunications extensively in conducting international business.

It would probably amaze the student that when the telephone was introduced, many thought it was a passing fad, and few saw its potential for business. People complained about the possibility of carrying on a private conversation with a person whom they could not see. To the Victorian-minded white European and American upper classes, it was odd to think of conducting business through such an arrangement.

Of course, people were wrong. The telephone emerged as a remarkable instrument for conducting business. Resistance did not erode that quickly, however. When interviewed for one of the Chicago newspapers in the 1920s, Edwin Booz was highlighted in a discussion regarding conducting business over the telephone. For a while, the telephone must have been an instrument for the very elite through which top-level managers, capitalists, and politicians could talk to one another.

The popularization of the telephone and the strong efforts made in the United States and elsewhere toward universal telephone service eliminated forever the idea of an elite type of communications system. Still, strategic types of telecommunications systems based on voice technologies kept emerging.

- The emerging airline industry developed the institution of the reservation center with dozens of operators cheerfully taking orders for flights.
- Public opinion organizations such as Gallup began to size up public opinion in the United States by using the technique of the telephone survey.

Business survived with the use of the telephone, telegraph, and telex through most of the Second World War and the immediate postwar period. The big change came when the computer started to radically alter business operations, first in the back room, then with the world

Exhibit 1.2

Evolution of Telecommunications Technology and Applications

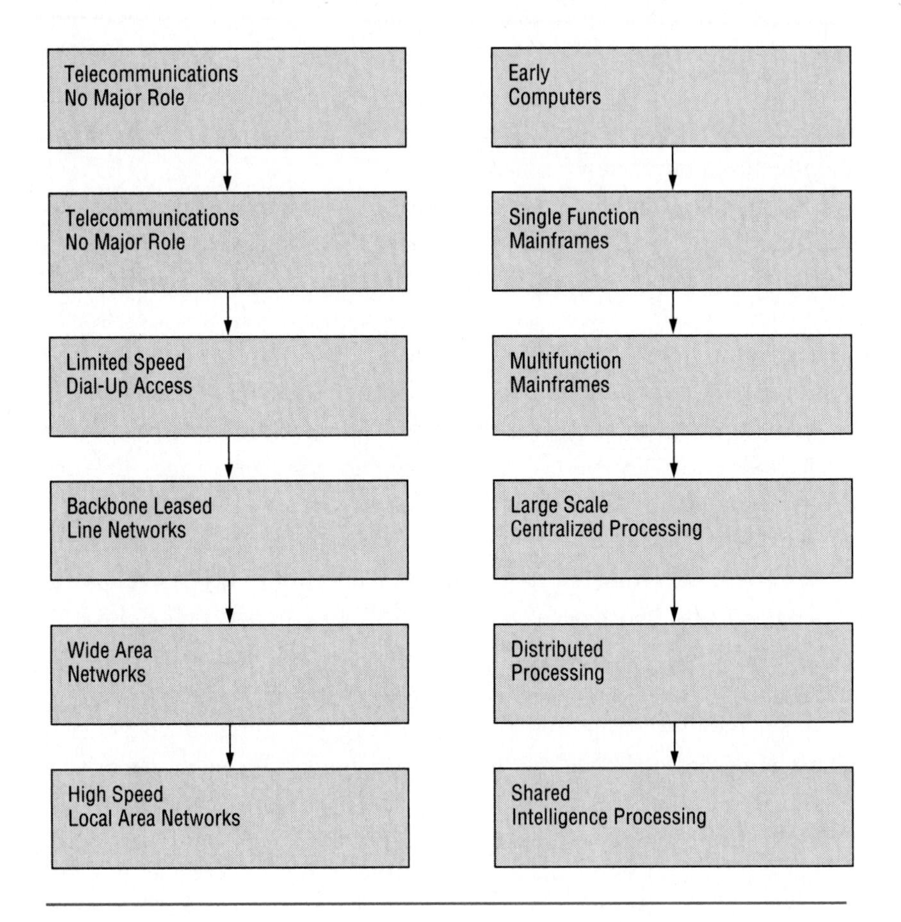

outside of the corporation. Exhibit 1.2 diagrams the progress of this technology.

The early 1970s saw the emergence of the first time-sharing computers, which made possible the linking together of remote terminals with the main computer center. This first type of linkage joined the computer to "dumb" terminals, which acted only as extensions of the main computer. These first linkages were made through fixed telecommunications lines. This is the period in which the IBM philosophy of a

Exhibit 1.3

Growth of Electronic Funds Transfers

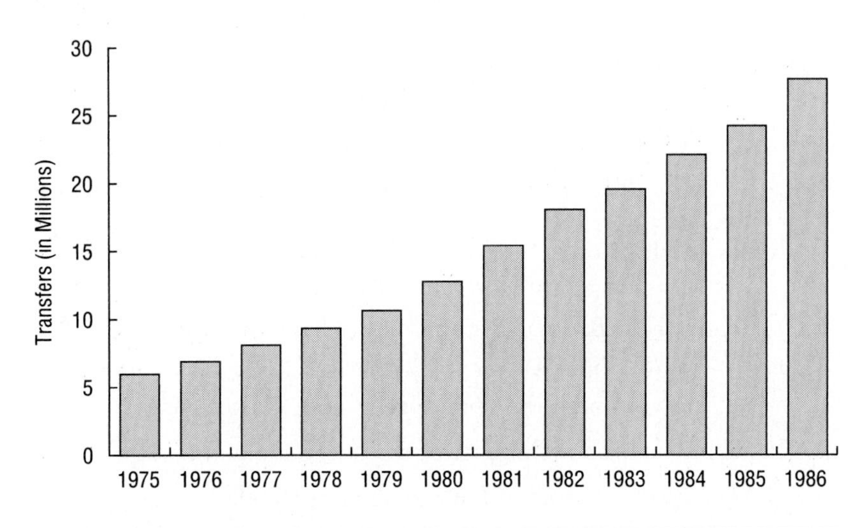

large, centralized computer system started to exert heavy influence on the configuration of the ensuing telecommunications system. Much of that design philosophy still remains today, although many alternatives have emerged as telecommunications made possible links between information systems through other components in addition to terminals.

The dramatic rise in the volume of information transmitted through the computer communications systems of the world is indicated by the increase in the number of money transfers through the New York clearing house (see Exhibit 1.3). It is reasonable to suppose that the number of transactions in the financial marketplace showed the same dramatic growth tendency.

Questions

1.1 Why do you think that the international arrangements governing telecommunications have survived through two world wars?

1.2 a. Why does the United Nations Charter mention cutoff of telecommunications as a sanction that can be carried out by the Security Council?

b. What would be the effect on the economy of the United States if all of its telecommunications were cut off?
c. How would people conduct business if telecommunications suddenly became unavailable?

1.3 a. Do you think that telecommunications gave European countries an advantage over less developed countries?
b. Can telecommunications give businesses advantages today?

1.4 a. Why did the international radio convention have to work closely with the international telegraph convention?
b. If television could be sent to you over your telephone line, would that regulatory structure have to change also?

1.5 Who was P. B. Strowger and why is he important?

1.6 a. What is the Morse Code?
b. Do you think the same type of code is used today?

1.7 What has been the impact of the telephone on business?

1.8 How have computers changed along with telecommunications?

1.9 Do you think the recent stock market crash in 1987 had anything to do with telecommunications?

Part ■ *I*

Technological Foundations

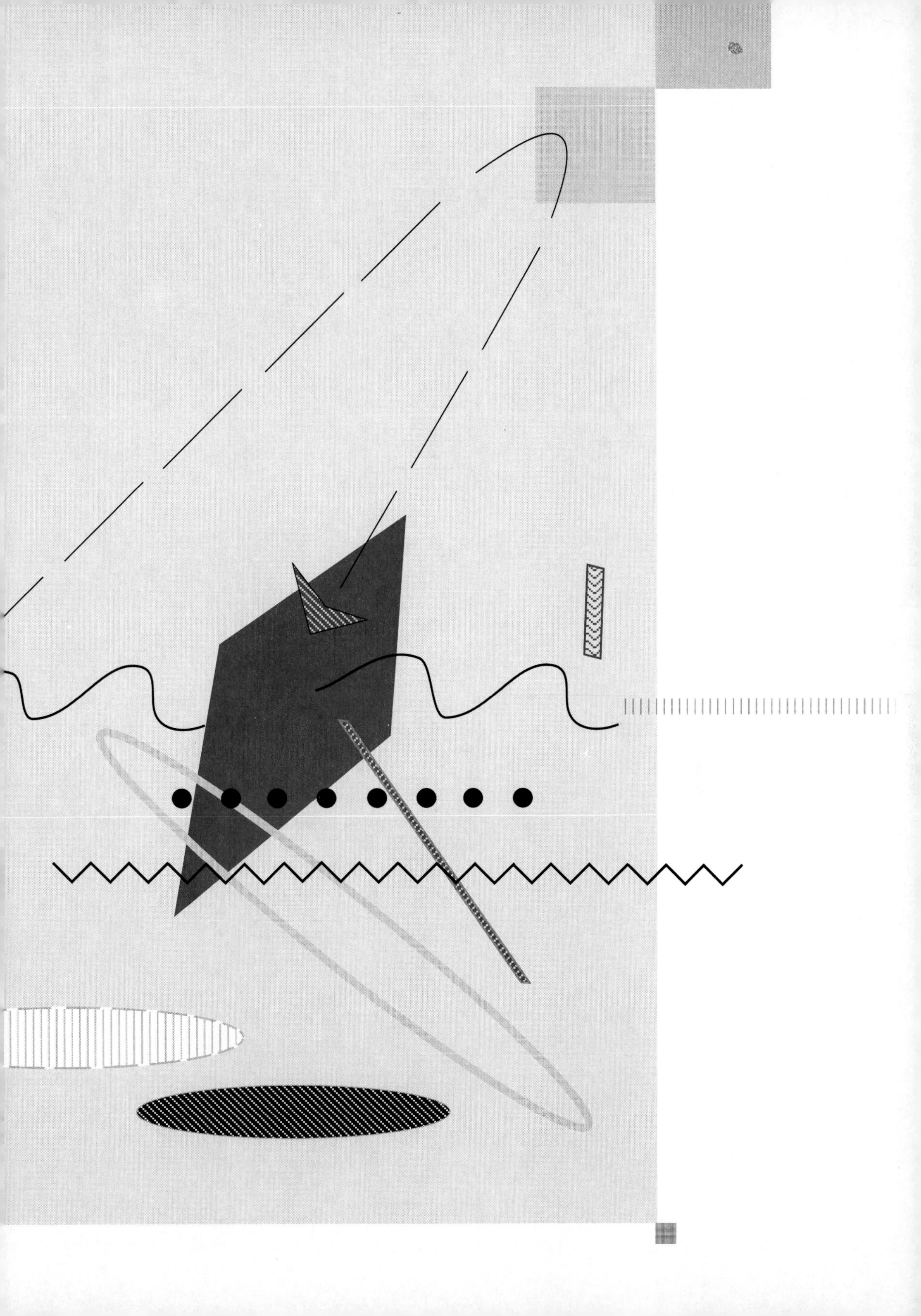

Chapter 2

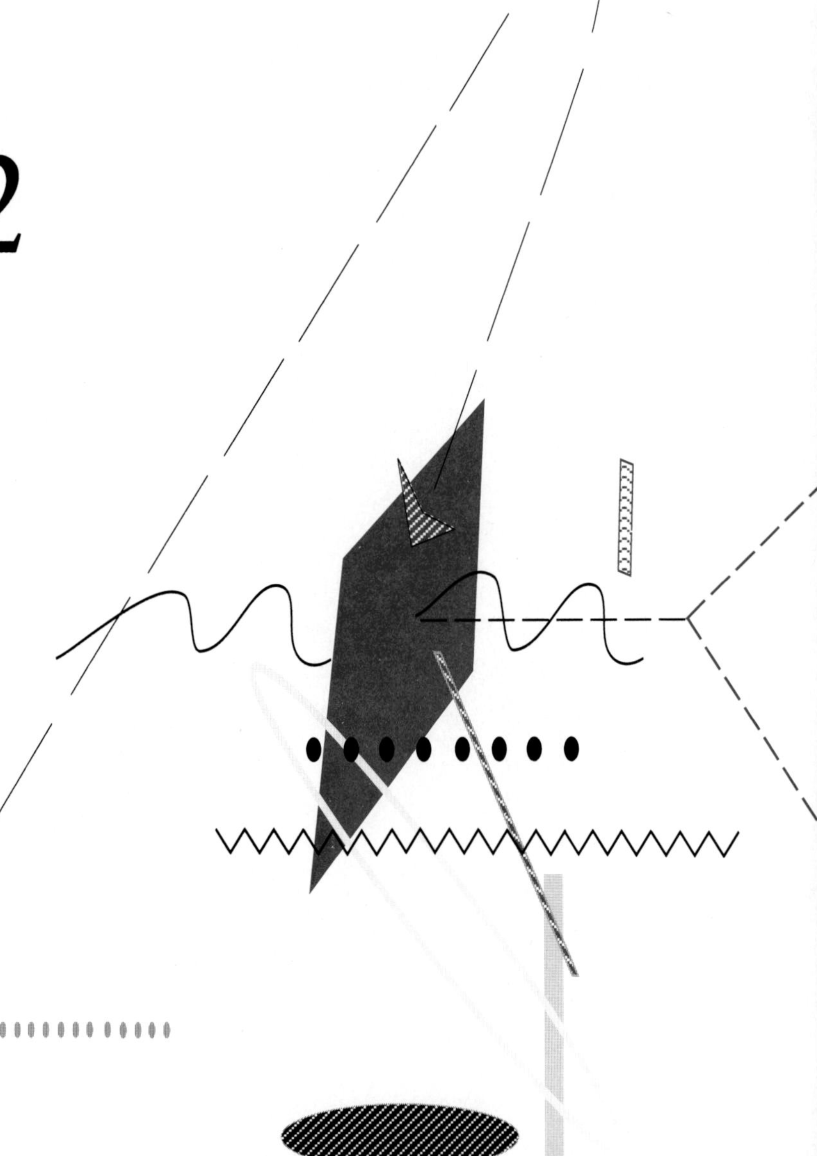

Types of Networks and Telecommunications Families

This chapter presents a way to classify the many different types of telecommunications according to their basic carrier media, the type of information they carry, and their business applications. It reviews the development of voice telephony networks and discusses the central role of the Private Branch Exchange in providing more advanced services to businesses. It also reviews the important customer service-oriented peripherals that perform voice processing applications for PBXs. It then discusses large-scale data processing networks and shows the characteristics of leased line point-to-point versus packet switched networks. It also reviews image-oriented networks and shows which factors drive their unique features, and it reviews teleconferencing systems and complex facsimile systems. Computer-Aided Design workstation networks as well as combined media networks are also discussed. Finally, a case study of McDonald's Corporation's ISDN trial network is discussed.

Classification of Networks

Telecommunications is a complex field because there are many different types of telecommunications systems, or **networks**. In its simplest sense, telecommunications involves the movement of information from one location to another by physical means. Telecommunications systems can be classified in several different ways: according to the type of information being telecommunicated; according to the media through which the information is being telecommunicated, or by their business applications.

Type of Information

Telecommunications systems link together many different types of people and machines, so many different types of information flow through the telecommunications systems. For example, two persons talking over the telephone are sending voice (or audio) information; in this case we may be looking at a **voice network**. If two computers are sending information (or data) between themselves, then we would probably be discussing a **data network**. If we are attempting to classify an office automation system sending pictures of documents, or video pictures of humans, then we would probably be discussing an **image network**. Perhaps we might examine a system which is simultaneously telecommunicating several different types of information; then we may be looking at a **combined media network**. For the most part, however, we usually only distinguish between voice and data networks.

Type of Media

Regardless of the type of information being transmitted through the telecommunications system, the media over which the information is moving can be different. The same information can move through a copper cable, a light tunnel of glass, a microwave relay, an undersea cable, a satellite transponder, an infrared beam, a laser, or several other media. In fact, it is common for a bundle of information to pass through several different media on its way from point A to point B. From the point of view of the information, the media channel may not be important—as long as the information arrives undistorted, it doesn't matter. However, from the point of view of cost, efficiency, flexibility, and many other technical factors, the selection of media is extremely important. One medium may be optimum for short distances, another for very long distances such as from one country to another. Therefore, the classification of telecommunications networks according to their media is a useful tool when solving technical problems.

Type of Business Application

A third useful type of classification for telecommunications networks is based on the function they serve for the business. The distinction between internal and external networks is fundamental. Internal networks are concerned with telecommunication of information within the firm. External telecommunications networks link the firm with outside elements, either other firms or customers (which in many cases can also be other firms). Note that because this system of classification is based on business strategy, it is useful to separate out telecommunications linkages that connect the firm to its customers. As we will see in the various case studies presented throughout this book, telecommunications-based linkages between the firm and its customers may be a strong source of competitive advantage.

This book does not attempt to be an encyclopedia of all possible telecommunications networks. This level of detail would not be appropriate to a business major or MBA. However, various high-level telecommunications network concepts are enduring and helpful in plotting business strategies. We present these selected networks in this chapter.

Voice Networks

Every firm has a telephone system, and therefore a voice network. The degree of private ownership of voice networks varies, as do the service options available from the service providers. In the United States, service is provided through a local telephone company in cooperation with a long distance carrier, usually AT&T. In Western Europe, telephone service is provided by the Postal Telegraph and Telephone agency (PTT), almost always a government-owned or -controlled organization. In the United Kingdom, companies choose between either British Telecom or Mercury, although most telephones are connected to BT. In Japan, most telephone service is provided through Nippon Telegraph and Telephone, although many other telecommunications companies provide a wide variety of services to cater to different segments of the population. In the rest of the world, virtually all telecommunications services are available through government organizations, usually PTTs.

Until the 1980s, when the telecommunications "monopoly" of AT&T was broken up, the United States had the best, most advanced telephone system in the world. Unfortunately, this is no longer the case, but it is nevertheless useful to discuss the evolution of services available to business with reference to what has happened in the United States. Most of the great developments in telecommunications were created in the United States up until the mid and late 1980s, after which the focus of innovation shifted elsewhere, primarily to France and Japan.

Central Role of the Private Branch Exchange (PBX)

In the early stages of business telecommunications, each telephone was unique, and had a separate number. In other words, if a business wanted to use several telephones, it would make arrangements to get several telephones from the telephone company. As the number of telephones being used by business, particularly in such areas as Wall Street and midtown Manhattan, increased rapidly, the strain on the central switches connecting these telephones together also increased. If a company had several telephones, and if employees from within the company wished to speak with each other, then their calls were channeled through the telephone company's switch.

In order to cut down costs, the private switchboard was created and each company with a great number of telephones was able to have an operator who would answer the telephone and direct the call. These switchboards were the forerunners of today's **Private Branch Exchange (PBX)**.

The switchboard solution made it possible for the company to purchase only a few lines between itself and the telephone company's switch, thus reducing costs. Each telephone no longer had to be connected to a separate telephone line. In addition, when one person within the company was calling another person within the company, the call could be routed through the switchboard without leaving the company to transit through the telephone company's switch. Naturally, this saved a great deal of money and was a generally effective solution. It also made it possible for a company to have a single telephone number, a great advantage to the rapidly expanding commerce in the period immediately after World War II.

Technology advanced with the Strowger switch which allowed dialing with rotary telephones. Rotary telephones are still used in many parts of the world. The clicks of the rotor on the telephone signal a series of switches and relays which transfers the call from position to position until the correct circuit is found. This technology was basically electro-mechanical in that it relied on magnetic coils and magnets to move the switches back and forth.

This development in switching was transferred from the telephone company switches to switchboards on-site in major corporations, transforming those switchboards into PBXs, and changing the role of the operator significantly. As a result, most incoming calls were routed directly to their destination without having to go through the operator. A company would have a single number, with only the last few digits changing to allow for the different internal telephones within the company. This is the familiar pattern we see today: a company might have several similar numbers, such as 880-9123, 880-9124, 880-9125, etc. The switching of the incoming telephone call to the 880-abcd number is done by the central switch of the telephone company. After the call reaches the PBX, the last digits would be read to direct the call within the company.

As the PBX evolved, it gained several features. The ability to place calls on hold, and the ability to forward calls to another "extension" of the PBX, were two of the most popular features of the new systems. It is interesting to note that these features helped to determine the sociology of interpersonal communications using the telephone system.

As the microelectronics revolution swept through the world of telephony, the PBX became digital. To the user, a digital PBX does not appear to be any different from an older style analogue or Stowager switch PBX, except that it offers more features (many of them rarely used). In the digital PBX, the switching of telephone circuits takes place through microelectronics, usually through tone dialing instead of pulse dialing. Since the digital PBX is based on microprocessors, it has many features which have never before been available to telephone users.

- *Programmability of Numbers for Moves*
  The digital PBX is very useful in allowing changes within an organization. Changing telephone numbers is done through software programming rather than by actually disconnecting and reconnecting telephone wires. This makes it possible for people to "take their telephone numbers with them" as they move from one department or office to another within an organization. For example, people who have printed up calling cards with their telephone numbers do not have to print new cards each time they change offices. This feature saves a great deal of time and expense in reconfiguring the telecommunications network.

- *Call Detail Reporting and Private Accounting*
  The PBX is a very powerful tool for keeping track of the calling patterns of individual callers. At the end of each month, for example, the PBX can provide a complete listing of each call made by each individual telephone, including the charges associated with those calls. In this way, it is possible to isolate the calling patterns of each individual in order to enforce expenditure patterns for long distance calling. On many PBX systems, the individual is given a unique number-code for making long distance telephone calls. Independent of the extension from which they are calling, the use of the accounting-code tells the PBX who is making the call. This information is combined into the end-of-the-month call detail reports for the individual. These features are very useful for the accounting function which must track costs in business, perhaps on a project-by-project or department-by-department basis. Call detail reporting is an important feature for establishing the true source of costs, and it helps a great deal in the management of calling patterns. Persons who make excessive long distance calls or personal calls can be quickly identified and notified that they must change their calling habits.

- *Number Access Control*
  Another useful feature of the PBX is its ability to discriminate between different telephones. Some telephones can be given complete dialing access to the outside, whereas other telephones may be restricted only to calls within the PBX system, i.e., within the organization. Another variation might be allowing local calls within the city, but restricting long distance calls. This feature is very useful for providing universal telephone service within an

organization without the risk of uncontrolled long distance telephone bills from persons who should not be calling long distance.

- *Speed Dialing*
  The PBX may allow the caller to program frequently called numbers so that they can be "speed dialed" using only one or two numbers. For example, 880-9678 might become #1. This feature is now available on many handsets, and also through most local telephone companies for numbers which are frequently dialed outside the PBX organization.

- *Automatic Callback and Camp-on*
  In order to save the time spent repeatedly calling a busy extension, the automatic callback option allows the user to place a call automatically when the desired extension becomes free. It becomes necessary to dial the busy extension only once, and the PBX will automatically call back the caller when the number that was formerly busy becomes available. On some systems, the PBX automatically places the call and calls back the original caller at the same time, thus completing the call at both ends. In a setting such as a university, for example, this feature might be very useful in making calls to heavily used numbers such as registration or a sports ticket office. The PBX therefore allows the caller to "camp-on" to the busy extension until it becomes available.

- *Call Forwarding*
  Most PBX systems provide an ability to have incoming calls forwarded to another extension on the same PBX within the organization. The user programs the PBX to turn on call forwarding and keys in the number to which the calls should be forwarded. This useful feature helps persons without secretaries to avoid missing important incoming calls because they are temporarily at a different location (such as a conference or meeting room) and away from their desk. In addition, some PBX systems allow calls to be forwarded after three rings. For example, a busy person who wished to ignore an incoming call might let the phone ring three times, after which the call would be forwarded to another party, perhaps a secretary, to take a message. In addition, PBXs may be programmed to forward incoming calls to another number if the number being called is busy.

- *Call Pickup*
  Call pickup allows a person to pick up an incoming call which is going to another person's phone. This is useful when the second person is away from his or her desk or out of the office. The

software features of the PBX allow pickup groups to be programmed into the system. Persons who are in the same pickup group can intercept incoming calls to other persons in their group.

- *Conference Calling or Party Calls*
  This feature allows more than two persons to speak together through the telephone. Usually one person calls a second person, places that person on hold, then calls a third person, and reconnects the second person. At that point, all three persons can speak with one another. Depending on the capabilities of the PBX, it is possible to add more than three persons to a conference call. However, as the number of persons connected to a call increases, it becomes increasingly difficult to manage the call. For very large conference calls, AT&T provides a conference calling service in which each of the parties involved calls a special number, and all parties are then connected together. For the purpose of limited party calling within a company, these features of the digital PBX are very helpful, but only a small portion of calls made are conference calls.

- *Hold*
  The hold feature on the PBX allows the user to temporarily suspend a call while doing something else. Under these circumstances, the telephone connection is not lost, and it is possible to reactivate the call with a simple flash hook (pushing the hang-up button for an instant). On many telephone systems, a melody or an announcement, perhaps an advertisement, may be played to entertain the caller who waits on hold. This feature helps to reassure callers that they are not forgotten.

- *Call Transfer*
  The transfer feature of a digital PBX allows the person receiving a call to redirect it to another extension within the organization. In some organizations with several locations, a call may be redirected not only to another party within the same building, but to an entirely different city, many times without the caller realizing that the call is being redirected such great distances. On Wall Street, incoming calls can be automatically redirected to overseas locations as easily as down the hall within the same organization.

- *Shared Modems and Data Circuits*
  Many PBXs also provide access to data circuits. For example, they allow one or more extensions to be dedicated to data communications. Persons wishing to get access to a data circuit might dial a particular extension to be connected to a modem. The advantage of this is that the cost of the modem can be shared between many

occasional users. With the newer digital technologies, it is possible to have both data and voice information traveling through the same circuits. As this occurs, the PBXs will make the transition to allow both types of information to flow through their circuits.

As useful as these many features of the PBX are, they are well-known and perhaps boring. A new type of technology that developed quickly in the late 1980s has made great improvements in customer service for many business organizations. These new technologies give callers different types of options: they might make direct inquiries of a computer, or leave messages for pickup when the person being called is away. A firm can achieve these new capabilities for its telecommunications system by adding new types of equipment to the PBX network.

- *Automatic Call Distribution (ACD) Devices*
  For a business setting up a call center to handle a large number of incoming calls, the ACD is indispensable in routing calls "to the next available agent who will assist you." Telemarketing applications, consumer service or information centers, answer centers, airline reservations centers, and other similar business operations typically use this technology. The ACD makes sure that calls are distributed evenly among the employees answering the telephone.

- *Voice Recognition Equipment*
  The use of voice recognition equipment has been very successful when a business is required to give a high volume of answers to highly similar questions. For example, when customers call up a bank and wish to know what their account balance is, or whether or not a check has cleared, a voice recognition system can "talk" to them. It may be more popular in Japan than in the United States because Japanese accents vary less. A more popular option is the use of voice response units.

- *Voice Response Units and Transaction Processing Interfaces*
  Voice response units take information from a computer and generate a human voice for the caller. When VRUs are connected to a transaction processing interface unit, a linkage directly to a computer system can be established. Using a touchtone telephone, the caller can ask questions of the computer, such as "what is my balance," and get answers via the voice response unit. From the point of view of business, this type of system may free up a great

deal of human resources for higher quality customized service for customers. (It can also be used simply to cut back on labor costs.)

- *Voice Store and Forward Systems (Voice Mail)*
  Voice store and forward systems act more or less like a personal answering machine, except that they are centered in the PBX. Messages can be received remotely, the same as with an answering machine, but messages can also be forwarded to other persons; some messages may be sent to several persons at the same time. This extremely friendly and flexible technology option is very useful for persons such as consultants or salespeople who have to travel a great deal and yet stay in close contact with others who also move around a great deal.

Voice Processing Technologies

Think to yourself: How many telephone calls have I missed in the last year? How many times have I called someone and not gotten through to them directly? How many times has the message I left for someone been distorted or lost altogether? How many times has a company from which I am buying services failed to get back to me as promised?

Now, think of this problem from the point of view of a business, particularly a business that transacts a great amount of its daily activity over the telephone. This, of course, includes almost all businesses. What does it mean if the customer calling you is unable to get through to the right person, or has difficulty in being satisfied? Fairly serious problems can arise from bad voice management:

- Customers get angry from being put on hold too often and keep calling back, further straining your system.
- Customer service representatives spend too much of their time answering simple questions because there is no other way for the customer to get the information.
- Customers get tired of never being able to get through to their salesperson and start placing orders with another company.
- Customers cannot report problems with merchandise or stolen credit cards, so the company never knows how its products are doing in the field.
- Secretaries are burdened with taking telephone messages, and routinely put other callers on hold while taking messages.

- Customers involved in impulse buying have time while on hold to rethink their purchases, lose interest and hang up, thus resulting in a lost sale.
- Customers may suffer injury, or even death, due to poor response time in answering an emergency call.

Several important technologies came into the market in force in the last part of the 1980s. These technologies—**voice processing technologies**—perform a wide variety of important functions that enable an organization to more effectively handle incoming calls. Utilization of the tone system for dialing instead of the obsolete rotary system made it possible for the customer to feed information into the system by pushing buttons on the telephone. This makes it possible for the system to ask certain questions of the customer, such as: "Are you calling about a lost card?" "Do you wish to know the balance on your account?" "Do you know the extension of the person you are dialing?" The callers can also answer those simple questions (known as "prompts") by pressing keys on their touchtone phones.

These technologies perform several important functions, as illustrated in Exhibit 2.1.

- Audiotext units give the caller an option to select among different recorded messages by pressing the different telephone keys in response to prompts.
- Voice response units act as an interface to the organization's information system. When the customer responds to prompts like "please enter your account number," or "press 1 if you want current balance information," the unit translates the information into a proper query for the information system, receives the response, then generates the voice response to the customer. In other words, the voice response unit reads to the customer the information from its mainframe contact.
- Voice mail units give the customer or an internal employee the chance to leave a message for a person who does not answer the phone. These units also support functions such as mailing lists, broadcast announcements, and forwarding messages with comments.
- Automatic attendant functions bypass the switchboard operator, giving callers the option of dialing known extensions directly.

Exhibit 2.1

Families of Voice Processing Technologies

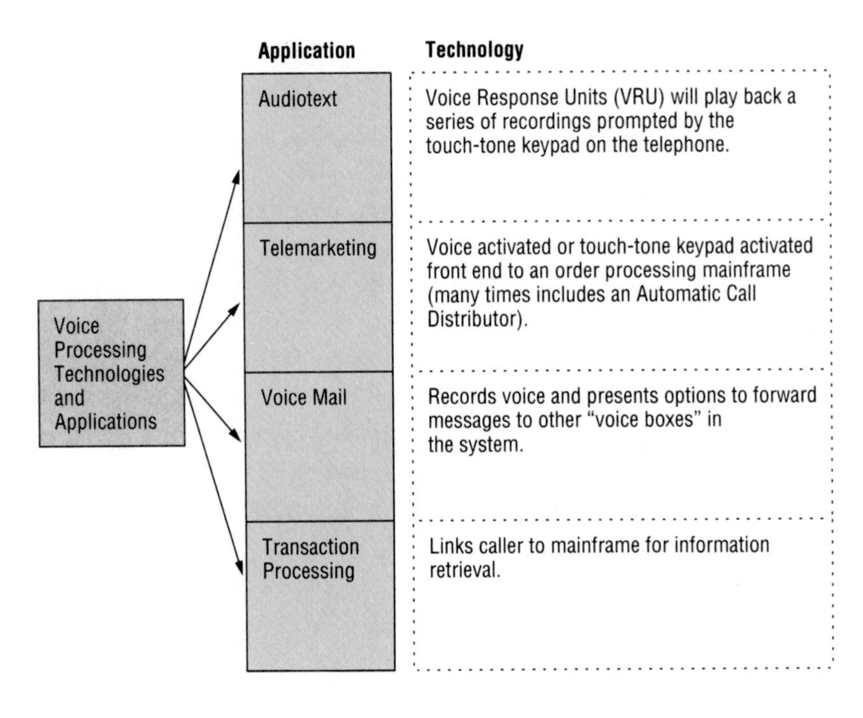

- The PBX acts as the central switching system for incoming calls into the organization. The other voice processing units typically interface very closely with the PBX. Many advanced PBXs have very sophisticated built-in voice processing features.
- Automatic Call Distribution units hold onto incoming calls, then connect them to the first available agent. Anyone who has ever called a large service organization is very familiar with ACDs. They usually play music while you are on hold.

These technologies, which are diagrammed in Exhibit 2.2, can provide very smooth ways of handling customer calls. For example, a customer may call for information that is available through a standard recording. After listening to the recording, the caller might want further

Exhibit 2.2

Basic Technologies of Voice Processing Systems

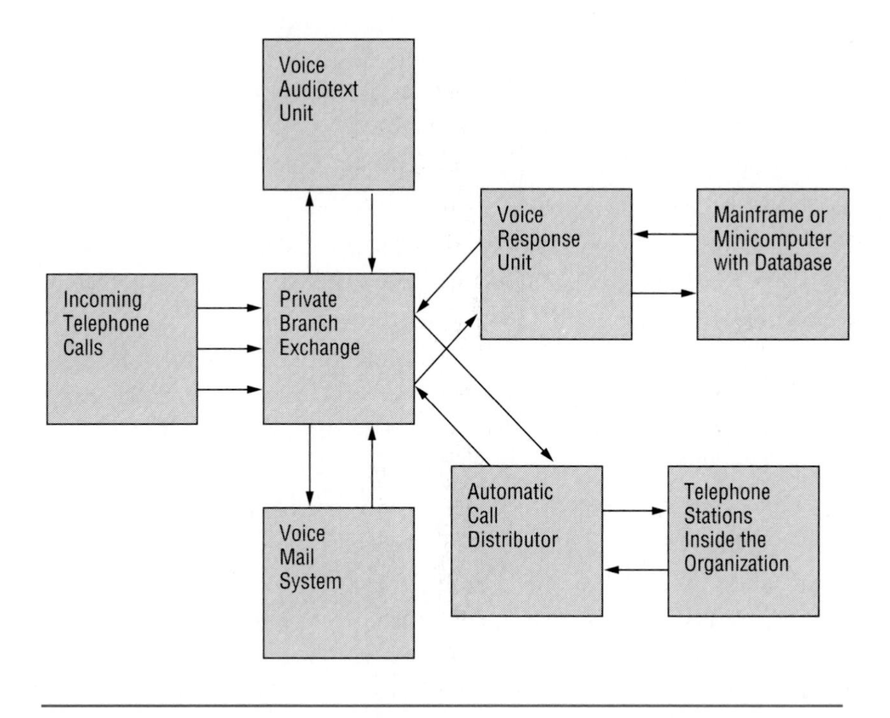

information. These techniques of handling customer calls provide important efficiencies because they:

- Reduce the traffic load on the console operators who answer the phone for the organization as a whole.
- Bypass company employees to give frequently requested information automatically.
- Allow the caller to leave messages without having to go through a secretary. Such messages can include details, instead of just a short note saying "please call" or "will call back."
- Automatically route calls to the first available representative, thus fully utilizing the time of the employee, and reducing to a minimum the waiting time for the customer.

- Reduce the number of double and repeat calls on the part of both the customer and the employee.

A customer might first call the organization to get some very general information. Upon first reaching the organization, such a caller has several options. The caller chooses among the various options by pressing keys on the telephone. A call requesting information on the new services that are available is forwarded to a voice audio text unit which gives the caller further options to listen to different messages. After listening to the selected message, the caller might decide to check his or her current balance with the company. The call is passed to a voice response unit that can provide a bridge to the corporation information system (a mainframe or minicomputer system). The customer might be asked to key in the account number or credit card number, and again the caller responds through the use of the telephone keypad. The voice response unit reformats the information and then sends it to the host computer so that it looks like a regular query. The voice response unit is typically emulating a terminal connected to the mainframe. The customer is given different options, such as "Press 1 if you wish to see your current balance" to which to respond. When the response comes back from the information system, the voice response unit reads it to the customer. This continues until the customer is satisfied.

But suppose the customer then has a question about the account. The system can automatically transfer the call to the first available customer service representative through the automatic call distribution unit. If the representative cannot resolve the question, a supervisor may be contacted. If the supervisor is not available, the customer can leave a message on the voice mail system. Exhibit 2.3 shows the progress of the customer's call throughout the organization's voice processing technology system.

Voice Mail

Voice mail is a type of telecommunications technology that enables callers to leave a recorded message in a central location. Exhibit 2.4 shows the user's options with a voice mail system. When the intended recipient of the message calls the central location, she is alerted to the presence of a message. She may then request that the message be

Exhibit 2.3

Progress of a Customer Call through a Voice Processing System

A. Customer places call.

B. Call enters into audiotext system to receive information recordings. Audiotext system allows the customer to transfer to a customer service representative.

C. Customer's call routed to first available service representative.

F. Customer not satisfied with conversation with representative, wants to speak with supervisor, who is out to lunch.

G. Customer's call transferred to voice mail system where customer leaves message for supervisor.

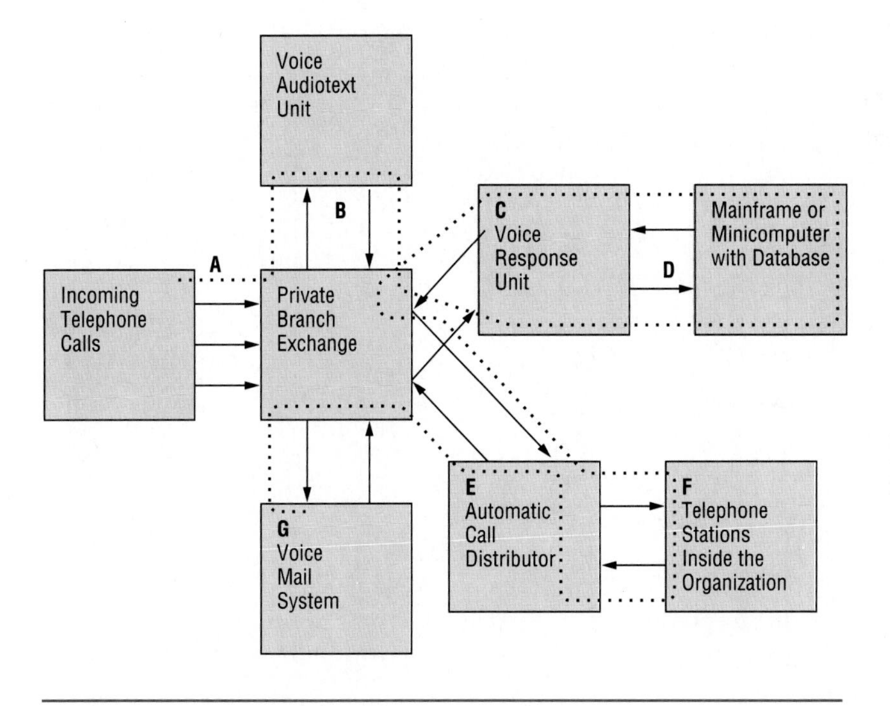

played back. After playback is completed, she may respond to the message by entering a recorded message. That response is then automatically made available when the original caller calls into the central location.

Voice mail has become popular because of three cost factors:

- It saves time by eliminating phone tag. This time savings can be very substantial in a telephone-intensive environment.

Exhibit 2.4

Options for Voice Mail Users

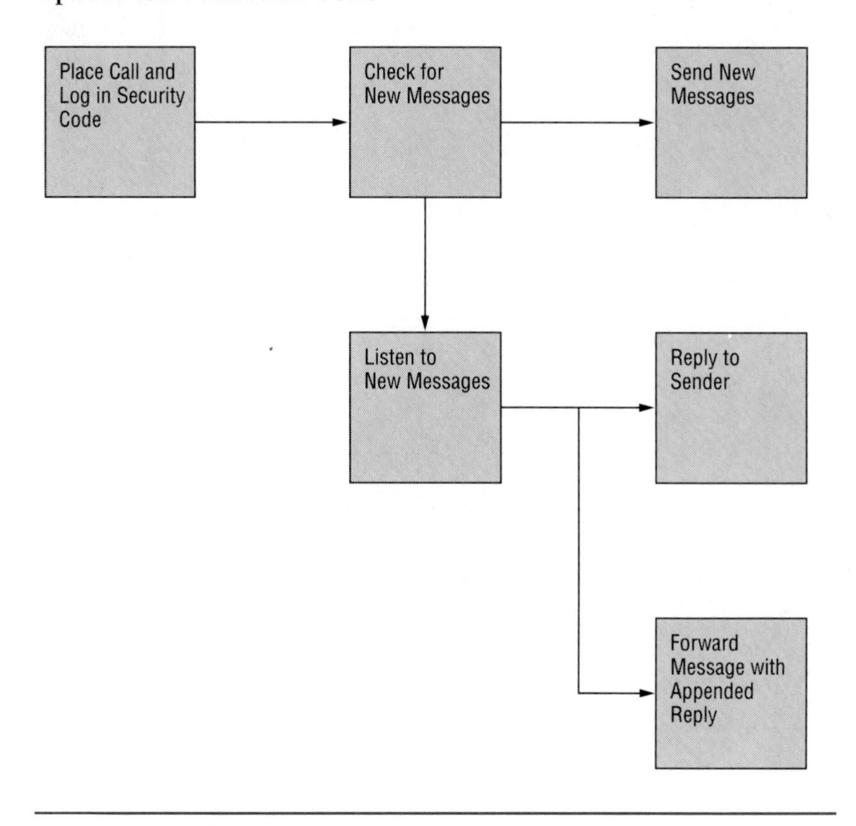

- It saves on telephone costs because it allows one message to be sent to many recipients simultaneously with only one telephone call.
- Under some circumstances it can eliminate the need for intensive secretarial support.

Booz, Allen & Hamilton Voice Mail • Booz, Allen & Hamilton, Inc., is one of the leading management consulting firms in the world, behind McKinsey and the Big Eight Accounting firms. It conducts business and consulting assignments in many parts of the world, including a few developing countries. Its type of management consulting practice ranges from information systems to planning market entry strategies or technological strategies for the world's major transnational corporations.

By its very nature, management consulting demands a great deal of energy, and the working environment is such that communication between different team members on an assignment is critical. Messages come and go virtually 24 hours per day. When assignments are located simultaneously in different cities, or even in different countries, problems of phone tag and time differences between different parts of the world start to hinder efficient operations of the consulting practice. Phone calls miss their intended recipient, and during the off hours it is difficult if not impossible to get messages to one another.

In order to solve this problem, Booz, Allen & Hamilton began utilization of a voice mail system. Each employee needing access to this type of tool was given an identification code to enter into the voice mail system before getting to use its features. Voice mail acts like a very large telephone answering machine. People send voice recorded "letters" to their colleagues by storing them on a host computer, then entering the correct destination codes. When the intended recipient calls the voice mail system, she is notified by a synthesized voice that a message is waiting. By pressing the correct codes into the telephone, the recipient receives the recorded message from the other team member, and is able to record a response which will immediately appear in the sender's "voice box" to be retrieved when that person calls in to the system.

Booz has observed that on some assignments, voice mail appears to offer great advantages, primarily in time savings from the need to return calls and play phone tag. It has been noted, however, that for some people, voice mail does not suit their way of working. For those persons, voice mail does not appear to offer significant advantages.

Large-Scale Data Networks

Data networks transmit information to and from computers over relatively great distances. This procedure follows several generic models:

- *Terminal to Computer*
  The most common type of data communication involves linking a terminal to a computer. Sometimes the distance between the terminal and the computer can be very great; other times it can be

only a few hundred yards or less. Depending on the complexity of the data communication system, and the distances involved, a terminal's message may actually go through several different networks before reaching its final destination at the host computer. However, most systems involve a relatively simple arrangement.

- *Computer to Computer*
  Most large corporations, including multinational corporations, have computer centers located in many different locations. For applications such as financial control, information must be telecommunicated from the remote data center to the headquarters data center. Under these circumstances, data files are sent from computer to computer. In addition, many computer networks employ a complex system of automatic updating of files located in many different computer centers hooked into the network. For example, American Express must update its card authorization files constantly so that they are available in many different parts of the world in addition to Phoenix, Arizona, the location of the headquarters data processing center.

- *Workstation to Workstation*
  With the rise in the number of high-speed workstations, some networks have been set up to link them together into giant networks, without the intermediation of a host computer. For example, oil companies such as Sunoco have networks like this in which Sun workstations used in analysis of seismic data are tied together in several different locations throughout the world.

- *Other Variations*
  As you will see in this book, many hybrid networks are possible that link together a great variety of equipment throughout an organization.

It is common for a company to have many different data networks, each custom tailored to a specific application. In many cases, companies could save a significant amount of money if they attempted to consolidate their networks, but there is a great deal of reluctance to do this because of the risk of change and the inconvenience of having networks temporarily out of service. In addition, the total data communications budget of many information processing organizations within firms is only a small portion of the firm's data processing expenditures, typically 5 percent or less. Under these circumstances, there is little incentive to make massive, disruptive changes in the telecommunications infrastructure that would save only a few tenths of a percent of what is already a small portion of the information systems budget.

Networks tend to grow in companies one by one, with each serving a new application as it is developed by the information systems function. It is usually easier to build a new network isolated from other networks than to modify an existing network in order to accommodate the new applications that are needed. Given this observation, it is not difficult to understand why companies frequently have many different data communications networks.

The telecommunications circuits which are the building blocks of networks are available for hire from the major telecommunications carriers such as AT&T, MCI, and Sprint in the United States, and from either PTTs or private companies in other countries. The speeds of the lines available for data communications vary greatly. For communications at the personal computer level, speeds of 300 to 2,400 bits per second (bps) are possible over dial up lines using inexpensive modems. If the telephone lines are directly connected, speeds up to 9,600 bits per second are obtainable as a matter of routine. If a type V.32 modem is used, then 9,600 bits per second are possible over standard dial up lines. If leased lines are obtained from the telephone company, then higher speeds are possible. Leased analogue lines can give speeds of 19,200 bits per second full duplex. (In countries outside of the United States, it may not be possible to get the same levels of performance over dial up telephone lines.)

American Telephone & Telegraph in the United States offers higher speeds through Digital Data Service, which operates at 64,000 bits per second. For 56,000 bits per second, AT&T offers a switched service. An even higher-speed service, T-1, can give 1,544,000 to 2,048,000 bits per second. By the late 1980s, 10,000,000 bits per second (10 mbps) circuits were available over private fiber optic links and microwave circuits.

Leased Line Networks

The most common way for a firm to link together its various locations through telecommunications is by purchasing **dedicated circuits** from the telephone company. These circuits are reserved only for use by the purchaser, and the telephone company guarantees the level of service and quality on the circuit. Depending on the size and geographic distribution of the telecommunications network being built, the business must make contact with one or more companies. If the circuits are to be located only within the same town, such as from a headquarters location to a plant or branch facility, then the company can purchase its needs from the local telephone company. If the circuits required are intended to move data and information from one city to another, then generally a long distance carrier must be used. Occasionally, depending on the city and the specific circumstances, it is possible to connect your organization directly to the long distance carrier without going through the local telephone company, but these situations are less common. When they do occur, however, the local telephone company is bypassed, and the total costs of the circuit become correspondingly cheaper.

However, when it is necessary to build circuits to move data and information from one country to another, the services of an **international record carrier** might be used. Companies such as AT&T International, MCI, Western Union International, and a few others are able to offer this type of service. At the other end of the line, however, your organization will have to strike a deal with the foreign PTT, not always an entirely easy matter.

The use of leased fixed circuits has some great advantages and disadvantages. On the positive side, these circuits are less expensive than using dial up—using the regular telephone system on an ad hoc basis for each data transmission. At lower levels of utilization, dial up may be less expensive. For example, if a subsidiary is required to file a single summary report, such as sales or payroll, only once a month, it is difficult to imagine that establishing a leased line would be less expensive. The calculations rely on defining a crossover point where the total expected utilization of the circuits with dial up exceeds the monthly cost of having the dedicated line.

Another advantage of the leased fixed circuit is that there may be a great deal of flexibility regarding the types of equipment which can be attached to the line. In liberalized countries such as the United States, it is possible to attach sophisticated equipment, such as band-width managers and multiplexers, which help to pump the maximum amount of information through the dedicated circuit, including voice traffic. Unfortunately, in many countries, it is not possible to get permission to use the most advanced equipment. Instead, the PTT would rather have you buy more and more circuit capacity in order to boost its revenues. This picture is changing, but slowly.

The utilization of leased line networks depends on these connections between the computer or PBX circuits and the circuits which are designed to carry the data and information. The capacity of the circuit (measured in **bandwidth**) can vary greatly. The largest circuits, for example, can carry dozens of telephone conversations and video image transmissions simultaneously. Smaller circuits, such as that from a branch store to a local warehouse, can be designed to carry a much smaller trickle of information. Circuits are generally priced according to the amount of information they can carry.

Most often, firms use leased lines for circuits with long life cycles. They are more or less permanent networks. A telecommunications planning department would never consider setting up a leased line network for a short period of time, perhaps less than a year, except under exceptional circumstances. It takes quite a bit of effort to get the network up and connected and running, and there is a corresponding amount of reluctance to change the network once it has been established.

As a result, **leased line networks**—also known as "point-to-point networks" because they always connect together only two points—tend to be permanent. This stability and strength can also be a shortcoming under certain circumstances. There are several disadvantages to leased line or point-to-point networks:

- *Difficult to Change Capacity*
  When the network or circuit is in place, it is generally difficult to make changes. For example, what if you wish to add an intermediate point to the network? What if you wish to increase the capacity of the circuit by 25 percent only a few days per week? What if one of your branch locations shuts down or moves to a new location? Under any of these circumstances, it is difficult to make the required changes. This is the essence of the difficulty. It is also a time-consuming process, taking weeks or sometimes months for each change. If you are managing a complex network, then the management and recordkeeping for such network maintenance can

also be complex and error prone. The section on packet switched networks discusses alternatives to leased line networks that are much more flexible.

- *Difficult to Add or Subtract Network Nodes*
  Since leased lines are fixed in a point-to-point configuration, they are not really flexible. If the network topology is in need of a change, then each new point-to-point connection must be established, and each old connection in need of elimination must be disassembled on a one-by-one basis. This makes it difficult to manage complex network changes. For example, some networks are so large, the U.S. Arpanet being one example, that at any single time no one knows the exact extent of the network. Every day hundreds of nodes are either taken off of or added onto the network. However, the Arpanet is not a leased line network in many respects. We will see later why there are better alternatives to the leased line network if your most important criterion is making changes and modifications to the network topology.

- *Not Economical If Usage Falls*
  Leased lines are economical as long as there is a substantial need for their services. If, for some reason, the traffic volume on the leased line falls, then they automatically become an expensive albatross hanging on the neck of the telecommunications function. In large, complex networks with hundreds of circuits spanning through many different cities and locations, it is not uncommon to find a substantial number of leased line or point-to-point networks operating below economic use which need to be re-evaluated. The problem is that the cost of changing to another type of delivery system for the firm's information and data can be high. This raises the switching costs involved in converting to a new system. Under these circumstances, organizations many times let their suboptimal leased lines stay in place.

- *Long Lead Times for Availability (Particularly in Western Europe)*
  Another problem which repeatedly surfaces in interviews with MIS and telecommunications professionals is the difficulty in getting leased lines. In the United States, firms must sometimes wait a few weeks or months. However, overseas the situation is much more serious. In parts of Western Europe, in Spain, for example, there may be a wait for two or three years before getting a leased line. Persons who have built most of their professional experience in the United States find it difficult to imagine waiting for such a long time for telecommunications service; this complacency can prove to be

a serious disadvantage when building networks overseas for multinational corporations.

- *Difficult to Get Maximum Utilization without Sophisticated Equipment*
  Since the service provider usually only provides the capacity, it is most often up to the company to attach the equipment that will allow it to take the fullest advantage of the line. Many times, this equipment is expensive and difficult to manage without further expense for telecommunications professionals. Under certain circumstances this can prove to be a disadvantage, but if the business is large enough, and if it can afford a specialized and highly trained staff of telecommunications professionals, then this really poses no serious problem.

Packet Switched Networks

Imagine that it is the 1950s in the United States, and the Cold War between the USA and the USSR is nearing its peak. In various locations around the country, the Department of Defense is building underground missile launching silos that will be capable of delivering hundreds of huge thermonuclear devices to the landmass of the Soviet Union, should World War III break out. On the Soviet side, military and political planners are worried about imperialist aggression against the Soviet homeland and as a result are taking similar countermeasures to prepare for nuclear attack from the United States.

In the midst of this situation, military planners and other groups in the United States are contemplating what would happen to the United States in the case of nuclear war. For example, what would happen if Chicago were "eliminated"? What would happen to the United States if Washington, D.C. or Los Angeles were eliminated? If several major metropolitan areas were eliminated, how would the United States continue to organize its production and logistics for the coming protracted war? How would messages be transmitted from one end of the country to another if several major metropolitan areas were eliminated? Suppose, for example, that major telecommunications networks were built upon leased lines that had major switching points in the destroyed cities. Would they not be highly vulnerable to attack against the switching center?

Out of this concern, scientists supported by the Department of Defense created **packet switched networks**. They were to be a type of network that would have no center or single point of failure. If several switching nodes of the network were destroyed, the network would be intelligent enough to redirect the information and data through other switching nodes that were still operating. This system of passing information from one node to another would be reliable, because each node would wait for a confirmation from the next node that the information had been received and transmitted on safely. If no confirmation was received, the node would automatically try another node. Rather than having information and data move from fixed points, such as with leased line point-to-point networks, the new type of network would be able to send information on many different routes, depending on how crowded various circuits were and on the reliability of various nodes. If the Soviet Union or China dropped an atomic bomb on Houston, then the data traffic which had been traveling through that area would be automatically sent through different channels.

This concern was the genesis of today's packet switched networks. Packet switched networks are linked together by many different switching nodes, so organizations that use the packet network can access it from many different locations. Packet networks have no logical center, although they do have operations control centers that control network maintenance and diagnostics in case problems arise with lines or nodes. Packet networks are easy to access because they are usually available through a local telephone call, which connects the caller with the nearest node. Recent changes in the regulation of local telephone companies has raised the possibility of their providing packet switched services in competition with the various major companies such as Tymnet, Telenet, and CompuServe.

The clever innovation of the packet switched network lies in its use of "packets" as diagrammed in Exhibit 2.5. When a long message is to be transmitted through the network, it is divided up into small units. Each unit of information is given a "head" and a "tail." The head of the information packet acts as a navigator through the network. As the information packet reaches each node of the network, the head tells the node where it is going. The node then passes the information packet on to the next appropriate node. The tail of the information packet tells the node that the information packet is concluded. Upon learning this, the

Exhibit 2.5

Packet Switched Networks

A	B	C	D	Message 1
A	B	C	D	Message 2
A	B	C	D	Message 3

Messages 1, 2, and 3 are combined in a packet transmission.

A	A	A	B	B	B	C	C	C	D	D	D

node waits for the head of another incoming packet, which it will guide to the next appropriate node.

The heads and tails on information packets help the network become highly flexible. For example, a long stream of information being sent from Manhattan to East L.A. (perhaps to confirm the purchase of high capacity shocks for a Chevy "low-rider") can be broken down into many smaller information packets, each packet taking a different route through the network to reach its destination. Some information packets might go through Chicago, others might take the more scenic route through Houston and Phoenix. It doesn't matter which route they take through the hundreds of nodes of the network, because each node will receive routing instructions from the head of the information packet and switch the message to the next appropriate node.

At the receiving end of the system is a device known as a PAD, or **Packet Assembler Disassembler.** This piece of equipment lines up all of the incoming information packets, which may have arrived out of order because they have taken very different routes through the network, and arranges them back into the original message. The message is then passed on to the receiving machine. At the beginning of the network, the PAD divides up the message into the different information packets and adds the heads and tails. At the receiving end, the PAD simply puts things back together so no person (or machine) can tell the difference from the original message.

What makes the packet switching network so efficient is that many messages can be transmitted over the network at the same time. Dozens, hundreds, or even thousands of users and machines connected to the network at the same time can all send and receive information back and forth over great geographic distances, including across international borders. Since each message's head tells the nodes where it is going, there is rarely a mix up. Suppose, for example, you are connected to a distant computer, on which you are editing a file; if you pause for a few seconds to think about what you are doing, during that small pause in time, the packet network can send thousands of messages over the same circuits. This makes packet networks highly efficient compared to leased lines. The only exception is when the leased lines are under constant use. In that case, of course, the two systems are equally efficient.

Although the technology for packet switched networks was developed in response to the threat of nuclear war, in the early 1970s the same technology was spun off from the military, the result being the development of **public packet switched networks**. These networks have grown in importance and in size. For example, a major network such as Telenet, owned by GTE, has more than 700 nodes throughout the United States alone. It is possible to get to a node of a packet switched network from any telephone and, except for very remote areas or small towns, the telephone connection to the node is usually a local telephone call.

In both Western and Eastern Europe, in Japan, and in many other countries, packet switched networks have been built to handle data communications in the same way. With so many networks in place around the world, it became necessary to figure out a way for messages to cross from one network to another and still arrive at the other end of the trip in good condition, without any distortions or errors in the information. The X.25 standard was agreed upon for this purpose. Exhibit 2.6 depicts the flow of information through such a system. As a result, people frequently talk about "X.25 networks" when they really mean packet switched networks. Using X.25, for example, it is possible for persons connected to different networks around the world to

Exhibit 2.6

X.25 Standard for Packet Switched Networks

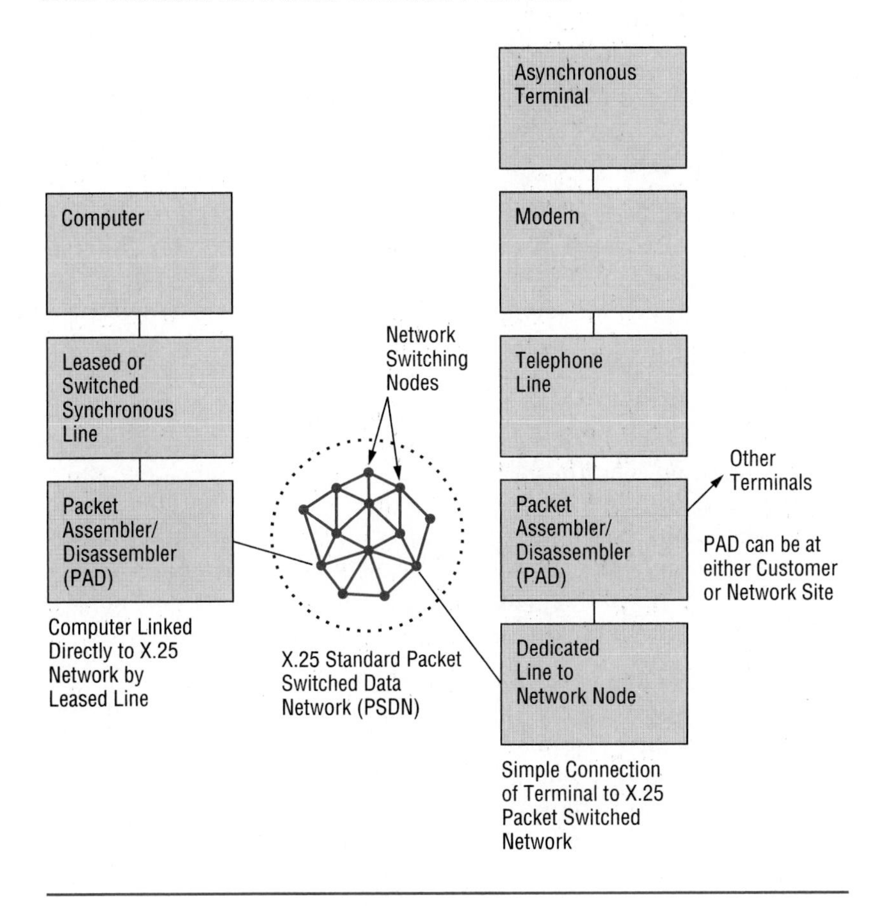

communicate with one another, or for computers to be connected together over great geographic distances. Packet network technology is truly a case of "beating swords into ploughshares."

In spite of all of the remarkable advantages of packet switched networks, they do have some disadvantages:

- *Speeds Are Limited*
  One disadvantage of public packet switched networks is the limited speed. They are optimum for medium speed applications from

1,200 bps to 9.6 thousand bits per second (kbps), and perhaps a little higher. However, in terms of the overall speeds of networks, these are in actuality in the low to medium range. For intensive applications involving applications such as linking together computers or sending very heavy traffic, the leased line alternatives are generally better since they support very high speeds with both lines and different types of multiplexers.

- *They Can Be Expensive*
  One great advantage to using public packet switched networks is that they are essentially a shared resource. Since many different organizations are able to use the resource, it can be relatively inexpensive to the individual user. This is because the excess capacity is paid for by other users who are on the network at the same time. The problem is that since many organizations use packet networks for a limited amount of time, the best way for the owner to charge the customer of the network is not by time, but by amount of information sent through the network. Many organizations experience growth in their use of packet networks beyond their initial expectations. When this occurs, they find that the fees for network usage balloon out of control. Packet networks are the most efficient option only within a certain range of usage. When that range is exceeded, other network alternatives are more attractive.
- *They May Deteriorate under Heavy Work Loads*
  Although packet networks are efficient in handling many different users and their streams of information simultaneously, this efficiency is good only up to a point. As more and more information packets start to flow through a packet switched network, the network performance starts to deteriorate. This is normal. However, the efficiency curve for packet networks takes a nosedive at a certain saturation level. The deterioration is not graceful. When a packet network reaches a saturation point, delays in transmission of information may become a serious problem.
- *They Do Not Generally Support Voice Information*
  The conversion of voice information to digital signals involves a processing of sampling. Generally, the higher the number of samples per second, the higher the quality of the reproduction of sound through the system. Unfortunately, this improvement requires more and more data to transmit the extra information. As more data are transmitted, it takes a longer time to pass through the network, unless the speed of transmission increases accordingly. Since many packet switched networks are limited in speed, the volume of data they can handle limits the quality sample rates.

Thus, the quality of voice reproduction at the other end is considerably poorer than regular telephone service. In addition, the nature of packet switched networks, the fact that many users are sending data simultaneously, means that the response time for the network can change unpredictably. Messages can occasionally be slowed by a second or more. Anyone who has ever spoken on an international call sent through a satellite with the incumbent delays in voice response, knows that significant delays in voice propagation are a strong irritant. In spite of their obvious advantages in many other areas, packet switched networks are not suitable for integration of voice information with data.

Virtual Software Defined Networks

As discussed earlier, leased line networks have some disadvantages in terms of flexibility. They are very good for handling intensive, stable applications, but if the organization is constantly undergoing change, they may become a burden because the administrative tasks of changing networks are severe. We have shown how packet networks answer some of the problems associated with the inflexibility of leased line point-to-point networks. However, it was also pointed out that packet networks have a few disadvantages, particularly when considering particularly large networks with high throughput and the need to integrate voice into the data traffic flowing through the network.

The **virtual software defined network** solves some of these problems. It acts like a leased line network, with all of the advantages that implies in terms of capacity, and yet has flexibility to change its configuration as needed by the user. The network can grow or shrink, speed up or slow down, or change its geographic coverage as commanded by the user. These changes are typically made by issuing change orders to the network control center of the service provider (i.e., the telephone company). After receiving the request for a change, the service provider initiates the change in the network. These changes are particularly easy when the network carries more digital information. Changes in such a newer network do not require as much physical modification as older networks.

In some variations of this theme, the telecommunications function within the corporation is able to initiate the changes in the network through network control software and programming available through terminals connected directly to the control center of the network provider. This type of service is not available in all areas or from all providers. A network can be modified in several ways:

- *Increase or Lower Capacity*
  Given the current utilization of a large private network, an organization may wish to either increase or lower capacity on a particular circuit. For example, if at the end of each month an organization expected that its data communications traffic would increase greatly between New York and Chicago, it would be able to increase the network to the needed capacity on demand. This is quite a different approach from purchasing high bandwidth and data flow capacity all of the time just so the organization would have the needed capacity at the end of the month. As a side note, this flexibility also makes the network as a whole more available to others, since organizations that adopted this approach would network resources only as they needed them.

- *Create a Temporary Instant Network*
  The ability to quickly set up and configure networks can be a major advantage for certain situations. If needed, a network can be dialed or programmed on the spot with the topology and characteristics needed by the user organization. Networks can be both established and also destroyed quickly, all through software. It is conceivable that organizations would be able to create their networks as needed either monthly or even weekly, maintaining no network in the intermediate period. This is a new concept, but there is little if any difficulty in adapting to this type of system. This might be particularly useful for applications such as inventory control or financial reporting that are not done on a minute-by-minute basis, but rather on a biweekly or monthly basis.

- *Remove or Add Nodes to the Network*
  We discussed the administrative difficulties involved in reconfiguring networks. In the past, the work associated with this operation was time consuming, bothersome, and error prone. With software defined networks, it is possible to accomplish this type of change quickly, in a programmed fashion, and without undue administrative difficulty. For companies with a heavy administrative burden of network topology changes, this feature of the software defined network is very helpful.

- *Perform Routine Network Control Functions*
  Another advantage of the software defined network is the enhanced ability to perform network management functions. (See Chapter 6, Network Control and Management.) The user organization is

given a very good view of the network as provided by the network provider. This capability helps monitor the structure of the network, including identification of trouble spots.

The easiest way to think of a software defined network is that it is a "network on demand." The user organization is able to dramatically reduce its interaction with the service provider, including automation and direct computer linkage with many of the critical functions. This means that, whereas in the past the network was something which was more or less carved in stone after it was set up, it has now become a resource which is available on demand and according to the specifications and characteristics needed by the user organization.

There are some parallels between software defined networks and the newer services such as switched 56 kbps lines. In the past, an organization wishing to get a 56 kbps line would have to make arrangements similar to, and as complex as, getting a special dedicated line installed. This was expensive, time consuming, and would end up locking the organization into a fixed, rigid network. In the late 1980s, service providers such as AT&T introduced switched 56 kbps service. Under this service, companies would have a 56 kpbs line hooked directly into their offices. They were then able to dial calls on the 56 kbps line to other points on the 56 kpbs circuits and were charged for the use of the line at a cost similar to regular telephone service, but more expensive. This type of circuit gives a company great flexibility. It allows dialing many different locations at different times, rather than locking the organization into the old idea of a single circuit. Companies such as Revlon found that this type of service was very useful for hooking teleconferencing sessions together on an ad hoc basis.

These variations in networking and other changes have emphasized the increasing flexibility of networks. Rather than having to choose between leased line fixed point-to-point networks and packet switching, other more flexible alternatives are emerging.

In the future, a company's choice will not be a question so much of the availability of different types of networks that will fit its needs. Rather, the more important questions will center around how those networks will be used to promote the business. This is a more difficult problem.

Image-Oriented Networks

Common belief holds that there are only two types of networks: voice networks that carry sound, and data networks that carry data. To the extent, therefore, that networks carry images translated into digital data, they are really no different from data networks. The applications being handled by **image-oriented networks** are, however, very different from either data-oriented or voice-oriented networks. In addition, the speeds and bandwidths required for image networks are generally greater than those for data networks. This is due to the nature of images and the need for data to store them. For example, a page such as this, stored in the data format of the word processor through which it is being written, requires only a small amount of storage space. The file has two components (called "forks"); one contains the letters of the text, the other contains the information required to format the page when it is printed out on a laser printer. The format file, which is generally invisible to the user who is not a programmer, sends the correct formatting instructions to the printer, which then combines these instructions with the text entered by the writer. For this page to be viewed by the reader, it must be "processed" by the microprocessor in the printer. Although the file is relatively small for a page, it is really in an intermediate form that is inaccessible to the reader.

Image networks are based on replacement of this intermediate form with the transmission and processing of whole images. If this page were stored in the data format, it would be no more than a kilobyte of information. However, if it were scanned into an image file, the resulting file would be considerably larger, probably several *hundred* kilobytes at least, and possibly more.

One of the greatest changes to have taken place in information and telecommunications technologies during the late 1980s was the emergence of technologies that merge image information into standard database information. Instead of handling paper documents, companies such as Citibank Mortgage started scanning documents as they arrived at the office, then passing the documents through the office electronically. This type of system was pioneered by companies such as FileNet, but was adopted as well by the giants, such as IBM, DEC, and Wang. In addition University Microfilms offered full scanned images of articles from key journals on their CD ROM system.

The emergence of these image technologies has greatly increased the speed and volume requirements of telecommunications systems. Scanned images occupy much more electronic storage capacity than compact data files. Depending on the resolution of the scanned document, up to a megabyte or more may be required to store the image, assuming no encoding is used. In order to get the rapid response rate required for working with images, the image should appear within less than a second. The Ethernet data transfer rate of 10 megabits per second is still somewhat slow for a local area network employing image technology.

The most common forms of image networks are teleconferencing networks, facsimile systems, office document image processing networks, CAD/CAM networks, and other specialized workstation networks. Each of these types of networks has specialized applications and is relatively rare, although very important.

Teleconferencing Networks • Teleconferencing networks allow the transmission of video images with accompanying sound. Persons in one geographic location can have full business meetings with persons in other geographic locations. The technology resembles that for the nightly newscast's remote interviews. Teleconferencing networks are usually dedicated and specialized networks, and their economics derive from their replacement of executive travel time.

Office Document Image Processing Networks • Image networks for office documents rely on taking photograph-like images of each incoming document and storing these images in the computer system. Instead of manually searching for the document among large stacks of paper files, information workers can access an image of the original document through the telecommunications system.

CAD/CAM Networks • Computer-aided design and computer-aided manufacturing networks are used in creating and distributing detailed design specifications for machining and stamping parts. For example, a few automobile companies have been moving toward global manufacturing and sourcing of components. Under these circumstances, design information must be distributed worldwide.

Specialized Workstation Networks (Publishing, Oil, Exploration) • Many businesses are also building specialized networks that link together high-powered workstations (such as Intel-80486-based machines) into specialized applications. These networks usually do not rely on central mainframes or minicomputers for processing. Examples of this type of network include distributed publishing applications and seismic analysis in oil and natural gas exploration.

Teleconferencing Networks

Teleconferencing takes place when groups at different locations communicate with each other using audio and video channels simultaneously. It is something which you have become accustomed to on the nightly news shows when persons in different locations around the United States and the world carry on conversations. What makes the use of teleconferencing different in business is that it is private, and that businesses rarely find it cost effective to employ the expensive technologies used by the major television networks.

In order to transmit information on television from city to city, the professional networks use very large circuits, with bandwidths (capacities) up to 6 million bits per second. Such circuits are very expensive. Private teleconferencing in a point-to-point fashion (rather than via satellite, which is point-to-multipoint communication) is done through the use of dedicated or dial up high-capacity circuits. The quality of the picture is not as high as full motion television because smaller bandwidths are used. Over time, different compression methods were developed to take the standard video image and transform it into the very minimum of data required to transmit the image.

For example, one of the earliest methods of compression scanned the picture only every second or so, instead of 60 times per second as is found in standard television. The result was a jumpy image such as you might see in early films of NASA space missions, like the mission to the moon. As the astronauts moved, the image would move in visible increments. This type of compression was called, appropriately, "slow scan."

Another refinement in compression came with the technique of transmitting only those portions of the image that change, instead of the entire image. Thus, for a teleconferencing image of a person sitting in front of the camera who simply opened his mouth, only the information from the image immediately around the mouth area and the mouth itself would be retransmitted. The rest of the information in the image, particularly information about stationary objects such as the background of the room or the table top and the things on it, would not be retransmitted. At the other end of the teleconferencing channel, the monitor would simply keep displaying all the received information until it received a message to change it. Since only a small amount of information was being transmitted each time, and since the scanning was slow, this also improved the compression.

As compression improved, the amount of bandwidth required to transmit a teleconference continued to decline, thus reducing costs as well. The key to this technology was the CODEC, which is a multiplexer that takes the video and audio images, samples and converts them into digital format, then transmits them to another CODEC at the other end of the line. Such a system works like that in Exhibit 2.7.

One of the startling developments in teleconferencing technology is that costs have declined rapidly for full motion, color videoteleconferencing. In the last five years of the 1980s, the costs declined in two significant dimensions. First, the cost of the teleconferencing equipment and room set up has declined by more than an order of magnitude. Second, the telecommunications charges for the required dedicated circuit have declined also by an order of magnitude.

In the early, experimental days of teleconferencing, the general plan adopted by most pioneers, such as Mobil Oil Corporation, was to build large dedicated rooms and facilities to handle the teleconference. These rooms typically had special insulation, several cameras, write boards, and many other features. Unfortunately, their cost could be as much as $1 million for each installation.

This was dramatically changed by the introduction of more sophisticated and miniaturized equipment. Equipment made by corporations such as PictureTel of Peabody, Massachusetts, eliminated the need to build expensive dedicated teleconferencing facilities. The new generation of teleconferencing equipment is set up on a mobile dolly which can be wheeled from office to office as needed. The bulk of the dolly is taken

Exhibit 2.7

Teleconferencing Network

The CODEC takes audio, video, and data signals and combines them into one stream of data for transmission.

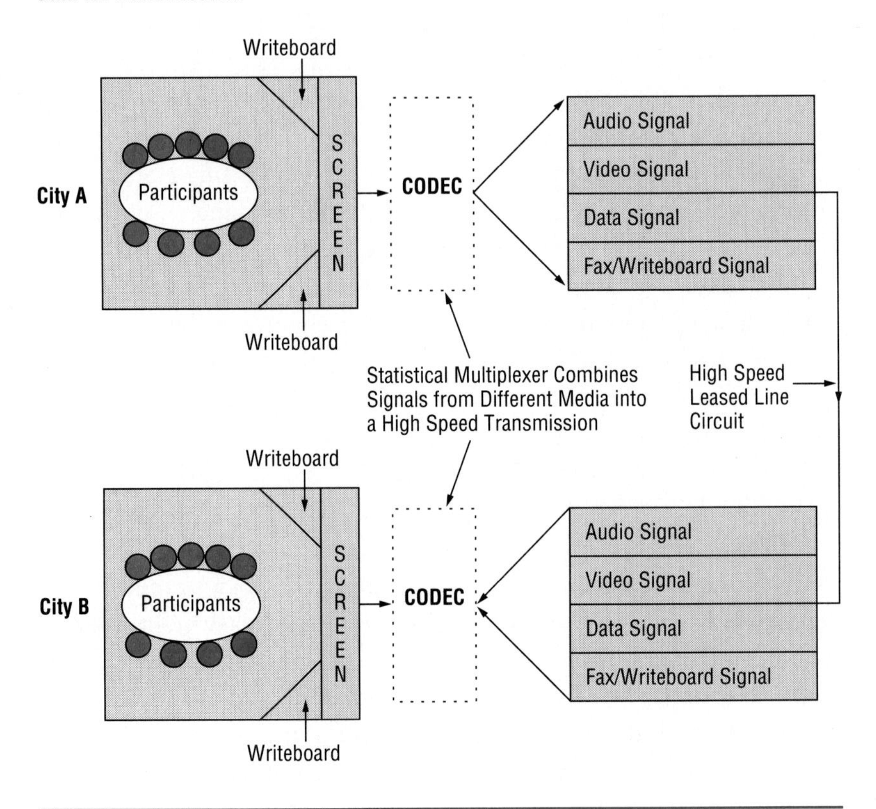

up by a large standard video monitor and the CODEC, made by PictureTel. The only technical requirement for this arrangement is a 56 kilobit circuit jack for the unit at each location where it is used.

Another important factor in the teleconferencing field is the change in bandwidth required for reasonable quality full motion video. Generally, as schemes for compressing the video and audio signals improved, the amount of required bandwidth decreased accordingly, as shown in Exhibit 2.8. Although in the late 1970s virtually no compression was possible, technology improved continually.

Exhibit 2.8

Teleconferencing Bandwidth Requirements

Full motion teleconferencing has come to require smaller and smaller bandwidths.

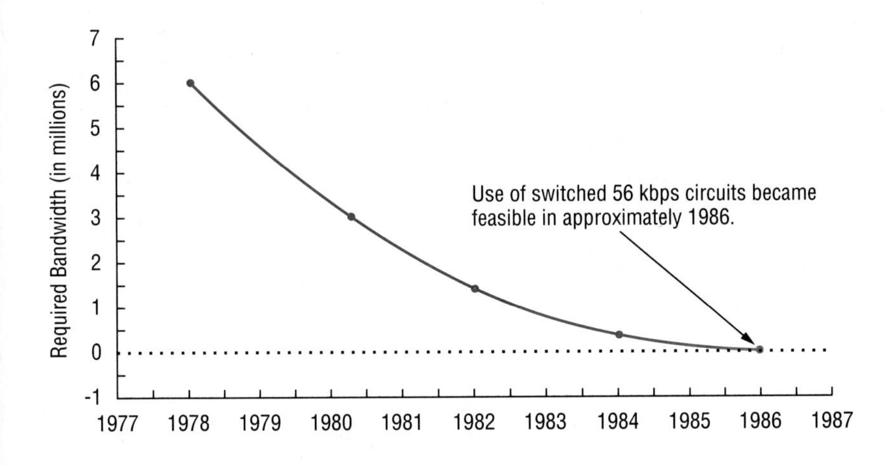

This has resulted in a great decrease in telecommunications costs, as Exhibit 2.9 shows. For example, a switched 56 kbps service provided by such companies as AT&T with its ACCUNET service costs as little as 65 cents for the first minute and 60 cents for each additional minute. For calls over more than 1,000 miles, the rates are 88 cents for the first minute and 80 cents for each additional minute. This means that the most expensive service costs $48 per hour and the least expensive costs $36 per hour. In other words, you could have a transcontinental full motion video teleconference for half an hour for about $20. That's cheap.

Besides being inexpensive, using switched 56 kilobit telecommunications channels for carrying teleconferencing signals is easy. The dialer simply dials the telephone number of the receiving teleconferencing station just like dialing a regular telephone call. The two sets establish communication and the teleconference begins.

This is very different from the past, when greater bandwidths were needed for transmission. Much greater bandwidths required a timed reservation for availability of the high capacity circuit which was going

Exhibit 2.9

Teleconferencing Costs

Video teleconferencing costs have decreased more than an order of magnitude in five years.

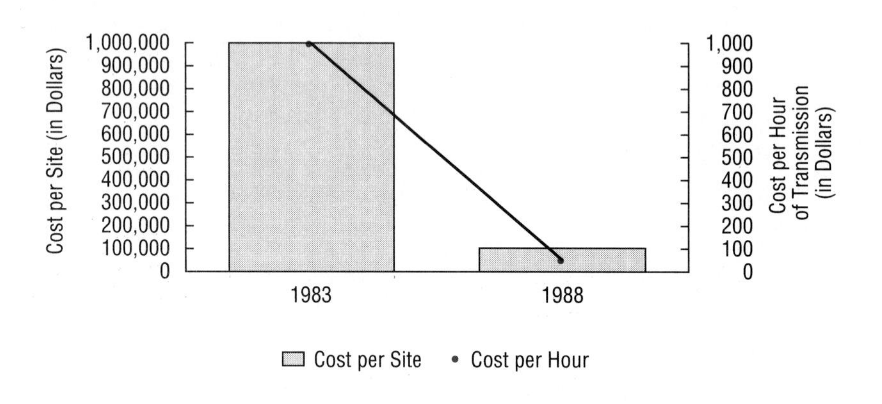

to be used. Some organizations had their own dedicated networks for this purpose. In any case, the cost was substantial. In a reservation system, teleconferences had to be scheduled ahead of time, and thus in many situations spontaneity and creativity suffered. In a dedicated network situation, the teleconferencing system could exist only as an adjunct to other data communications applications, not by itself.

Teleconferencing is a relatively new and emerging technology in terms of its installed base and the fact that it is still not widely available. People are still not accustomed, in general, to working in a teleconferencing mode, but one can expect this to change in the future as compression technology gets more sophisticated and the basic electronics become very much cheaper. Exhibit 2.10 details likely payoffs that will help induce more firms to make the investment.

Another factor which must be noted is that teleconferencing will be very much improved by a new technology called **ISDN (Integrated Services Digital Network)** which will allow each telephone line to have essentially two channels and a control channel without having to add more wires. See further discussion of ISDNs later in this chapter. This

Exhibit 2.10

Teleconferencing Payoffs—Cost Estimates for an Average Business Trip

Total Time Lost Each Way: 7.75 Hours
Total Time Lost Round Trip: 15.50 Hours
Equivalent to two business working days.

Activity	Hours	Percentage Lost	Lost Hours
Productivity Costs			
Scheduling/Coordinating	0.50	100	0.50
Personal Preparation	1.00	100	1.00
Trip-Related Preparation	1.00	100	1.00
Early Departure	1.00	100	1.00
Travel to Airport	0.75	100	0.75
Preflight Time	0.50	100	0.50
Average Flight Time	3.00	50	1.50
Postflight Time	0.50	100	0.50
Surface Travel	0.50	100	0.50
Check In/Out, Miscellaneous	0.50	100	0.50
Travel Costs			
Tickets (round trip)	$500		
Ground Transportation	100		
Hotel (two nights)	150		
Meals (two days)	100		
Travel Department Charges	50		
Miscellaneous	50		
Total Expenses	$950		

will mean that teleconferencing information might possibly be sent together with an audio channel.

Facsimile Systems

Facsimile systems have been rapidly replacing courier services to send instant images over long distances. Exhibits 2.11 and 2.12 show the growth in the number of terminals. The simplicity of facsimile technology is that it most commonly uses standard telephone lines. Like many key technologies today, facsimile systems were originally

Exhibit 2.11

U.S. and Canadian Facsimile Terminals in Use by Groups

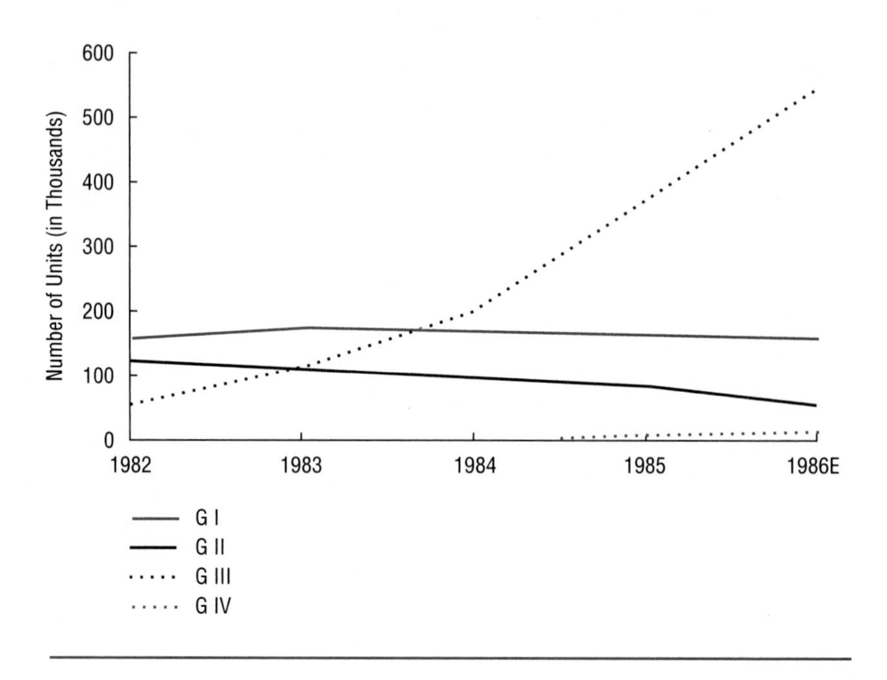

invented in the United States, but the center of innovation in the world of facsimile systems has shifted to Japan and East Asia. In the 1990s, a new generation of digital facsimile machines will become more popular. These digital machines will rely on packet switched networks or ISDN channels to transmit the information that describes the image. As this development proceeds, the network complexity of facsimile systems will increase, and many new features will be added.

Facsimile systems work by the telecommunication of scanned image information. This information is enough to create a copy of the image at 200 dots per inch at the remote location. **Scanning** is the process whereby an image is illuminated for the purpose of defining whether it is dark or light. Characters printed on a plain piece of paper are dark, and the surrounding untouched paper is considered to be light.

Exhibit 2.12

Forecast of U.S. Market for Facsimile, 1987–1991

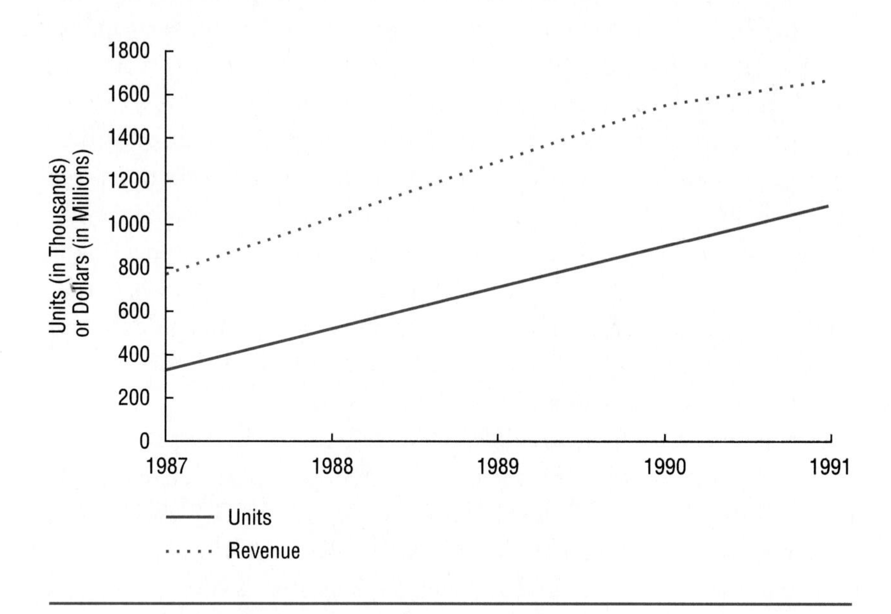

The information that describes the lightness and darkness is then modulated into sounds through a modem and sent through the voice telephone circuit to a receiving facsimile machine at the other end of the line. This second machine takes the information and prints the corresponding pattern on paper. Most facsimile systems use thermal paper since it is very inexpensive compared to other means of printing. More expensive facsimile machines print images through a laser driven system. These images are far more stable in terms of shelf life than are images printed on thermal paper.

In its most simple sense, facsimile transmission is no different from data transmission using modems. Just as the data of a regular computer transmission are decomposed into a series of 0s and 1s before being sent, the digital images of the light and dark areas of an original page are decomposed into audio signals for light or dark shades.

The facsimile machine has penetrated the market extremely rapidly. It is a highly useful tool in all areas of business, and this has helped power its growth. In addition, most facsimile machines are made by the major Japanese companies and their fierce competition with one another has driven the costs rapidly downward. Facsimile machines that cost more than $3,000 in the late 1980s now cost between $300 and $700 in the stores of discount dealers in Manhattan.

As facsimile machines have grown in complexity and sophistication, new features have been added, including broadcasting, delayed timed dialing, and storage. The broadcasting feature allows the facsimile machine to send the same message to many different machines. It will attempt to contact other machines one at a time and transmit its message. Delayed timed dialing and redialing allow a machine to keep trying until a circuit becomes available or wait until off-hours early in the morning to get through.

Standard facsimile machines are called Group III machines from the name of their class of transmission standards set and agreed upon by the International Telecommunications Union CCITT groups. For the most part, these machines rely on the regular telephone lines for transmission of their materials. When a connection is being made, the chances for success are the same as getting a telephone call through. Generally, this is fine for transmissions between developed countries. However, in less developed countries and many other countries such as the newly liberated nations of Eastern Europe, the telecommunications circuits are very poor. As a result, one frequently finds it difficult to get through with a telephone call. For example, although it is now possible to direct dial Moscow, it is usually very difficult to find an available circuit for transmission of facsimile data. Under these circumstances, facsimile store-and-forward systems are very useful.

Facsimile Store-and-Forward Systems • When one encounters repeated difficulty connecting a facsimile machine to its destination because of unavailability of circuits or because the target machine is constantly busy, it is possible to use **facsimile store-and-forward systems.** In using a store-and-forward system, the sender transmits the facsimile message to an intermediate destination. The message is then stored temporarily at the intermediate destination. The store-and-forward system then automatically retries to send the message over and over until an open circuit is found. It then transmits the message, and returns a confirmation to the original sender that the facsimile has been sent, at what time, etc.

Any person who is familiar with the world's **telex** system knows that it offers a similar service. A telex machine relies on making a direct connection with the machine being contacted. That connection must be held during the entire time of the message transmission. Telex is particularly popular in developing countries. In fact, telex is probably the world's greatest and most pervasive data telecommunications system; it is found in many of the most remote areas of the world. Telex contact with the most remote corners of the globe is particularly vulnerable to difficulty in transmission because of unavailability of circuits. For example, getting telex transmissions into communist China is very difficult because of the poor telecommunications infrastructure there and also because it is one of the most common means of carrying on the telecommunications necessary for international commerce. The circuits are frequently jammed with traffic.

The world's telex system operators (the PTTs in most countries of the world) long ago recognized the problems and built store-and-forward systems into their networks. If you have difficulty getting through to a remote telex, you can send the telex to an intermediate "mailbox" where it will be held until the master message switching computer is able to establish a connection and get the message through. The message computers usually wait until the early morning hours to try to send the messages. After the message is sent through, an acknowledgement is forwarded back to the originating telex machine. Sometimes the delay is quite substantial; other times the process is relatively fast. After a certain number of unsuccessful tries, the message will be aborted.

Facsimile systems use the same type of arrangement. The difference is that the store-and-forward computers store the image of the paper being transmitted, instead of the simple letter-by-letter code of the international telex system. The unsuccessful user gives up trying to transmit the facsimile and instead sends it to the store-and-forward number. There it is stored until a line becomes available. Considering the very rapid growth in the number of facsimile machines around the world, facsimile store-and-forward systems are sure to grow in importance. It is even conceivable that eventually organizations will keep "electronic mailboxes" for receiving all of their facsimile messages.

Facsimile Interface Data Conversion • Up until the early 1990s, facsimile systems remained separate from the mainstream data processing centers of the firm. Facsimile for the most part was run through the telephone system, and no computer technology was involved in processing or distributing the messages. In many applications, critical information was sent via facsimile to a recipient point where it was manually entered into an information system. If the frequency of movement of information was low, then it was not economically viable to create a special telecommunications network to support the limited traffic. Under these circumstances, transmission of the needed information via facsimile was economical. As a result, the general pattern would be that a collection point in the information flow of the organization would be set up. This collection point would be the facsimile receiving station. From many different locations, portions of the organization would send their information via facsimile and this information would then be collated and fed manually into the computer information system.

A **facsimile interface device** is a piece of equipment which accomplishes this process without human intervention. When the facsimile arrives at the collection point, the image from the page being transmitted is interpreted by the interface device. This interpretation consists of taking the image of the paper being sent and translating this image information into hard data. These data are then fed directly into the computer information system.

The efficiency of the facsimile conversion interface unit is that it eliminates the step in the process taken up by a manual operator. This results in a substantial speeding up of information processing in the organization as a whole.

Since the interface unit is designed to handle a very specific transaction, the forms it receives via facsimile must be standardized in nature. If they are not standardized, then it is impossible for the interface computer system to properly interpret and translate the information into the specific data requirements for the computer system. This is because the data entry forms opened up from the computer system are not only highly specific, but they require that the information be in a very specific format and in a predetermined order. If any of these conditions are not met, then the information entry into the computer information system will be aborted.

On the other hand, the chief advantage of the facsimile interface unit is its ability to automate a manual process. In addition, this automation makes it possible for the system to handle much greater flows of information at the same cost than would ever be the case in a manual system. It must be reprogrammed when a new form comes into use, and this adds an element of inflexibility to the system. It is therefore best used when the application is not expected to change much. These systems point to the increasing merger between the world of facsimile and the world of information processing.

E Mail Interfaces (MCI) • Another merger between data processing telecommunications networks and facsimile is found in the ability of some public electronic networks to take information from **electronic mail systems** and transmit it directly to a facsimile machine. MCI Mail, for example, offers this type of capability. A sender who addresses a message through the electronic mail system can specify the type of message. It can be a regular electronic mail message, a telex message, a cable, or a facsimile. If it is a facsimile, then the system asks for the addressee and the telephone number of the facsimile machine which is the target destination. After the message is sent into the electronic mail system, that system acts much like a facsimile store-and-forward system. It dials up the target destination facsimile machine then transmits the message. After the message has been received, an acknowledgement is then sent back to the sender.

There are a few drawbacks to this system. First of all, it is not possible to handle much graphic information beyond the simple typefaces available through the electronic mail system. No pictures, no blue prints, no CAD/CAM information can be sent. If, however, the information being sent through the facsimile system is only textual in nature, then the convenience of the electronic mail interface is very great. It can create distribution lists and have the electronic mail system send the same message to more than a single organization. In addition, it can keep a record on store for future transmission or modification.

Finally, since the message is stored in the electronic mail system, it can be sent through other media as well, for example through telex, regular electronic mail, or through a letter format in which it is printed out at a distant location then delivered through the regular mail service.

Personal Computer Facsimile Systems • Another merger between the data processing world and the facsimile world is taking place in the personal computer. In particular, expansion options for personal computers are allowing the personal computer to emulate a facsimile. This makes it possible for the personal computer user to transmit a facsimile directly from the personal computer to a distant facsimile machine. The Apple Macintosh series, for example, cannot only perform the required emulation, but it is able to capture the different font designs and graphics that can be prepared with the Macintosh.

Under the old way of operation, the user would first create a document, then print it out, then spend the time sending the information through a regular facsimile machine. With the direct facsimile system coming out of the personal computer, these other steps in the process are eliminated.

The Promise of Digital Facsimile • Up until approximately 1990, there were three families or generations of facsimile equipment. They are called Group I, Group II and Group III. Group III, the most modern and most popular form of facsimile, is the only type of facsimile standard which coincided with the move by the giant Japanese manufacturing firms such as Matsushita to enter the market. This brought a precipitous decline in price, and a correspondingly rapid increase in market penetration throughout the general population.

From the consumer's point of view, the major differences between Group I, Group II, and Group III facsimiles lies in the amount of time it takes to transmit a standard size letter. Group III is the fastest, and Group I, the first and oldest generation of the technology, is the slowest. Roughly speaking, Group I took about 6 minutes per page and most commonly used a rotating drum to which a single piece of paper at a time was attached. This rotating drum would send the information on the page past a light sensitive scanner which would slowly traverse the horizontal dimension of the rotating drum, much like a needle on predigital vinyl records which began to disappear in the late 1980s and early 1990s.

The newer Group II facsimile standard was faster, taking approximately 2 minutes and using stationary scanners in some models. In addition, a type of *compression algorithm* was used to require less and less information to transmit the same page image through the network. Group III facsimiles offer more sophisticated features such as polling, automatic pagination, and better compression algorithms. The Group III facsimile machine is able to transmit a page in less than a minute, has multiple page feeding, and other features not found on the earlier models.

Since many Group III facsimile systems operate at 9,600 bits per second over the standard twisted-pair analogue telephone line and already use a sophisticated compression algorithm, they are more or less at the technological maximum. It is difficult to standardize on a more efficient compression algorithm that can be easily handled by the inexpensive microprocessors which are found in the control unit of the facsimile machine. In addition, it is difficult to push the speed of the telephone line much higher because of the necessary errors which would occur, requiring retransmission of individual pieces of information. In many parts of the world, the regular telephone system through which facsimile messages are sent will not support 9,600 bits per second speeds. Under these circumstances, many of the facsimile machines automatically adjust their speeds downward until the optimum speed is found.

The next generation of facsimile, called Group IV, is no longer based on analogue telephone circuits. These machines rely on digital transmission of information. At the beginning of the 1990s, the growth of Group IV facsimile was almost nonexistent. Consumers, including business organizations, were busy installing Group III facsimile machines throughout their organizations. On their technological growth curves, Group IV facsimile systems were at the beginning, whereas Group III machines were still in the process of reaching maturity as a market.

Group IV facsimile works through digital networks. When the machine is connected to a digital network, its speed is much higher, and therefore the amount of information it can transmit is much greater in a shorter period of time. For example, compare a Group III facsimile system transmitting information at 9,600 bits per second to a Group IV facsimile machine which is capable of transmitting at the standard Ethernet speed of 10 mbps—10 million bits per second. This is approximately 1,000 times faster!

The much higher speeds possible with digital Group IV facsimile systems will change entirely the economics of transferring images through networks. Whereas in the past, the great amount of data required for transmission of images combined with the slow data communication speeds tended to limit the application of facsimile technologies; with the higher speeds promised by transition to digital networks, these barriers no longer exist.

This raises the concept of the complete merger between facsimile and image-oriented networks used in the office. These networks are built around the rapid creation, processing, and movement of images through networks.

Office Document Image-Oriented Networks

As office work has grown in complexity and volume of information processed, particularly since the 1950s, the number of forms that must be processed has increased a great deal. In spite of the massive improvements in data processing and the large economies of both scale and business scope made possible by these technologies, many facts about office life have not changed much: the file cabinets and mountains of forms, correspondence, and paperwork have not gone away—in fact they have increased greatly. The result is a dichotomy in the efficiency of the modern office: on the one hand, data processing, letter and form creation, and functions such as mailing bills are highly efficient; on the other hand, the way in which many offices handle filing of incoming information has changed little since the 1940s. Actually, a historian of bureaucracy would observe that drawers with files of papers were around during the Middle Ages. The last big revolution in office storage of information was the change from rolls of paper or scrolls to the storage of flat paper. Offices have developed this split personality: creation of information and paperwork is highly mechanized and efficient, but storage and retrieval of paperwork is primitive and little changed from the Middle Ages.

The application of information technology, particularly image-oriented networks and systems, to the problem of paperwork in the office has resulted in creation of a new generation of systems. In the **image-oriented system** the flow of paper through the office is entirely different.

When papers arrive at the office, they are scanned. The type of technology used for the scanning process is very similar to Group IV digital facsimile. Just like facsimile, the digital scanning technology uses a compression algorithm in order to reduce the amount of storage space needed for the image information. Reduction of the amount of data required to reproduce the original image also greatly facilitates the telecommunication of the information throughout the information system network. What is interesting about the implementation of image-oriented networks is that they store not only the information in the incoming paper, but also the image of the information. This means that signatures can be verified, and pictures or diagrams can be included easily in the files without the special software that would normally be associated with such a compound document.

After the digital image information is recorded in the information system, often stored on large capacity optical disks, it can be recalled as needed. Typically high resolution workstations are connected to this type of system as terminals. The high resolution of the terminal display is needed in order to reproduce the actual picture or image of the original document, which has been filed to be recalled only in case of emergency or legal dispute. For general office functioning, the image of the document rather than the original document is used.

The telecommunication of the image of the original document from the optical storage bank to the high resolution workstation takes place over a high-speed local area network, generally operating in the Ethernet range of 10 mbps. It is interesting to note that the transmission of the image is done in much the same way as digital facsimile. You can think of the optical storage bank as little more than a giant store-and-forward system, only with the ability to act as a library for users requesting to see the information as needed.

Using this networked workstation approach to handling the volumes of paperwork in the office, it is possible to forward letters and documents in image form from one person to another within the same organization. This step also eliminates the bane of the interoffice mail system, the loss of large numbers of documents in the process of being moved from one office or department to another within the same organization. "I'm sorry sir, but we are unable to locate your record—can you call back in a few days." This is the typical result for the customer of this type of primitive system. On the other hand, using the image-oriented network approach, the representative who answers the telephone is able instantly to call up the records of the customer, and to reproduce digitally the details of any letter which has been written to customer service from the customer.

Using gateway technologies and a technology called "bridges," it is also possible to link together office local area networks from one location to another. Under these circumstances, the accessibility of the network and all of its associated information, through the optical disk tank, of employees located in different geographic areas gives the corporation a way to extend a paper processing bureaucracy far beyond its original scope, which had been confined to a single geographic location. For example, office workers with image network stations located in one city can access more information about their customers because they will have telecommunications linkages to the central image-base of the corporation.

CAD/CAM Networks

The 1980s saw the gradual introduction and proliferation of **computer-aided design (CAD)** and **computer-aided manufacturing (CAM)** technologies throughout most of the major manufacturing corporations in the world. CAD/CAM technologies have greatly increased the efficiency in manufacturing design and prototyping because they allow drafts of components and engineering designs to be done electronically. Designs can be printed out on high-quality plotters and studied. If revisions are made, they are entered in the electronic file, the result being that the revision cycle as a whole is radically reduced since a minimal amount of redrafting is necessary.

A revolution in manufacturing design came in the 1980s when it became possible to link the design data from CAD equipment to the programs that drive computer numerically controlled (CNC) equipment such as metal lathes, and in particular machining equipment. The effect of this was that the experimental designs in the memory of a computer terminal or workstation could be outputted directly as metal component prototypes. If this process is contrasted to design in the 1950s, when each redrafting of a design was done by hand, and when each prototype was first hand-carved to specification or hand lathed piece by piece, the remarkable change is apparent.

In terms of telecommunications, the first groups of networks linked the large design terminals, such as those made by Evans & Sutherland, into mainframes, which at the time were the only machines of sufficient size to perform the intensive calculations necessary to reproduce three-dimensional modeling in systems. These networks were essentially just terminals linked to mainframes. As the technology changed rapidly in the 1980s, the processing power of personal computers, and then workstations, increased greatly. By the end of the 1980s, it became a relatively simple matter to have a workstation with a processing power of 2 million instructions per second (abbreviated MIPS) located on the desk of an individual design engineer.

With such powerful workstations, much of the computational effort was transferred away from the mainframe and toward the workstation. However, very large projects, such as the design of a jet engine, an aircraft, a car, a turbine generator, etc., still required the use of the computational power of mainframes and supercomputers.

Nevertheless, the networks supporting high-powered workstations continued to proliferate. Using **telecommunications protocols** such as TCP-IP, it became possible to move large files quickly and efficiently from workstation to workstation with relative ease. (Telecommunications protocols are agreements on the form of data transmission that keep the information from being distorted as it moves quickly through the world's telecommunications infrastructure.) The gradual pattern was a trend toward very powerful workstations doing the bulk of the "number crunching" (a popular slang term for intensive computation typical of three-dimensional modeling). This combined with quick telecommunications transmission of completed files back and forth between different workstations, each with a different information worker.

This development in the technologies underlying computer-aided design coincided with the shift towards greater internationalization on the part of many corporations. For example, Ford Motor Company introduced the World Car, a term that described its strategy of striving for similarity of parts for its car throughout the world, and a coincident strategy of manufacturing each part in the cheapest location in the world. The problem with this strategy was the need to keep track of these developments at a central location, where basic design and archiving of manufacturing design data took place.

This type of strategy, which was not unique to Ford, generated a requirement for a type of global coordination that had never existed before in world economic history. How could it? Until the last two decades of the 20th century, no communication technology would allow precise transmission of detailed design data across such vast distances in a timely, efficient manner. To move a single workstation-generated three-dimensional modeling file of a complex object around the world in the 1940s or even the 1950s, would have taken days if not months to transmit, and one-third as much time again to reassemble into a usable format. Any time before the first half of the 20th century it would have taken half a year or more to transmit the file. It is an interesting tautological question whether the move toward global manufacturing drove the developments in international networking of high-powered computer-aided design stations or the other way around.

Regardless of the causal relationships, multinational corporations such as Ford and many of its suppliers have set up global CAD networks to link together their engineering design centers located in different parts of the world: North America, South America, East Asia, Australia, and Western Europe. These systems allow the global coordination of engineering design specifications. For example, once a specific design is made of a component for a car being produced in several locations in the world, such as the Ford Taurus or Escort, these design specifications can be immediately telecommunicated to each of the different design centers around the world within seconds. At each of the design centers, these designs can be integrated into the manufacturing process. It is clear that in the handling of manufacturing changes, which are very common, particularly to rectify defects, this ability to move information around the world is very critical. In addition, the ability to quickly move design information around makes it possible to simultaneously carry out design activities in different parts of the world. It is no longer necessary to have all design taking place in a single location.

Telecommunications has effectively made the entire world a single location. This is a new age, and global linkages of telecommunications are only the beginning.

It is interesting to note that many of the global engineering networks being built are more or less separated from major corporate networks. In order to build the networks with the high capacity for moving CAD data around the world, firms have set up dedicated networks. This will change in the future as general backbone networks gain capabilities, and as technologies of network management allow further integration of heterogeneous network functions.

Other Specialized Workstation Networks

The change in the nature of CAD/CAM applications made possible by linking together workstations around the world is only the first in a series of applications that will be built upon this type of telecommunications infrastructure. In the oil exploration business, high-powered workstations are being linked together from different parts of the world in order to handle the complex computations required for analysis of seismic data. The companies that handle this information efficiently can gradually gain some advantage in the market, although many other variables obviously affect the ultimate success, not the least of which are basic selection of the correct tracts for exploration and even luck. In addition, the publishing industry is finding that publishing layout and typesetting stations can also be linked together over great distances to produce publications.

The long-term effect of these developments in linking together of high-powered workstations is the construction of new networks that bypass the traditional information processing infrastructure of the firm. Instead of being tied together at the center with a large mainframe installation, the emerging networks will not have logical or architectural centers. Instead, they will be true *networks* in which each node is a workstation.

Combined Media Networks

The worlds of computer, telecommunications, and video technologies are merging, but it is still difficult to see exactly what types of options will be available to the user and systems designer. By the late 1980s,

Exhibit 2.13

Integrating Teleconferencing Technologies

Image technologies will integrate full motion video into the workstation display.

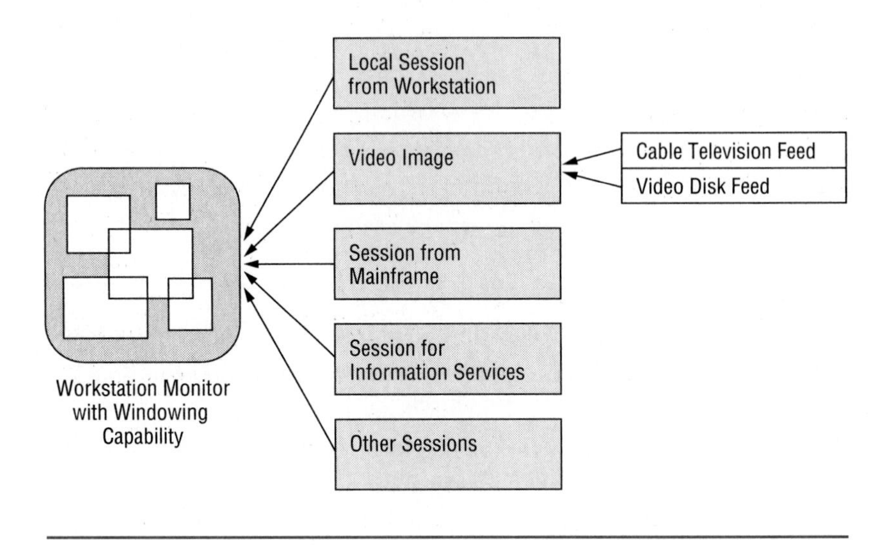

designers had managed to have a full motion video channel feed piped through a digital image processing card directly onto the screen of a personal computer or workstation. The video image was scalable into a window that would be one among many on the screen. The inputs to the workstation monitor could come from several different types of telecommunications channels at once, as illustrated in Exhibit 2.13.

The applications being developed for this type of technology include:

- In the finance sector, political developments or statements available on television appear to influence prices with which traders actively involved with the market must cope. Under those circumstances, having a video feed directly available in the workstation would help the trader make up-to-the-second trading decisions.

- In video training applications, students will be able to view moving images on their workstations as they are doing research. Not only will they receive static information, such as is available through an

Exhibit 2.15

ISDN Developments

1926	Japan introduced step-by-step automatic switching systems.
1950s	Pulse Amplitude Modulation (PAM) switching systems Pulse Code Modulation (PCM) switching system trend
1959	H. E. Vaughan originated the concept of integrated switching and transmission technology
1968	PCM Working Party of Study Group XV of CCITT—collaboration of personnel representing transmission and switching signified the start of integration in the CCITT
1969	Study Group Special D; Study Group XVIII
1971	Working Party 2 of Study Group XI established to discuss Integrated Digital Network and Integrated Services Network
1972	Beginning of studies of digital networks
1981	CCITT issued its first statement of direction concerning ISDN
1984	CCITT published "Red Books" including approved ISDN Recommendations called the I-Series McDonald's request for proposals (RFP) for new telecommunications system
1985	McDonald's selects Illinois Bell's ISDN
1986	McDonald's ISDN begins operation
1989	Publication of CCITT's "Blue Books" containing updated ISDN standards
1990s	High-capacity broadband ISDN becomes available commercially in certain countries

University (the training campus), and the new Home Office opened in 1988.

The corporate Home Office in Oak Brook had over 20 separate voice and data systems in three layers of networks, illustrated in Exhibit 2.16, that were becoming increasingly expensive to maintain. The separate layers of coaxial cable needed for each network, known as "spaghetti wiring," were almost unmanageable.

These convoluted information systems still were not able to meet the growing need for data. McDonald's employees needed a combination of information—voice, text, and image—from different sources and for different applications to do their jobs. The regional and district offices fed information on sales and inventory back to the headquarters daily. In addition, they tracked the product mix and promotion information. They needed to access or deliver information that was not

Exhibit 2.16

McDonald's Telecommunications Network before ISDN

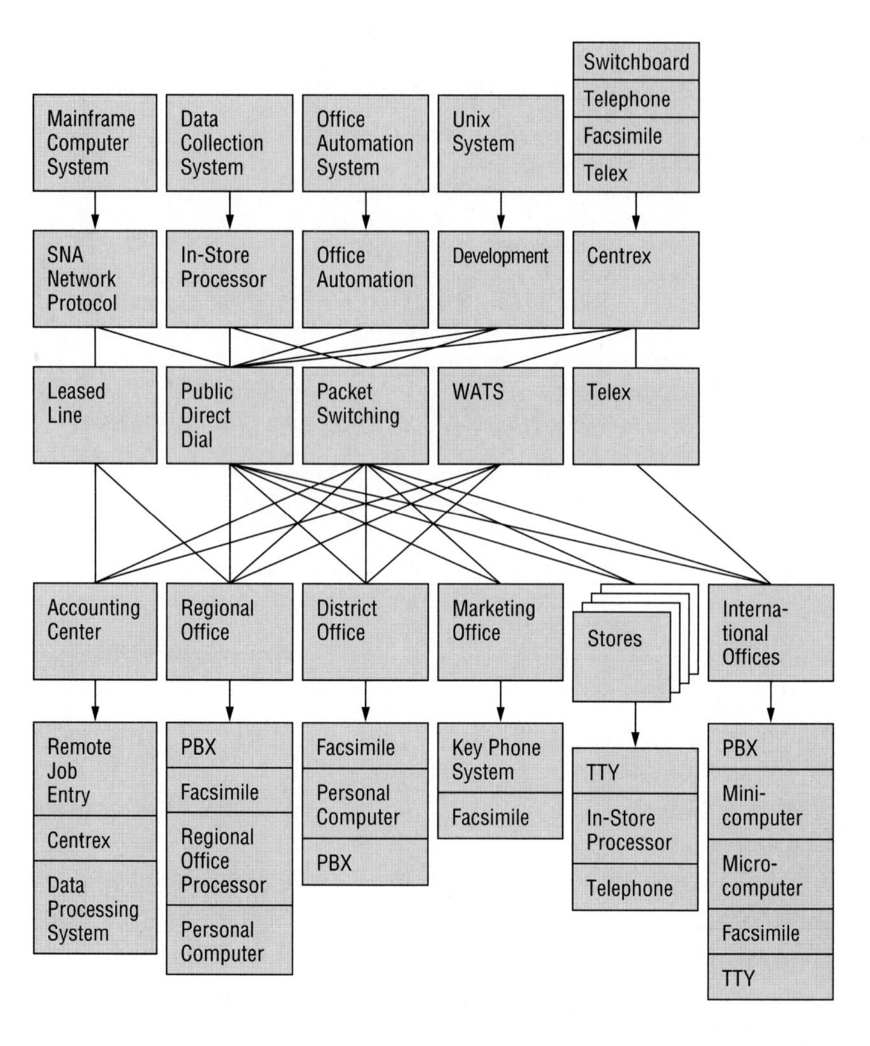

time sensitive and there was not enough traffic between any two locations to justify an expensive dedicated network.

As the company grew, the employee offices became increasingly expensive and complicated. McDonald's moves 50 percent of its headquarters staff of 1,200 each year. It takes several weeks to plan and implement the rewiring of coaxial cables this requires at an estimated cost of $1,000 to $1,500 per move.

Office automation capabilities had not grown with demand for them. Employees had complaints about telephone tag and cumbersome message taking which slowed productivity. Personal computer communications were highly inefficient because there was no method to transfer data from one machine to another except to take diskettes manually between machines. Furthermore, although there were many mainframes and applications, personal computers could not access multiple hosts with different applications. It was a virtual Tower of Babel.

On December 16, 1986, Illinois Bell and McDonald's switched on ISDN lines with data, voice, and video capabilities. Over the same interface, the ISDN combines office automation, access to IBM host computers, networking of personal computers, facsimile workstations, electronic publishing, and a comprehensive voice network. The integrated communications system provides easy access to information stored in databases on many different hosts and applications running on many different processors. The system is illustrated in Exhibit 2.17.

If the McDonald's case is indicative of a larger trend, then ISDN will provide a competitive advantage to businesses that seek a complete global communications network.

An important advantage of ISDN is that it uses the twisted-pair wiring that is already installed in the walls of most buildings. The problem of spaghetti wiring is solved, preserving the investment in the installed base of wiring. Ultimately, wiring costs are greatly reduced. The elimination of coaxial cable and the reuse of old wiring are key to the success of ISDN.

The ISDN system functions as a virtual private network to act as a transfer point between the mainframe computers and users. Because the end-users at McDonald's are geographically dispersed worldwide, they have three basic ways to send data to the mainframes:

- Messages from IBM 8100s in the field, and from regional Unix-based systems via IBM's SDLC (Synchronous Data Link Control) links.

Exhibit 2.17

McDonald's Telecommunications Network after ISDN

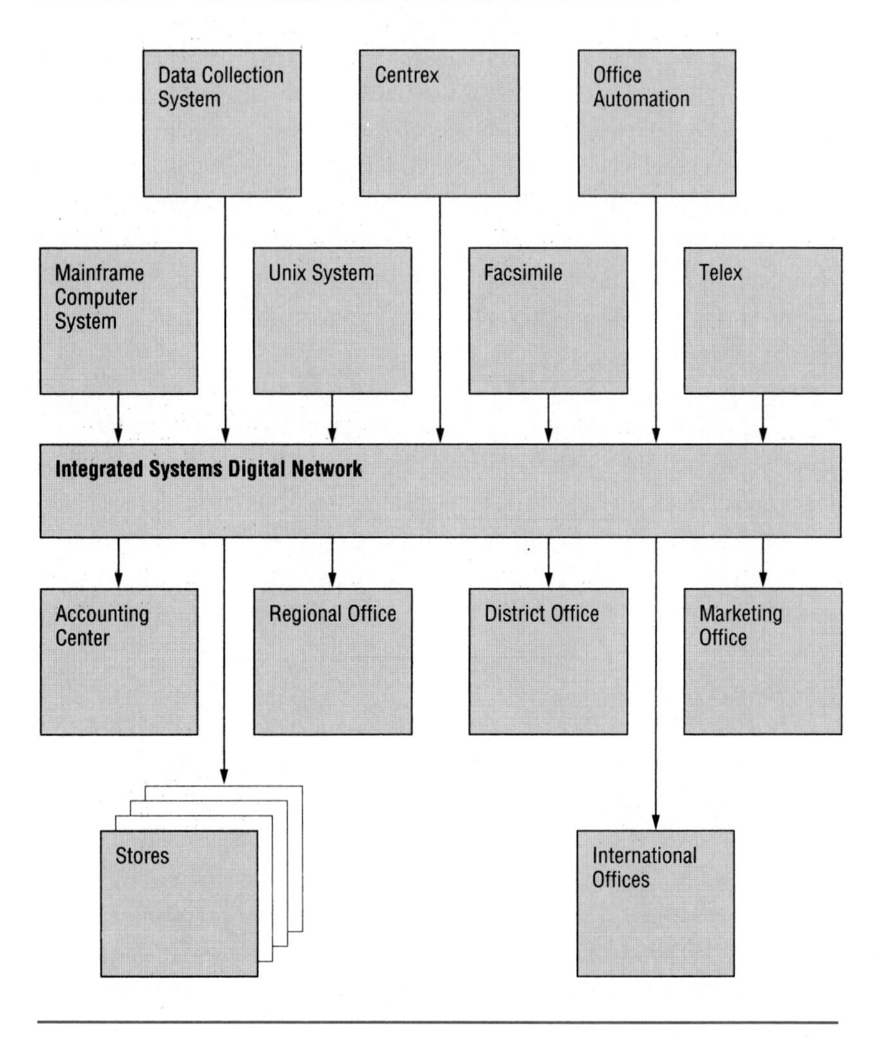

- Sales and inventory information uploaded from many different types of terminals to Amdahl 5890 and IBM 3090 mainframes located in Oak Brook.
- International transmissions of data via CCITT X.25 packet switching networks to Tandem front-end processors attached to the mainframes.

With ISDN, McDonald's has reduced its equipment and station rearrangement cost. The integrated system reduces the cost of moves as well as the actual number of telephones, terminals, and modems. ISDN's use of the twisted-pair wire facilitates the movement of employees in the corporate Home Office. The same move that would have taken several weeks and cost over $1,000 can occur with little or no expense and administrative difficulty. Since the Home Office has approximately 1,200 employees with an annual move rate of 50 percent, the average cost per move of $1,250 gives total annual savings to McDonald's on this factor alone of $750,000.

The new office automation system takes advantage of the capabilities of ISDN. McDonald's conversion from the coaxial-based office automation system to ISDN has brought several advantages. One key benefit is the ability to easily move users from one location to another. The system includes the classic office automation functions, including word processing, graphics, spreadsheets, database management, and electronic mail. In addition, resource management is simplified, since the entire network is controlled by the communications manager.

Summary

Telecommunications systems can be classified into several different categories according to the *type of information* they telecommunicate, the *media* through which they telecommunicate, or their *business applications*. The most widespread and fundamental network is the system of voice telecommunications. It has gone through many changes over time, culminating in the ability of businesses to have their own Private Branch Exchanges (PBXs) which act as private switching centers with many specialized features made possible by software programmability. Call detail reporting, speed dialing, call forwarding, conference calling and other features are common for PBXs. As the PBX world expanded, various peripherals were added such as automatic call distribution devices, voice recognition equipment, voice response and transaction processing interfaces, and voice store-and-forward (voice mail) equipment. A result of these voice processing technologies has been the ability to provide greatly enhanced customer service.

The second major type of telecommunications system is the data network. Data networks come in several varieties including those connecting terminal to computer, computer to computer, workstation to workstation, and other variations. Data networks are usually composed of leased lines which are fixed, point-to-point, dedicated circuits. Their advantages include stability and speed and the possibility of interfacing with very sophisticated equipment—and they are private. Their disadvantages lie in their inflexibility in changing nodes, speeds, or other characteristics. In addition, in many parts of the world they are difficult to obtain from the telecommunications authorities. Packet switched networks have emerged as an alternative to leased line networks. They break up each information stream into many information packets, each of which has a head and a tail by which the packet switching nodes direct the messages to the correct destinations. Packet switched networks can be used by many different users from many different locations, and they are extremely flexible. Their disadvantages lie in their limited speeds, expense for high volumes, their tendency toward rapid performance decay at high volumes, and inability to carry voice transmissions.

Virtual Software Network is a variety of leased line network that allows much greater flexibility in that it allows one to create instant networks and to increase or reduce capacity as needed. They also have much more sophisticated network control features. Image-oriented networks are used for telecommunicating complex images such as teleconferencing, office documents and facsimile, computer-aided design information, and other specialized information. Teleconferencing networks are used widely in business communications, and have decreased rapidly in cost. Facsimile networks have become more sophisticated with features such as facsimile store-and-forward, interface data conversion, electronic mail interfaces, and the development of Group IV digital facsimile standards.

Higher-order image-oriented networks such as office document image-oriented networks use a facsimile-related scanning technology to handle the massive amount of paperwork in today's bureaucracies. CAD/CAM networks allow global distribution of design and engineering expertise.

In addition, the spread of ISDN technology throughout the voice network will add many capabilities which will be easily accessible to organizations, making possible very comprehensive re-evaluations of telecommunications networks, as shown in McDonald's case.

Questions

2.1 What are the differences and similarities between voice, data, and image networks? What are the similarities?

2.2 What is a Private Branch Exchange and what does it do?
  a. Why was the PBX invented?
  b. How has it changed over time?

2.3 What is call detail reporting and what advantage does it have for business?

2.4 Explain automatic callback, camp-on, and call forwarding.

2.5
  a. What is an ACD and why is it used?
  b. In what type of industry is the ACD popular and why?

2.6 What is a VRU and how is it used?

2.7 Explain the operation and utilization of voice mail.

2.8 What is a leased line point-to-point network?
  a. What are the advantages and disadvantages of leased line networks?
  b. Explain how availability of leased line networks changes in the international environment.

2.9 What are packet switched networks? Why were they invented?
  a. What is a packet?
  b. How does it get switched?
  c. Explain what a PAD does.
  d. What are some of the advantages and disadvantages of packet switched networks?
  e. Why do packet switched networks have difficulty in supporting voice traffic?

2.10 What is a virtual software defined network? What are its main advantages?

2.11 Define image-oriented networks.

2.12 What does a CODEC do?

2.13 Approximately how much bandwidth is needed for full motion teleconferencing and how has this changed over time?

2.14 How might one justify the cost of building a teleconferencing network for an organization?

2.15 What is a facsimile store-and-forward system?
   a. How is it used?
   b. Why would one wish to use it?

2.16 What is an electronic mail facsimile interface? What benefits does it give the user?

2.17 a. What distinguishes a CAD/CAM network from a traditional design and manufacturing system?
   b. What is a typical application for such a network?

2.18 What are combined media networks?

2.19 a. Define ISDN and highlight a few major events leading to its development.
   b. Summarize the development and functions of McDonald's ISDN system.

2.20 What is a typical application for an office document image-oriented network?

2.21 a. What is the difference between Group I, II, III, and IV facsimile technology?
   b. Which type of facsimile is the most widely used today, and why?
   c. What is holding back the proliferation of digital facsimile technology?
   d. Can you explain what a facsimile interface data conversion system does?

2.22 What do the terms *2B plus D* and *23B plus D* mean?

Chapter 3

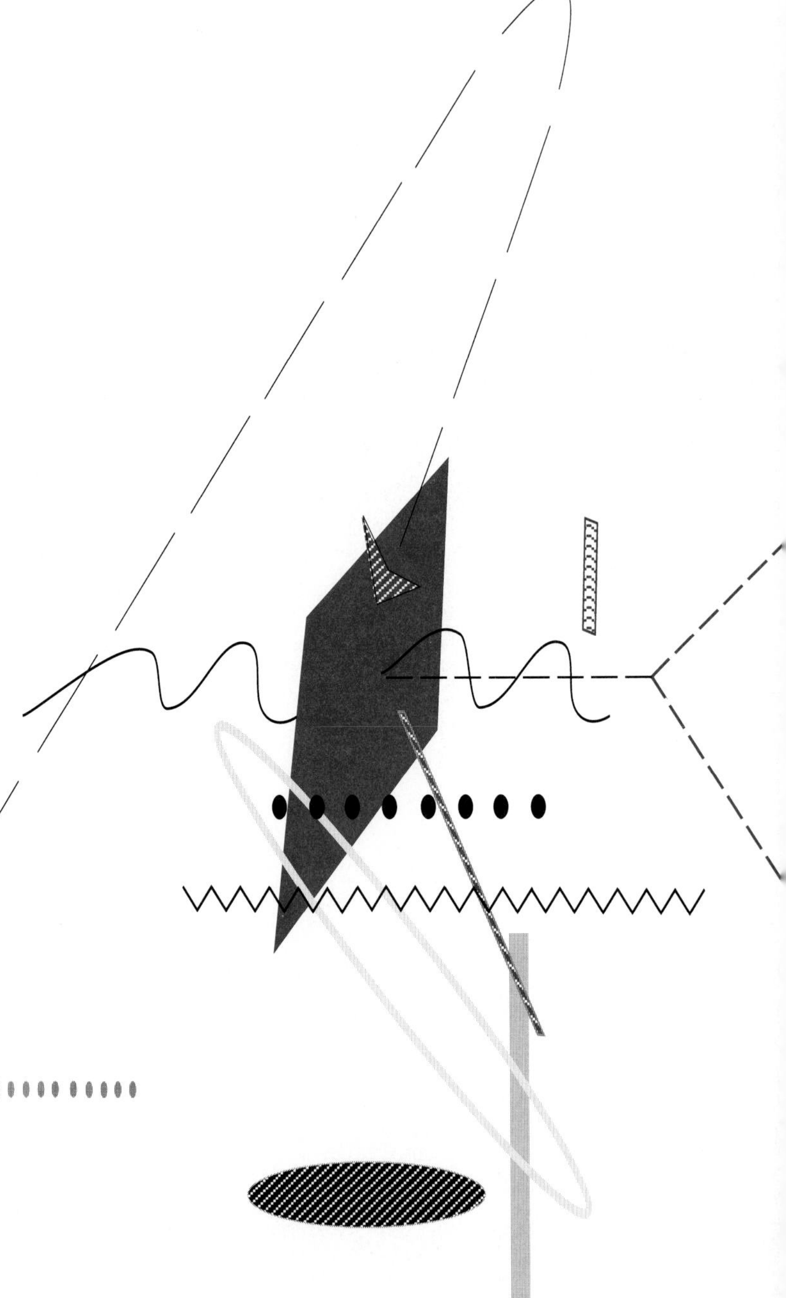

Getting the Maximum Use of Your Circuits

Switching and Multiplexing

The development of multiplexing technologies has allowed organizations to make maximum use of their data circuits, particularly for leased line point-to-point networks. Multiplexers interweave many streams of information into a single, high-speed transmission. At the other end of the circuit, a waiting multiplexer decodes the incoming stream and divides it back into its original component parts. Multiplexing technology has gone through several generations and has become more and more complex, particularly after the deregulation of telecommunications in the United States during the mid 1980s. The technology now combines switching and dynamic reassignment of bandwidths across complex circuits containing voice, data, and image information linking together the world's financial centers.

Bandwidth and Speed Are Sold in Different Sizes

When the telecommunications function of an organization wishes to purchase point-to-point or multidrop circuit capacity from a service provider, it finds that circuits are available in "models" or sizes classified by speed and capacity. Different models are available ranging from low-speed data lines, such as those associated with old-fashioned news wire services, to large-capacity circuits capable of linking together mainframe computing systems. A range of options are available—at a price.

In the United States, the most commonly purchased circuits include 56 kbps, T-1, and to a lesser extent, T-3 circuits. The terms *T-1* and *T-3* refer to bundles of circuit capacity which are much faster than individual circuits. T-1 circuits run at approximately 1.544 mbps and T-3 is a circuit composed of a bundle of T-1 circuits. In other words, a T-3 circuit has approximately 27 times more capacity than the popular 56 kbps circuits; that is to say, these circuits have three times the capacity of a T-1 link. Using these types of high-speed, terrestrial-based circuits yields practically no delay in response time whatsoever. Using satellite systems or packet switched networks inevitably involves propagation delays.

It is unlikely that an organization will have an application which uses T-1 and T-3 circuits by itself. The entire rationale for purchasing these large circuits is to divide them up into smaller parts and use them for many different applications simultaneously. For example, an equities trader may have ordering or processing centers in both New York and Chicago. This organization cannot tolerate any delay in response time whatsoever caused by the network. At any one time, there are many applications being processed between these two centers and seconds, or even fractions of seconds, can cost millions of dollars (or more). Under these circumstances the organization would buy one large circuit instead of many smaller circuits. The technology that allows the organization to divide up such a large-capacity circuit is the **multiplexer**.

Multiplexers are a critical technology in any large-scale computer operation, and the fact that they allow bundling together of many circuits into one helps to simplify the task of network management.

Several Types of Multiplexing

Multiplexing is a technique of getting more information to flow over the same wire or other media. Before multiplexing, it was possible to send only one signal at a time through a circuit. After multiplexing, it has become possible to send two, three, or even dozens of clear, distinct channels over the same circuit simultaneously. Of course, the use of multiplexing greatly increases the value of an existing circuit. Laying wire and installing circuits is expensive. It is much cheaper to place a "black box" at either end of a circuit to add capacity, instead of having to lay new circuits.

But what is the black box and how does it work? How can it be used effectively in business operations? Several types of techniques are used to cram more and more information down a single circuit. These types of multiplexing evolved over time, and have made such rapid advances that now ISDN and other multiple channel systems can be transmitted over the twisted-pair wire found in the average telephone installation.

How to Use Multiplexers

Multiplexers multiply the amount of information that can be carried down a circuit, and also divide up the amount of available transmission capacity into different pieces or channels. If one is using, for example, a high-capacity leased line between New York and Chicago, in order to make full use of it, the maximum amount of information should be carried at all hours of the day.

Many users or circuits are stacked on top of one another for shipment down a high-speed line, which carries bundles of information streams. You can think of a multiplexer as a black box that takes a fistful of raw, straight spaghetti strands and fits them into a garden hose. Each of the individual circuits gets transmitted down the single, large hose.

For example, suppose you have leased a T-1 circuit from New York to Chicago. What would be the possibilities for multiplexing? How many different types of information and data could be accommodated by the channel?

Typically, a T-1 circuit with a capacity of 1.54 million bits per second (mbps) can carry approximately 23 telephone calls using pulse code modulation. One of the important tasks of a multiplexer is to

Exhibit 3.1

Multiplexer Technology and Circuit Capacity

A T-1 circuit can carry varying numbers of voice calls using different types of compression standards.

PCM Pulse Code Modulation
ADPCM Adaptive Differential Pulse Code Modulation
DSI Digital Speech Interpolation

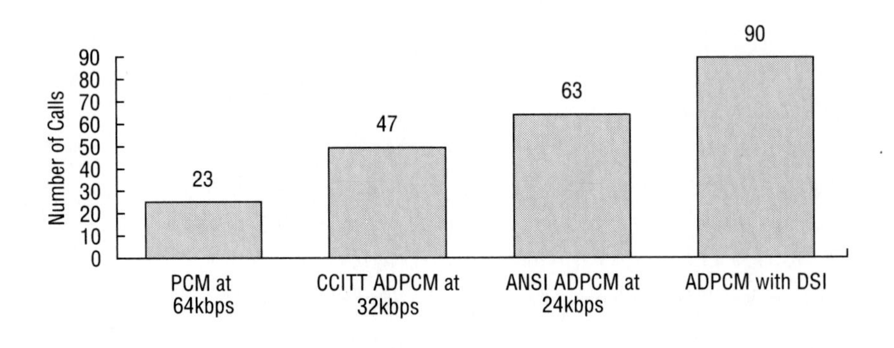

provide the sampling and coding necessary to convert analogue data to digital data. Pulse code modulation typically samples the voice circuit at 64 kilobits per second (kbps). Other types of coding for voice information further compress the signals up to about 90 voice calls on a single T-1 circuit. Exhibit 3.1 details circuit capacities with various types of multiplexing.

In a sense, it is ironic that so much effort was spent on compressing so much information into circuits at a time when circuit capacity was booming. Of course, there were a great many reasons for doing so during the period from the mid 1980s until the early to mid 1990s. During that period of time, the telecommunications system in the world was migrating toward higher speeds and integrated digital services. Furthermore, the availability and cost of leased lines in most locations tended to make sophisticated multiplexing cost effective.

Why was it cost effective? Because it allowed the user to combine many circuits into a single high-capacity circuit instead of buying the additional capacity from the telecommunications provider. As long as the price for several circuits was more than the price for a single circuit plus the cost of the multiplexing equipment, then the use of the multiplexer was more efficient.

Although in the beginning, multiplexing started primarily as a data communications strategy, by the late 1980s it was being used to link together large backbone data communication networks with the additional capacity of carrying voice circuits and full motion video teleconferencing. In addition, many of the very advanced multiplexers had the capacity to interface with the telecommunications systems of other areas of the world, such as Japan and the European Community. This enabled the user to have a sophisticated multiplexing environment spanning the globe without too much difficulty in interfacing with the domestic networks in different countries.

Time Division Multiplexing

Time division multiplexing was a more advanced step in getting more information to travel down the same channels. Given that a wire or circuit is already transmitting several frequency channels, time division multiplexing can get more information down a single channel. Time division multiplexing can be thought of as a highway. Imagine a procession of United Nations delegates heading from Kennedy Airport to the United Nations, and at the same time a funeral moving down the highway, and in between a few Federal Express and United Parcel Service delivery trucks. All of these vehicles enter onto the expressway at slightly different times. During their movement down the expressway, they mix in with each other, so that the Federal Express trucks are cutting into traffic with the UPS trucks, breaking ahead of parts of the delegation and funeral. All of the available spaces in the highway are taken up for the trip into the city. As the various groups reach the city, they take their respective exits, and the funeral and the United Nations motorcade reassemble in their strict orders for the remainder of the ride.

In time division multiplexing, any pause in the transmission of information down a channel is filled up by the addition of any other waiting information being held by the multiplexer for transmission. Just as several different motorcades can travel down a highway, even though they get a little mixed up during the ride, a time division multiplexed circuit can intersperse several streams of information and data on a single circuit. It is the job of the multiplexer to keep track of the information as it rides through the circuit so that at the end of the trip, it is correctly reassembled for movement to the correct destination.

Frequency Division Multiplexing

Frequency division multiplexing divides up the channel into several frequencies and then transmits a different channel of information and data on each frequency. It can be thought of as a radio band in which different radio stations are being broadcast simultaneously over the air. The listener simply tunes to a new channel to get a different station. For data channels being used for computer communications, the black box at each end of the wire would listen to several frequencies simultaneously and treat each as a different data channel.

Encryption

Highly confidential information must be **encrypted** during its movement over telecommunications circuits. When information is left open for inspection, a person tapping the channel would be able to learn very important secrets. For example, when money is transferred from one account to another, or from one bank to another, the amounts can be very great. Someone who could intercept the telecommunications channel which is transmitting the transfer information might be able to learn the codes necessary for the automatic transfer to take place. Once the codes were learned, it might be possible to place the same format information and codes onto the circuit and thereby have money transferred to a special illegal account, perhaps in Switzerland. The computer itself cannot tell the difference between illegal and legitimate money, particularly if the instructions for funds transfers appear in exactly the same format with exactly the same security codes as a legitimate transfer.

Much valuable information is transmitted through telecommunications channels: top board strategy information, drug formulas, marketing plans, design specifications and secrets, etc. The answer to the security dilemma this creates is to encrypt the information so that no one can read or understand it without a special deciphering code. **Encryption** is a technique to automatically change each piece of information to something else so that the original meaning of the message becomes scrambled. Without knowing the special key to descramble the information, the eavesdropper is unable to make sense of the information. Encryption is used widely in banking and electronic funds transfer networks. It was pioneered by the military for use on the battlefield and in times of peace for all communication that might be picked up or monitored by enemy forces.

The accepted standard for encryption in the United States is the Data Encryption Standard (DES), originally developed by the National Bureau of Standards. DES is a mathematical technique to allow for rapid changes in the keys for encrypted messages. Incorporating DES into your telecommunications channels is not particularly difficult. Encryption is available many times as an integral part of sophisticated multiplexers and modems.

Experimental Color Multiplexing on Optical Fiber Circuits

Color multiplexing is another technique to further maximize the amount of information and data which can be sent down a particular channel. However, color multiplexing is possible only in fiber optics, where the change in color of the laser light wave carrying the information can multiply the amount of available channel capacity. In this system, different colors are transmitted through the optical link, each carrying a different band of information and data. Color multiplexing has been developed by Bell Labs in the United States and also at various research facilities in Japan.

Bandwidth Managers and Switching

Using sophisticated switching and bandwidth managers such as that provided by IDNX™, it is possible to integrate voice data and images over the same circuits as shown in Exhibit 3.2. The bandwidth manager can dynamically change the circuit configuration to constantly alter the mix between data, voice, and image information. In addition, it is able to constantly test itself for faults and reconfigure the network in case of a problem.

The bandwidth manager is a sophisticated piece of technology with many telecommunications functions. To a certain extent, it merges

Exhibit 3.2

Central Role of Bandwidth Managers

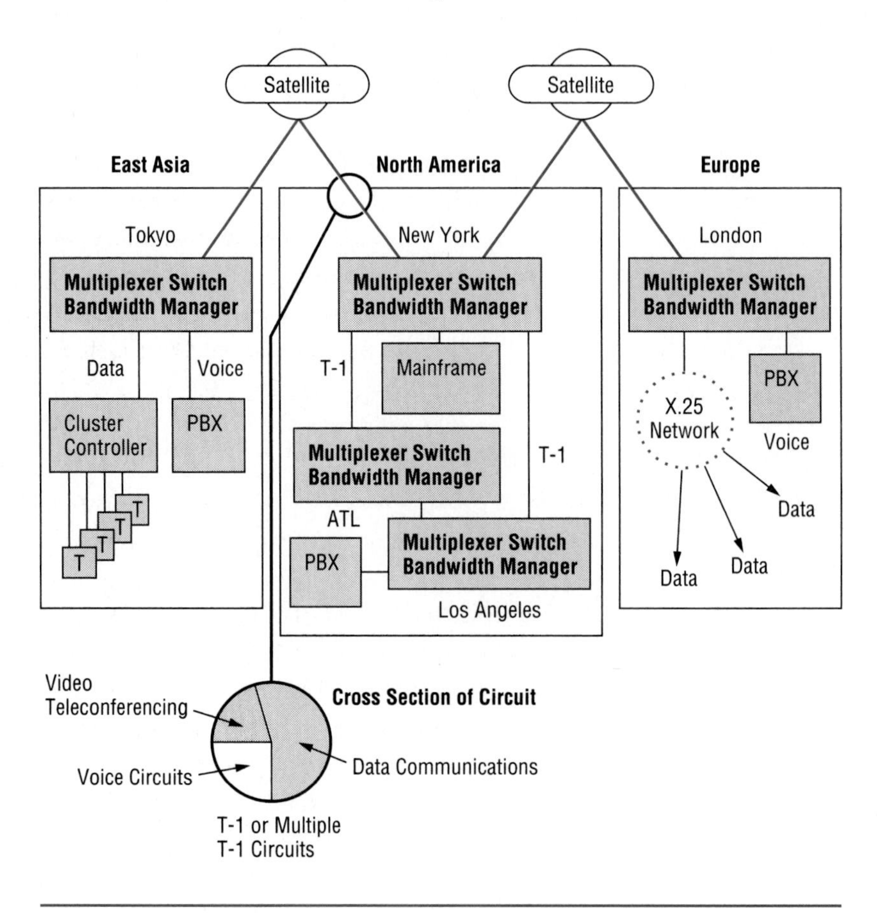

together several important functions which in the past have been handled by different types of equipment. For example, it acts as a multiplexer, but in such a way as to allow virtually any type of traffic to join the data stream. It acts as a switching node, giving some of the flexibility found in packet switched networks; it can act as an interface and gateway to other types of networks that are external to the organization. Finally, a bandwidth manager can dynamically reconfigure a network and change the distribution of different applications.

All of these features combine together to make the bandwidth manager a powerful and flexible technology for the telecommunications function.

Integrative Multiplexing Functions

The bandwidth manager's sophisticated multiplexing features allow all of the major types of information to be sent through the circuits. In a fully loaded bandwidth managed circuit, video teleconferencing, PBX-generated voice communications, high- and low-speed data transmissions, including local area network bridges, digital and GIII facsimile, and other types of data traffic could all be compressed, woven together, and pushed into high-speed circuits.

A great deal of efficiency can be generated by combining voice and data over the same circuits. If this can be done without disturbing the data processing activities of the organization, this integration allows a great deal of long-distance voice telephone traffic to be absorbed into the leased point-to-point circuits accessed through the bandwidth manager. To decide whether to tie two offices of the organization in different parts of the country together with a voice "tie line," the telecommunications function must compare the costs advantages between the two alternatives. Sometimes the trade-off may not be worth it.

Another important cost cutting feature of the bandwidth manager is the ability to take a variety of circuits, both high-speed and low-speed, and combine them together into a single circuit. Obviously, if this eliminates many single leased circuits in favor of a single circuit, there may be a great deal of saving. In addition, any time the single circuits are underutilized, this is money wasted; the bandwidth manager automatically fills the space in the circuit with something else. This is the second type of saving effect of network consolidation.

In addition to these features, the bandwidth manager also compresses the data streams and thereby saves even more space in the circuit. Altogether, it produces a rather substantial consolidation of networks into a single high-speed data highway. The cost savings depend on the extent to which various networks are being replaced or merged together.

Switching Functions

In addition to providing the multiplexing functions which allow many different streams of information to flow through the same circuit, bandwidth managers allow the network to make quick decisions about backup circuits and rerouting in case there is a difficulty with the network. This function is very similar to the function performed by packet switching nodes on a standard public packet switched X.25 network. If one portion of the network experiences a failure or degradation in its performance or simply becomes crowded and over-used, the bandwidth manager is able to sense this and make an adjustment.

Suppose, for example, an organization has three major bandwidth managers located in Manhattan, East L.A., and on Peach Tree Street in Atlanta. If all of these bandwidth managers are connected together in a network, then a stream of data moving from Manhattan to East L.A. could go through two routes: directly or via the Atlanta node of the network. If for some reason the traffic from Manhattan to East L.A. on the direct circuit were too heavy, or if the circuit started to fry out (deteriorate in quality), the bandwidth manager could push some of the traffic through Atlanta. Since this happens automatically, the users would more than likely not notice a difference. However, the telecommunications network management would be informed by the bandwidth manager that the Manhattan to East L.A. circuit was starting to crash. This would give them time to diagnose the problem and take corrective measures before the network actually crashed and burned.

If the organization is quite large, with nodes in many different locations around the world, the bandwidth manager gives the network more flexibility to take corrective action. As the number of nodes increases, the number of possibilities for rerouting correspondingly increases. Three nodes give two routes, four nodes give five reroutes, etc.

Interface and Gateway Functions

Since many organizations have large telecommunications connections both within and external to their information processing infrastructure, at some point their networks must connect to other telecommunications circuits that may not be private. Remember that it is economical to have private circuits only on lines of the network that are sufficiently heavily used to justify the expense of leased point-to-point lines. For other parts of an organization's telecommunications network, either the economics or the complete inavailability of private circuits (frequently the case in Western Europe) make it necessary for the internal, private part of the organization's network to interface to other networks.

Bandwidth managers provide this capacity by allowing the backbone network to interface with several other types of networks. One of the most popular interfaces is between the bandwidth manager backbone network and public packet switched X.25 networks. In Western Europe, for example, an organization may have one major bandwidth manager switching node in London or Brussels, responsible for providing the international **gateway** to other regions of the world. However, for the European Community portion of the business, the organization may build an X.25 network interface which will allow traffic from the backbone network to jump off onto the public X.25 network for its travels around European subsidiaries.

In addition, the bandwith manager might provide an interface to a major PBX at the remote location. Although the PBX is itself a major processor of voice transmissions, its connection to the bandwidth manager allows certain parts of its voice traffic to travel down the backbone private network, thus saving expenses, because the organization is already paying for this leased point-to-point circuit. This allows the PBX to use many of its advanced features, such as call forwarding, all the way across the network. In effect, a local call going into the Wall Street office of a firm could be automatically forwarded to a London extension because the PBX could simply route the call over the internal circuits being managed by the bandwidth manager.

Dynamic Circuit Allocation

Bandwidth managers are able to change the allocation of the circuits they control. For example, during certain parts of the day the circuits may be used predominantly for voice traffic with a smaller amount of data; in the evenings when most of the information workers have left work, the circuits may carry massive data transfers between computers for updating of files and clearing of the day's transactions. Under these circumstances, the bandwidth manager reallocates circuit capacity; the portion of the leased lines used for information worker voice traffic during the daytime hours can be borrowed and handed over for use by the data processing night shift workers to transfer large amounts of data from one part of the country or world to another. When the morning comes, the circuits can be shifted back to their original configuration as the information workers begin to use the voice circuit capacity during their working day.

The fact that the bandwidth manager allows this type of reallocation of circuit capacity without contacting the service provider gives a great deal of flexibility to the user organization. In many arrangements, this type of change in circuit allocation would be done by the service provider, but for a fee. However, the bandwidth manager cannot only make the changeover on a regular basis and without extra charges from the service provider, but it can also be *programmed* to do this on a regular basis, thus further automating the network control functions.

Summary

When the telecommunications function in an organization wishes to purchase leased line circuits from a service provider, it finds that the "models" available are based on the speeds of the circuits. There is a noticeable amount of variation between the types of leased lines available in different countries, but in the United States the popular lines are the 56 kbps, T-1, and T-3 circuits. T-1 circuits can carry 1.544 mbps, which is approximately 25 times greater than 56 kbps lines. A T-1 circuit can carry 23 to 90 or even more simultaneous telephone conversations depending on the type of compression standard used. Adaptive Differential Pulse Code Modulation with Digital Speech Interpolation offers the most compression. Most circuits use simple Pulse Code Modulation, which samples at 64 kbps allowing about 23 voice channels on a single T-1 circuit. T-3 circuits can carry approximately three times more than T-1 circuits. Multiplexing allows information to be compressed and transmitted down the circuit.

Time Division Multiplexing divides a circuit up into time slots, then fills those slots with a mixture of information from many different streams. Frequency Division Multiplexing treats the circuit like an FM radio band, with each stream of information being a different "station."

Encryption of data using the Data Encryption Standard developed by the U.S. National Bureau of Standards encodes sensitive information so it cannot be understood by anyone who does not have the correct key.

The most advanced form of technology for getting the most out of leased point-to-point circuits is the bandwidth manager, which functions as a multiplexer, fast packet switch, interface gateway, and furthermore has the capability of dynamic circuit allocation both for failing circuits and for changing the proportions of applications traveling through the channel.

Questions

3.1 What is switched 56?

3.2 What is T-1 and how does it differ from T-3?

3.3 What is the basic function of a multiplexer?

3.4 a. Using PCM, how many voice calls can a T-1 circuit carry?
b. How many calls could the same circuit carry using ADPCM?

3.5 Explain the conceptual difference between time division and frequency division multiplexing.

3.6 What is DES and why would one use it?

3.7 What is a bandwidth manager and what are its major functions?

3.8 What is dynamic circuit allocation?

Chapter 4

Emerging Network Alternatives

Fiber Optics, Cellular Radio, and Very Small Aperture Terminals (VSATs)

How are information and data sent from one location to another? Of course, of the many technologies involved, none is as important as the media and wiring that link the global system together. This chapter discusses the different technologies being used for transmission of telecommunications signals and gives some historical perspective on these technologies. Of course, the student should pay careful attention to the part on fiber optics, the clear wave of the future. In addition, the role of satellites and cellular radio technologies is examined. They provide interesting and useful alternatives to traditional telecommunications systems based on leased lines obtained from the telephone company.

Traditional Media Systems: Copper Cable and Microwave

The way in which information and data are delivered through the telecommunications system is in many respects independent of the medium used. Originally, copper wire was the medium of choice. At the time of the invention of the telegraph in the 1850s, cable was the medium of choice. Gradually, particularly in the 1970s and 1980s, other telecommunications media became more popular for various economic and technical reasons.

The electromagnetic cable is currently the most common telecommunications medium. Most of the world's wiring is done with copper cabling, since this is the preferred medium for voice telecommunications. The early undersea cables were also made of copper. The typical linkage for home telephone service employs copper twisted-pair cabling, except in advanced countries such as Japan, where glass fiber optics are available for direct wiring to premises.

The electromagnetic cables are out there right now, and they will certainly last well beyond the year 2000. They form the foundation of the telecommunications empire infrastructure. Other technologies have emerged, however, which offer interesting alternatives to copper wiring. From the point of view of business, they create several possible strategic plays.

Microwave Relay

Microwave relay was developed by the Bell System to handle remote telecommunications without the expense of laying land cables. It works by a transmission and receiving dish—usually a shaped, metal plate—capturing the line-of-sight broadcast of a tightly focused microwave, as shown in Exhibit 4.1. Particularly good for remote areas of the country that have relatively flat terrain, microwave relay stations are scattered around the world, linking together remote places with metropolitan areas.

In highly congested areas, microwave can face the problem of overlapping beams. One beam can spill over onto the receiving dish of another relay system. Under these circumstances, a problem can occur. Another difficulty with microwave is that it may be disturbed by bad weather; but it is not compromised too much by even relatively heavy rain.

countries and internationally. The newer generation of undersea cables use fiber optic technology because of its greater capacity.

For the telecommunications planner, the type of carrier system used outside the corporation has little relevance, except under very specific circumstances. For example, a leased fiber optics circuit may provide much faster response time than a satellite circuit, and this may be very important for a company that depends on instantaneous response time, such as a foreign exchange trading system. For other applications, the faster speed may make little difference.

However, in some environments, fiber optics may make an important contribution to efficient telecommunications operations: in manufacturing networks and in local area networks, for example.

Fiber Distributed Data Interface (FDDI) Standard, ANSI X3T9.5

As the late 1980s passed by, it was quite obvious that telecommunications networks were expanding their strategic importance, but still woefully inadequate to handle the more sophisticated processing and data transfer being demanded by the next generation of users. Users who had lived with simple data transmission and command line driven interfaces to information systems for years had begun to taste the simplicity, elegance, and even beauty of graphical interfaces, WYSIWYG processing of documents, voice and image integration of documents, and the ease of networking.

Users expected better networking of their new applications and the only way the installed infrastructure of telecommunications could handle these new applications was to greatly increase the bandwidth possible on each line.

Fiber optics, with its great speeds proven in the laboratories, offered a real benefit for the telecommunications planner who was being asked to provide subsecond response time for a new generation of highly complex and bandwidth intensive applications. The development had been known for years and a standard was being developed. The **Fiber Distributed Data Interface standard**, known as FDDI, was labeled X3T9.5 when it was issued by the American National Standards Institute (ANSI).

The promise of FDDI was that it could eliminate the need for multiple wiring systems to support computer systems, local area networks, PBXs, and other communications devices. It would do this by providing a single carrier and architecture for the data streams from all of these devices.

The strategy many organizations were adopting involved installing FDDI fiber cables through their physical plants in order to give network managers physical layer infrastructures that would both support current requirements and provide a path for future developments in fiber optics.

The FDDI was designed and intended to be the fiber-based, high-speed local area network of the 1990s. What are the speeds involved? The typical speed for an Ethernet standard local area network was 10 megabits per second. An FDDI network based on optical fiber could provide speeds up to 1,300 megabits (1.3 gigabits) per second. In other words, it ran approximately 130 times faster than the Ethernet standard, which had formed the bulk of the installed base for local area networks. Before the introduction of FDDI, customers that required those very high speeds were required to invest in expensive specialized equipment from boutique vendors. Since these technologies were not based on widely accepted standards, the risks of incompatibility were high enough to discourage most ordinary consumers of information technology. FDDI promised to change all of that by making available a standardized product that could be purchased essentially off the shelf.

The driving force behind the adoption of the FDDI standard was the change in bandwidth requirements for applications. Bandwidth requirements were increasing dramatically because applications at the end-user level were becoming more complex. These applications were handling more data and more creative work, frequently based on sophisticated 32-bit workstations, many of which were not directly compatible with other machines in the organization and on the network. In addition, a strong movement was pushing to allow multivendor operations. Users were attempting to build networks that would share all types of information technology resources, and it was no longer either economically wise, or even technologically optimal, to rely on a single vendor for strategic solutions in building an entire network.

These two historical forces generated the pressure to increase bandwidth, then: requirements for faster response time and re-

Exhibit 4.4

Growth of Generations of Information Technology

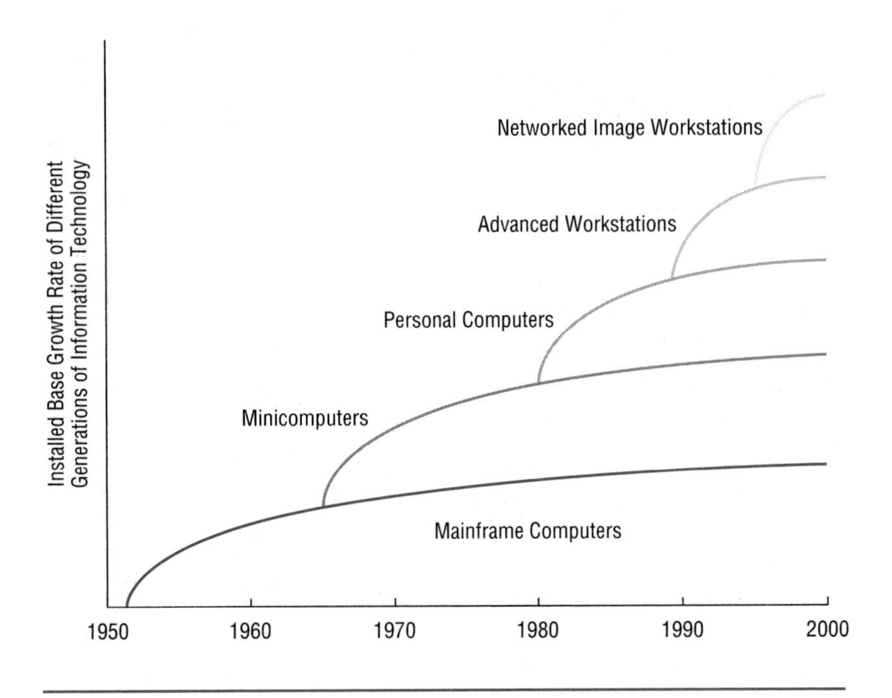

quirements for more data transmission across the network. These two forces were in turn driven by the changes in generations of information technology, as illustrated in Exhibit 4.4. As the 1990s arrived, the type of workstation being used was heavily dependent upon large bandwidths. By the end of the 1980s, most of the desktop media were involved in transition from graphics to color graphics to 3-D color graphics and finally to full-motion graphics and simulations, all on desktop equipment. In addition, workstations connected to minicomputers and mainframes, including supercomputers, also were required to transmit more and more information and data to their hosts and back.

Another factor involves the role of the local area network and the workstation. In the past, high-speed channel communications took place directly between different mainframe computers. As the role of the workstation increased, along with the advancement of high-speed 32-bit microprocessors in the workstations, the relative role of the mainframes and minicomputers diminished. Workstations took on much or all of the processing themselves. Local area networks linked the different workstations together for file sharing and transmission. In essence then, the local area network linking the workstations took on the role of the high-speed channel linking previous generations of mainframes.

High-speed applications such as desktop and professional publishing systems, CAD/CAM, and other graphics-intensive programs needed faster response times. In addition, technologies that had not become highly popular because of the specialized telecommunications required, including video teleconferencing, were migrating down in cost and bandwidth requirements.

The applications that will become more efficient with FDDI are many, as Exhibit 4.5 shows, the most important being:

- Large-scale file transfers between sophisticated workstations will be made more efficient by the high speeds possible through FDDI.
- Workstation local area networks handling graphics-oriented applications, such as computer-aided design, video applications, desktop publishing, and architecture design, will be able to move large amounts of data.
- FDDI will provide efficient bridges between already installed low-speed local area networks such as 802.x (Ethernet and Token Ring), thus providing the ability to interconnect isolated workgroups without losing network performance.
- FDDI will eliminate the need for local storage on workstations. If data and software can be quickly and efficiently moved around on the fiber optic network, then it is possible to keep all data and software at one central location and let inexpensive workstations access the information through the network. This might be useful for environments such as college campuses.
- Very high-resolution graphics communications, such as coaxial thermography transmissions (CAT scans) or nuclear magnetic resonance (NMR) images, could move over local area networks in a hospital environment with FDDI systems, giving doctors the ability to quickly scan through an image-base of information on conditions and references for diagnostic purposes.

Exhibit 4.5

FDDI Applications

Various applications require either faster response times or larger bandwidths.

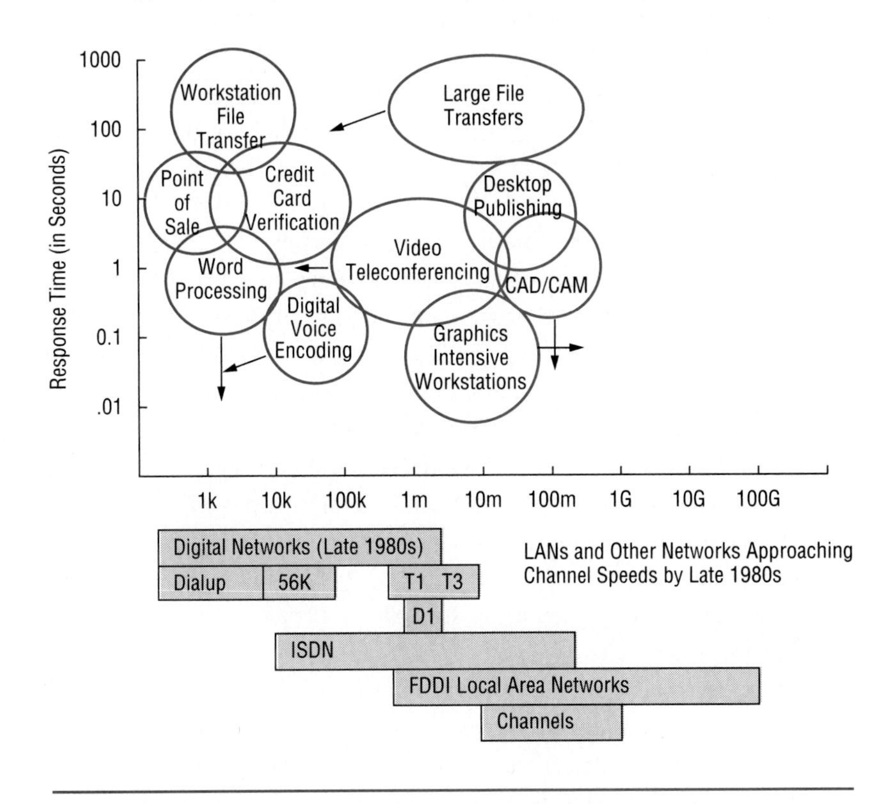

- FDDI gives fast access to files and database management systems in sectors where competitive advantage comes through subsecond response time to inquiries of large complex databases, i.e., finance. This gives traders and researchers the ability to access more and more data for any given decision point, possibly the critical difference in putting together deals.
- FDDI facilitates point-to-point digital imaging transmission applications such as the transmission and processing of images from satellites or from radar. This type of processing system requires movement of a very large amount of data through a telecommunications network.

- FDDI will provide a good way to build a secure network, since fiber optics have inherent security features (including inability to tap and redundancy).

The FDDI standard provides both packet switched and real-time data communications. The Link Access Protocol allows different workstations or pieces of equipment on the network to receive messages intended for them by the Token Passing method that requires each unit of information to be examined by each station before it is either accepted or passed on through the network. FDDI networks are arranged as dual counter-rotating rings. One ring is the primary ring, the other is the secondary ring. The secondary ring provides redundancy, since it transmits the signals in the opposite direction from the primary ring. As of the late 1980s, the transmission speed being recommended for FDDI was 100 megabits per second, with up to 1,000 physical connections being allowed on each local area network and as much as 2 kilometers between stations. Along the ring of the fiber optic cable are located various stations. Each station is a physical node on the ring with a unique address, and each is capable by itself of transmitting and receiving information. In addition, each station has the capability of taking incoming information and transmitting it on to the next station on the network.

Since each station is connected to both the primary and secondary rings, transmissions can follow the reverse path should problems occur, as Exhibit 4.6 illustrates. This "healing" ability will prove to be a very great advantage for users since the importance of the network increases with its capacity and the amount of traffic it is carrying. Since an FDDI local area network will be a major backbone system operating at the highest possible speeds, it is critical it have a built in redundancy that reacts in milliseconds instead of minutes or even seconds.

The Intelligent Building

The Cho Dong Insurance Company is one of the largest in South Korea. As it planned its landmark building, one of the tallest buildings in Korea covered with sleek, copper-plated glass, it needed to build in extra capacity for all types of telecommunications. Cho Dong decided to wire its entire building with fiber optics during the construction phase, even before the telecommunications requirements of the tenants were known.

Exhibit 4.6

FDDI Backup Ring

The secondary ring is a backup in case of failure on the primary ring.

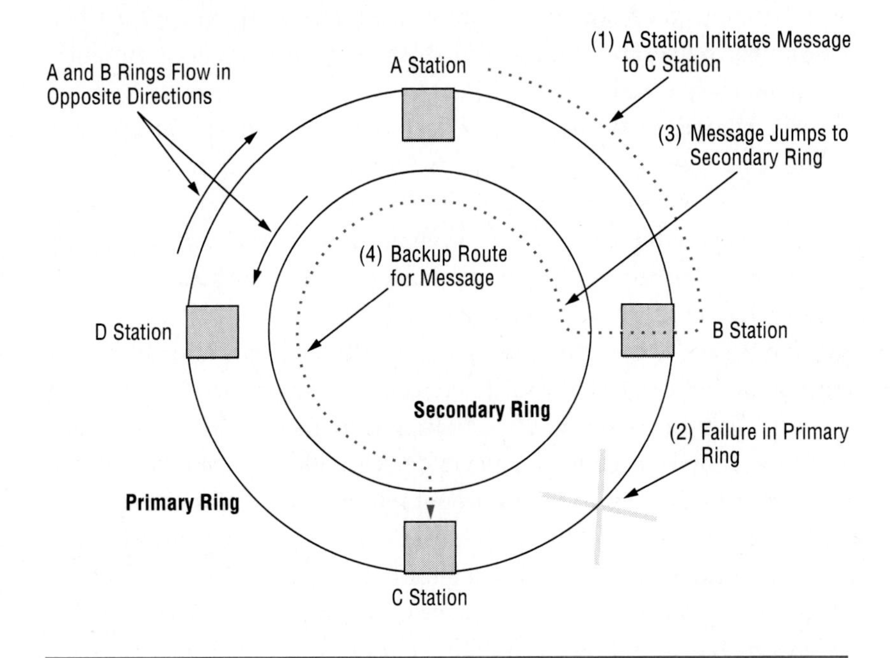

Cho Dong knew that it would be able to use the additional capacity as needed, and that avoidance of large telecommunications costs later would be a major attraction to possible tenants.

The Cho Dong building is indicative of a wider trend toward the intelligent office building. Real estate developers view the installation of a strong telecommunications infrastructure, many times including fiber optics, as a strong selling point in leasing available space to possible tenants.

Optical Fiber Submarine Cables

From the middle of the 19th century as telegraph cables started to spread out around the world, usually following the paths of the railroads, telecommunications has expanded greatly. At first, in order to cross over waters, cables were hung in the air or placed on bridges. Then a material called gutta percha proved to be a good enough insulator to protect the cables from the harsh conditions found underwater. Before the turn of the century, the trans-Atlantic system was beginning to operate, and the global infrastructure of telecommunications was being built.

Compared to what is available today, those first systems seem rather primitive, but they worked at the time and truly revolutionized business and history in general. For the first 150 years or more of telecommunications, practically all major innovation, commercialization, and application of telecommunications technologies was driven by developments in the United States.

This development of international telecommunications circuits, particularly undersea cables, has continued to improve immensely. By the early 1990s, the world's telecommunications authorities were busy laying very high speed fiber optic undersea cables, which promised the general public access to international telecommunications that would be almost as cheap as domestic. As the center of the world's economy, technological innovation, and cultural leadership shifted from the Atlantic to the Pacific, a great deal of emphasis was placed on telecommunications there. Exhibit 4.7 illustrates the network that developed.

Japan was linked directly for the first time to the U.S. mainland without routing circuits through Hawaii. This linkage made it possible for the world's financial community, which was focused in Tokyo, New York, and London, to link together through the United States with very high-speed circuits. The cheapest way to move telecommunications traffic from Tokyo to London and Europe was through the United States. In addition, the "dragons" of East Asia were also linked into the undersea fiber optic networks, including Hong Kong, Korea, Taiwan, the Philippines, Guam, etc.

Between the United States and Japan more than 20,000 circuits carried telephone calls from New York to Tokyo that sounded as though the person were next door. Webs of telecommunications networks were busy binding the world community even closer together. In the financial community, the 24-hour syndrome took hold in the late 1980s with firms monitoring the market as the opening hours moved

Exhibit 4.7

Submarine Optical Fiber Cables in the Pacific Basin

The number of circuits is calculated with a 64 kbps telephone circuit as a unit. Multiplexers can increase the cable capacity by about five times.

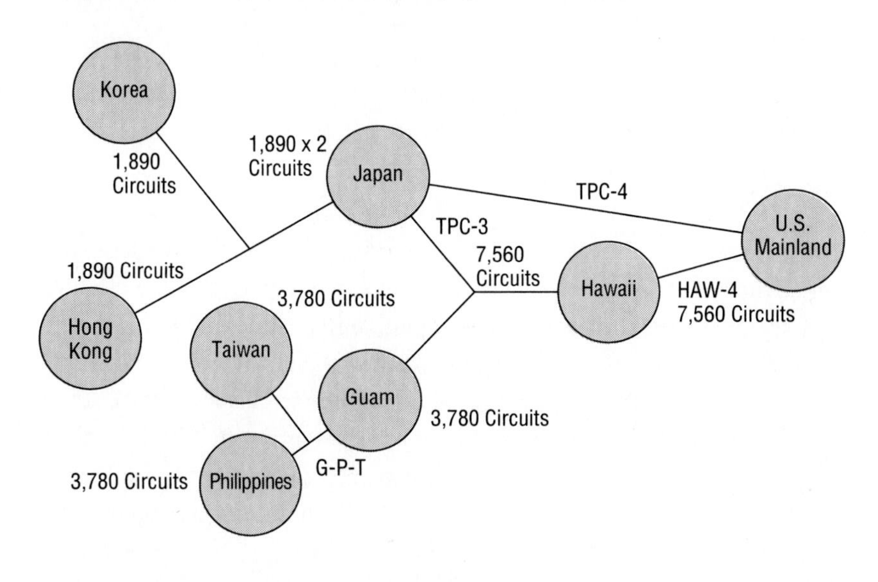

around the world. It became possible to cover positions in the Asian markets while New York or London was closed. The control and coordination of manufacturing operations on a global scale became even simpler as the vast developments in the capabilities of the global telecommunications infrastructures increased the capabilities of multinational corporations to operate on a truly global scale.

Satellite Networks and VSAT

VSAT stands for Very Small Aperture Terminal. Why "very small"? Before VSATs were introduced, building a ground station could be a very expensive proposition. Older generations of satellites used C-band frequencies for transmission of information. This made large satellite dishes necessary to receive and transmit information. Older generation satellite dishes were 10 or more meters across, and very expensive to

Exhibit 4.8

Satellite Frequencies are Different for the Uplink and Downlink

Band	Uplink	Downlink
C	5.925–6.425 GHz	3.7–4.2 GHz
Ku	14.0–14.5 GHz	11.7–12.2 GHz
Ka	27.5–31.0 GHz	17.7–21.2 GHz

install and maintain. VSATs use another frequency called the Ku band. Since the Ku band of frequencies are higher than the C band, as Exhibit 4.8 shows, the dishes can be very small. This single fact is one of the major reasons VSAT technology has taken off in the marketplace.

Another important factor in VSAT evolution has been the deregulation of the telecommunications marketplace in the United States. Deregulation has made it much simpler for organizations to build their own telecommunications networks and still maintain interconnection with regular telephone service. Size has been the driving factor, however. In fact a Ku band satellite dish is so small it can be sent in the mail or through a parcel service and assembled on site with relative ease. This is a very different situation from the manufacture, transportation, and set-up of a 20- to 30-foot-diameter satellite earth station.

VSAT satellite technology supports interactive data and video broadcasts. It typically employs a star configuration arranged in a point-to-multipoint architecture (see Exhibit 4.9). It is based on several major components:

- *Master Earth Station ("Hub")*
  The master earth station is located at the center of the network. The master antenna is typically larger than the dishes found at the remote sites, generally 4.5 to 7 meters in diameter.

- *Satellite Transponder*
  The transponder is the receiving and transmitting equipment located on the satellite. It receives the signals coming up from the ground station and then retransmits them back down to earth, but at a different frequency, where they are received by other satellite stations.

Exhibit 4.9

VSAT Satellite Technology

VSAT systems use a major hub with many remote sites.

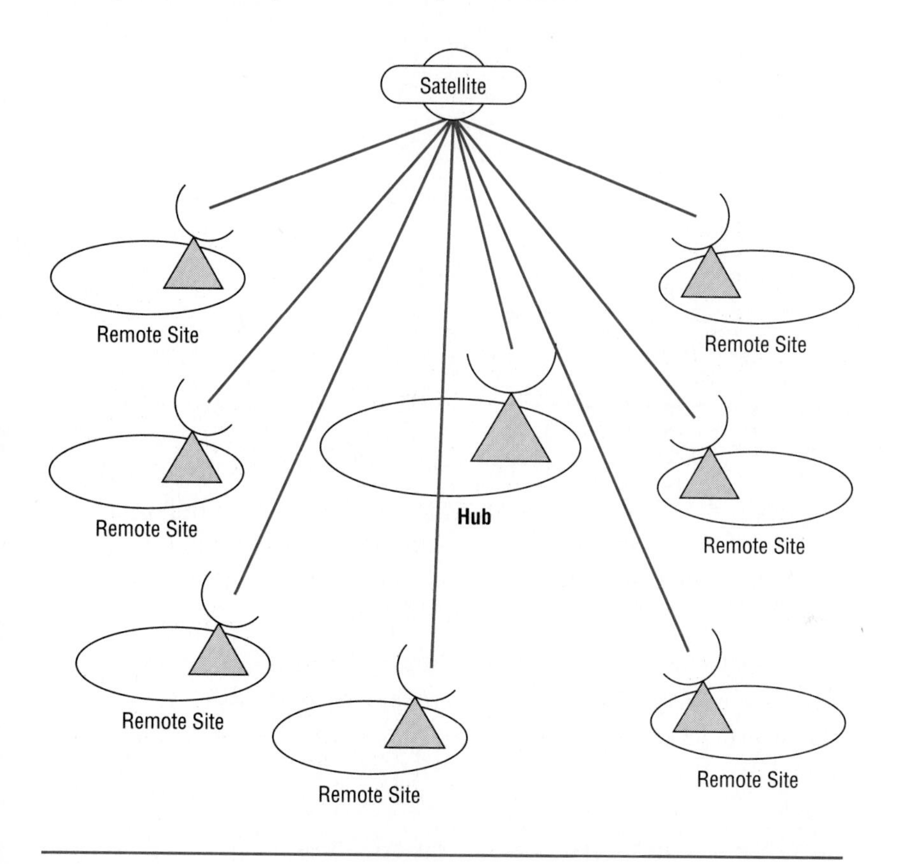

- *Remote Sites*
  The remote sites of the satellite network typically have small antennae 1.2 to 2.4 meters across. The sizes of the dishes depend to a great degree on their locations. Dishes closer to the horizon away from the satellite must be larger to better capture the incoming signals.
- *Computer, Voice, and Image Communications Interfaces*
  Each location has a considerable amount of equipment connected into the satellite network, including communications control

Exhibit 4.10

Growth in U.S. VSAT Sites

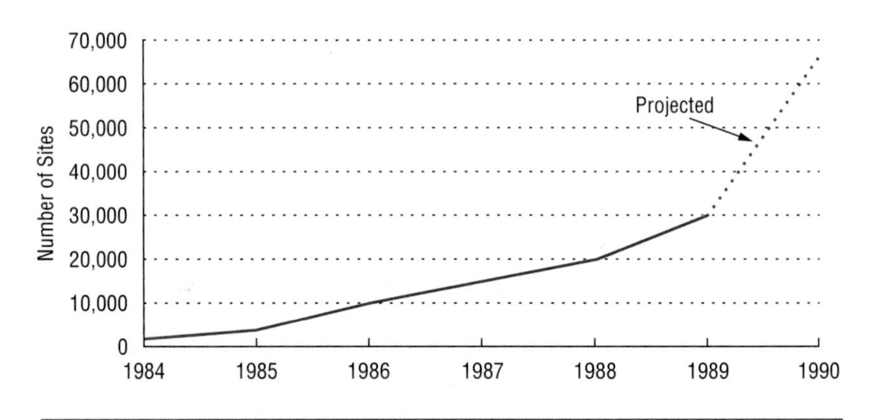

Source: Data courtesy of Scientific Atlanta.

devices, information processing technology, multiplexers, modems, etc. Also, video equipment changes video signals to high-speed digital transmissions. X.25 packet switched networks have PADs (packet assembler/disassemblers).

The availability of VSAT technology has driven its rapid growth (see Exhibit 4.10). Many firms from all parts of the economy use it to enhance their telecommunications capabilities, as shown in Exhibit 4.11.

Satellite Architecture versus Leased Lines

The basic network architecture for satellite systems is very different from that for leased lines. In fact, the critical difference between use of satellites and leased lines for telecommunications is the architecture of the networks. The satellite system uses a fixed, central hub as the control point. Each of the remotes can be placed anywhere within the footprint of the satellite, with very few exceptions. In contrast, each point in a terrestrial telecommunications network must be wired physically into the telephone system. As a result, a general pattern of a high-speed backbone network with multidrop lines connecting various lower volume locations has evolved. In a generic national network, the connections between major metropolitan areas would be served by

Exhibit 4.11

U.S. VSAT Users

VSAT users in the United States span all economic sectors.

Company	Type of Business	Number of VSATs
Psicor	Medical services	2
Toyota Lexus	Automotive sales	150
EDS	Systems integrator	300
Southland Corporation	Convenience stores	1,500
Wal-Mart	Department stores	1,500
Edward D. Jones & Co.	Financial services	2,000
Target	Upscale discount retail	420
Hartmarx	Specialty stores	500
Chrysler Corp.	Automotive manufacturing and sales	6,100
Kaman	Manufacturing	2
Pandick, Inc	Financial printers	9
GE Information Services	Business information services	500
Service Merchandise	Catalog retail	300
Computer Power	Telecommunications services	500
Pittsburgh International Teleport	Telecommunications services	8
Holiday Inn	Hospitality	2,000
Houston International Teleport	Telecommunications services	500

Source: Data from Hughes Network Systems.

high-capacity leased lines such as T-1 or multiple T-1 (see Exhibit 4.12). For linkages with smaller cities, lower volume leased lines, such as 56 kilobit per second lines, would radiate out from the points on the major backbone network. In turn, the smaller cities might use multidrop 9.6 kilobit per second lines to connect to small villages and hamlets.

A multidrop line is a leased line connecting together several points over one circuit. Like a leased party line, it provides an inexpensive alternative to multiple leased lines. Multidrop 9.6 kbps lines connect many sites to the medium-sized nodes of the network.

Ease of Reconfiguring a VSAT Network • When comparing the generic leased line network with the generic satellite network (such as Exhibit 4.13), it is easy to see that the VSAT network is easier to reconfigure. To add a communications node to the leased line network

Exhibit 4.12

Leased Line Network

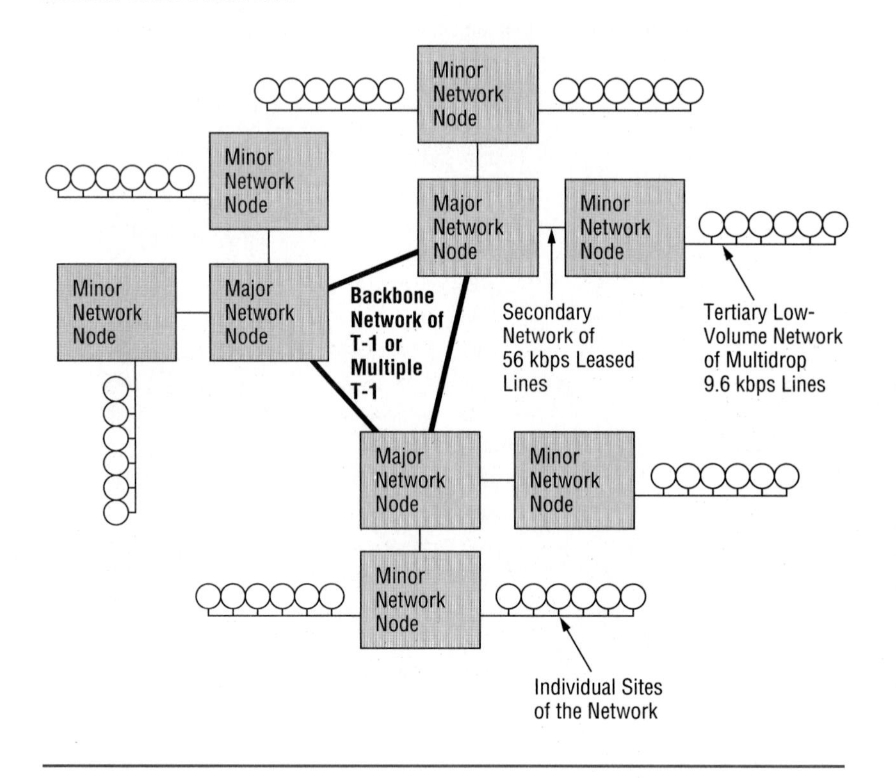

requires buying another leased line; adding a communications node to a satellite network requires only installation of the satellite dish and ground station. Addition of leased lines requires an application process and a delay in getting the line from the telephone company; to add a satellite remote station, just set up the equipment and turn it on. In the most severe case, the time difference between leased line set up and satellite ground station set up could be 6 months versus 6 hours. The implication of this is that organizations experiencing rapid growth or other change may find it better to choose satellite technology, although there is no hard-and-fast rule on this point.

Satellites Criticized for Unreliability • One of the most common criticisms leveled against use of satellites centers on their reliability.

Exhibit 4.13

Components of VSAT Network

A private VSAT network uses a large hub with many remote sites.

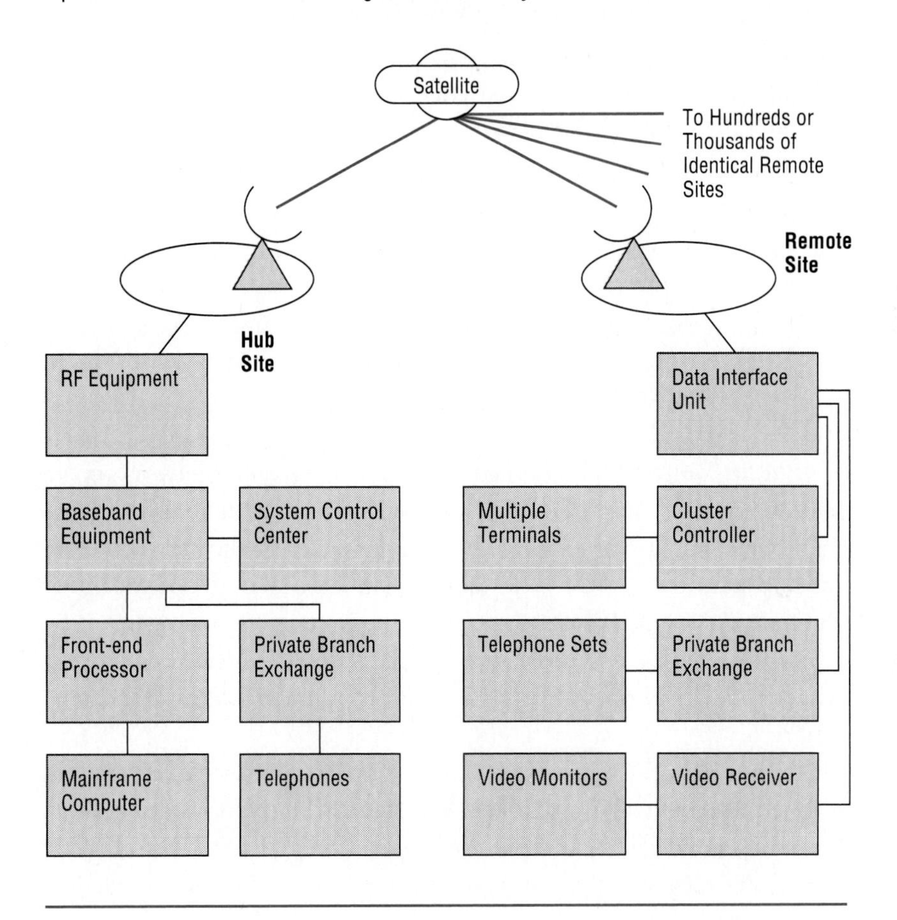

Satellite networks have several single points of failure. If the central hub malfunctions, the entire network will fail. Furthermore, if the satellite itself malfunctions, the entire network will cease to operate. In addition, reception of the signal can be disturbed by bad weather and other atmospheric effects.

To a certain extent, all of these concerns are very valid. However, they were more valid in the past than in the early 1990s. This is because satellite technology vendors have steadily improved the reliability and security of the data telecommunications carrier systems they are offering. In several ways, advanced, third-generation satellite networks can achieve very high levels of reliability and security:

- Packet switched X.25 networks can be used as backup in case of failure of either the entire network or any specific portion. These can be transparent to the user.
- A back-up hub can be kept on standby in case of a central hub failure, connected by a leased or dial-up line.
- Transponder failure on the satellite is backed up by quick switchover (within 3 minutes or less) to a spare transponder on the same "bird."
- Remote site failure can be prevented by equipping each site with a fail-safe switch which immediately links to a public data network in case of failure.

Bandwidth and Network Management • Satellite networks come in different sizes. Within a single network, it is possible to utilize more than one channel. For example, you might engineer the network from the hub station to have two outgoing channels and only one incoming channel. The critical factor in satellite networks is the arrangement by which the hub receives information from the remote sites. For example, what happens if two remote sites send information to the hub at the same time?

In practice, this problem is handled by a technique called **contention management**. After sending each signal, the remote station waits for an acknowledgement from the hub station. When two conflicting signals are received, the hub station refuses to send an acknowledgement to either of the conflicting remote stations. Each station then waits a random, independent period of time and tries to signal again. This type of contention scheme is called **"random aloha."** Random aloha has the advantage that it is relatively simple and inexpensive. Random aloha starts to get a high conflict rate, however, with channel loadings as low as 10 percent. At 18 percent, the system becomes unstable because so many retransmissions are required.

There are other schemes to control collisions, each slightly more sophisticated.

Slotted Aloha • This system is the same as random aloha, except that the times when the remote transmissions take place are timed at predefined intervals (slots). **Slotted aloha** remains stable up to approximately 36 percent channel capacity.

Reservation Access • The **reservation access** method is more sophisticated. It does not rely on conflicts between signals. Instead, a parallel channel carries signals from the remote to the hub when it is ready to transmit. The hub then sets a time for the transmission and informs the remote. The remote terminal in effect makes a reservation with the hub. Reservation access is useful because each reservation tells the hub the size of the incoming message, so it can reserve only the space the message requires. Of course, the reservation requests themselves are sent and received via the aloha methods, so even before the actual transmissions get to compete for time to transmit, the reservations can be colliding and queuing for retransmission. Also the response time is slower because the transmission first has to wait for transmission of the reservation and confirmation (two trips across the circuit) before being sent itself.

Stream Access • The method gives each remote a specific time slot in which to transmit. It is appropriate to applications, such as remote oil field monitoring, that generate a regular flow of data.

Adaptive • Adaptive methods for channel management are found in systems which can automatically switch from one type of scheme to another. For example, Scientific Atlanta equipment automatically switches to a more sophisticated channel management scheme when too many collisions start to occur.

VSAT Application Limitations • Choosing the VSAT option for a network is generally feasible only under specialized conditions. VSAT technology is not for every application. Discussion of some of the factors to consider follows.

Centralization and Decentralization • In the absence of a highly centralized information processing function, it makes little sense to operate a large hub. If an operation, such as retailing operations, is highly centralized, however, this is a positive factor for satellite technology.

Response Times • Satellite transmission of signals has inherent built in delays in propagation. Depending on the loading of the network and the type of contention management scheme being employed, response time can deteriorate severely. In applications where response time is critical, satellite operations are generally not feasible. For example, in a foreign exchange trading environment where extremely fast response time is a component of competitive advantage, satellite technology would not be appropriate.

However, for batch transmissions of information, satellites may be highly suitable. For batch transmission, a few seconds here and there are not critical or even noticeable.

Telecommunications Costs • Satellites may offer a poor man's network under certain circumstances. They may be 20 percent or more cheaper than traditional leased and dial up lines. For organizations that work on very low profit margins, and are therefore very sensitive to administrative costs, satellites may offer a reasonable route.

In some cases, however, telecommunications as a whole may be a small part of information processing expenses. Data processing expenses may, in turn, constitute a small part of overall administrative expenses. In such a case, that saving of 20 percent to 30 percent on telecommunications costs may make little real difference. For example, in securities or foreign exchange trading systems, telecommunications cost is not very important, relatively speaking. Such organizations are perfectly willing to purchase extra circuits as backups in case of failure. It costs more, but the costs mean little in the face of the possible financial consequences of a failure in the telecommunications system.

Voice Networking • Since VSATs are not ideally suited to voice networks, it does not make sense to use them in a business that is highly dependent upon voice communications, such as telemarketing. Most users include voice in VSAT systems for occasional use for internal calls where it is not critical to maintain call quality. The propagation delay in using satellites is also very important in understanding the problems with voice applications.

Volume of Data Transmission • VSATs are not suited to very heavy data transmission. A VSAT network cannot provide the capacity of a large SNA type T-1 or multiple T-1 backbone network. However, an organization that has relatively light data transmission needs will find that the contention problems will not cause too much loss of competitiveness. Satellites remain a good option for such firms.

Toyota's Hughes VSAT

Toyota has long been recognized as a producer of economy cars around the world. Partially to respond to the increase in "yuppies" in the United States market, Toyota made the decision to upgrade the cars it was producing for the U.S. market. In 1989 it introduced the Lexus, a more upscale car with either 6- or 8-cylinder engines, which was designed to compete in the luxury market with cars such as BMW and Jaguar.

To successfully introduce the automobile, Toyota wanted to have a completely integrated information and telecommunications system for its dealer network throughout the United States. After careful consideration, Toyota decided to build a Hughes Network Systems VSAT system to link its headquarters in Torrance, California with each of its dealers through integrated video and data communications. Toyota decided to build the dedicated hub for the satellite system at the headquarters. By the fall of 1989, approximately 100 Lexus dealerships were scheduled to open.

Each of the dealerships will have a Hughes Personal Earth Station™ on-site to send and receive signals. Toyota and Hughes decided to install the VSAT network in three phases:

- Build the uplink and hub control point at the Toyota headquarters in Torrance, California
- Build VSAT connections to each of the Lexus regional offices throughout the United States
- Build VSAT linkages to each of the individual dealerships located throughout the United States

"The satellite communications system is an example of our commitment to servicing the Lexus customer effectively and efficiently," noted Jim Perkins, Lexus senior vice president. "Through this network,

Lexus customers can have their vehicles serviced at any Lexus dealership and be assured that the dealership has access to a complete history of their car. Usually, such information is available only at the dealership where the vehicle was purchased."

By using the VSAT system supplied by Hughes, Toyota was able to completely eliminate expensive leased telephone lines and local connections. At the same time, Toyota was able to vastly improve service quality. Another great advantage of using VSATs is that, as the Lexus dealership network expands, the communication network can be easily reconfigured just by adding more VSATs. The firm need not go through the wait time and paperwork of leasing more telephone lines.

Edward D. Jones & Company's 2,000+ VSAT Sites

Founded in 1871, Edward D. Jones & Co. provides personal investment counseling to conservative investors in 37 states. With over 1,300 offices nationwide, the firm is one of the largest in the investment industry catering to individuals. In order to conduct its business, Edward D. Jones & Co. must process buy and sell orders, introduce products to the market, train internally, and provide a great deal of financial information to its numerous offices. At the same time, like all financial services companies, Jones faced pressures on margins and continued pressure to cut costs. Jones decided to use VSATs to link together more than 2,000 remote terminals installed at its facilities throughout the United States.

The headquarters in St. Louis houses the central hub. The system, supplied by Hughes Network Systems, was designed to transmit interactive data and video communications. This will help accelerate the speed of financial transactions for Jones. Jones plans to use the video capability for production and program introduction along with training and a few direct sales applications. According to John Bachman, a managing partner at Jones, "this new system will not only move all data via satellite, it will [also] enable us to transmit live television programming into each of our local branch offices." The VSAT stations support both synchronous and asynchronous data telecommunications, as well as voice telephony and broadcast-quality video.

Miscellaneous Technologies: Infrared Line of Sight

The future for glass fiber is very bright indeed. Eventually it will replace a large part of the installed backbone networks. In the developing world, it has emerged as the medium of choice. Other options remain, however.

Infrared line of sight uses a highly focused beam of infrared (not laser) light transmitted approximately a mile to a receiver. It is often used to bypass the local telephone companies in the United States. In such an application, two relatively close buildings of the same company could set up an infrared linkage between them, avoiding the expense of going through the local telephone company.

Infrared transmissions may deteriorate in bad weather conditions. This is not a widely used technology, but it does offer a cheaper alternative to the complexities and costs of the local telecommunications authority. It can also link buildings together without crossing public rights of way.

Summary

The basic infrastructure for telecommunications transmission is based on the copper wire. This technology is being replaced by newer types of technologies. The most important new technology is fiber optics. Fiber optics allows much more information to be transmitted at higher speeds over smaller media than copper wire. These higher speeds are necessary to accommodate the data communications requirements of progressive generations of technologies, from mainframes to terminal telecommunications to networked image workstations with very high computational requirements and large file transfers.

The newest emerging standard for fiber optics is the Fiber Distributed Data Interface standard. By the early 1990s, FDDI will allow the speeds of local area networks to approach data communication speeds associated with mainframe computer channel speeds of the 1980s, that is, from 1 megabit per second to 100 gigabits per second.

Another important technology shaping the development of international business is the continued proliferation of undersea fiber optic submarine cables which are quickly linking together the world's financial centers. Cellular radio technology has also proven to give businesses more strategic options under some circumstances, particularly in the area of logistics. Satellite technology also serves as another important alternative to the traditional terrestrial system built upon copper cabling and microwave transmissions. However, current satellite technologies are good only for specific types of applications: when the growth of the company is great, when data speeds are relatively low, and when interactive response time is not very important.

Questions

4.1 What are the basic alternatives to copper cable for carrying telecommunications signals?

4.2 What advantages does fiber optics have over copper cabling for telecommunications?

4.3 a. Explain the basic principles behind cellular radio systems.
b. What is a hand-off for a cellular telephone?

4.4 a. What does *FDDI* stand for?
b. What are the data communications speeds promised by FDDI?

4.5 Give approximate data communications speeds for each of these media: dial up, T-1, T-3, ISDN, and FDDI.

4.6 a. What is the difference between an uplink and a downlink?
b. What is the difference between the C, Ku, and Ka bands?
c. What is a hub and how is it used?

4.7 What are some examples of use of cellular radio telecommunications?

4.8 What are some ways in which VSATs have been used to get competitive advantage?

4.9 Do VSATs work well for every application? If not, then what are the best types of applications for VSATs?

4.10 Why is it easy to reconfigure a VSAT network?

4.11 What is the difference between random aloha, slotted aloha, and reservation access?

4.12 Discuss the Toyota Lexus case.

Chapter 5

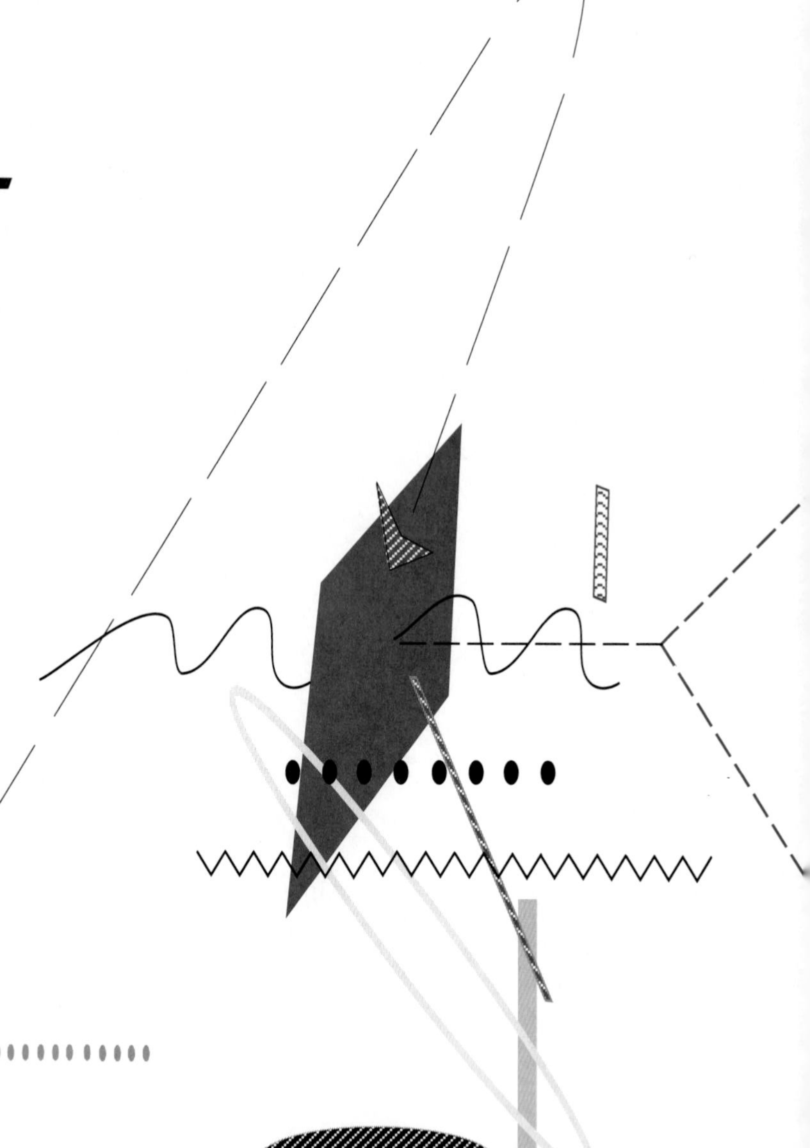

Vendor Solutions to Computer Networking and Information Processing

IBM, Digital Equipment, and Cray Research

The essence of telecommunications applications in business is the linking together of complex information systems. Two major vendors—International Business Machines Corporation (IBM) and Digital Equipment Corporation (DEC)—have done much to establish the dominant approaches to telecommunications networking. As a gross generalization, one could say that the IBM approach is centralized and the DEC approach is distributed, but these distinctions are changing. A third type of networking involves supercomputers. Although supercomputers are not yet widely used in business applications, apart from heavy engineering and design, a review of Cray Research's approach is included here since it leads the way towards a type of highly centralized solution that will emerge naturally for certain business applications in the 1990s.

IBM's Systems Network Architecture (SNA)

IBM's **Systems Network Architecture** originated in 1974 when the distributed function allowed each device attached to a mainframe computer to become independent of the application it was using. Prior to this, each application had a code for an attached device, and the device was therefore trapped with the application. SNA allowed a device to work with more than one application. The device became independent.

By 1975 and 1976, the **Virtual Telecommunications Access Method (VTAM)** was working on IBM's massive 370 machines and communications control moved away from the mainframe into a nearby programmable front-end controller, where it remained into the early 1990s. By 1976, with distributed processing already in place, multihost networking expanded the numbers of hosts that could be linked together. Network management features were added in 1978, then advanced networking in 1979. With connectivity enhancements, multiple network connectivity was added in 1983, enabling IBM users to link together multiple SNA networks.

In the early 1980s, the personal computer was introduced by IBM, and communications between smaller "peer" machines became important and peer-to-peer networking started to develop. This peer-to-peer communications was achieved without the hierarchical nature of VTAM with a central processor at the top of the system. SNA continued to develop, as well, with 1984 bringing expanded network size, enhanced network management in 1986, etc. By 1988, SNA supported high-bandwidth transmission. At about that time, Open Systems Interconnection (OSI) began to grow in importance, and IBM users were faced with operating widely varying networks around the world.

Under the IBM system, it became possible to link 370 machines through their communications processors in many different configurations, such as tree, multidrop, and star arrangements, as illustrated in Exhibit 5.1. The type of configuration used was highly dependent upon the geographic locations of the various data centers and the historical growth of the data processing establishment. For example, large mainframes may be connected together through this system within a

Exhibit 5.1

Front-End Communications Processor in IBM Network

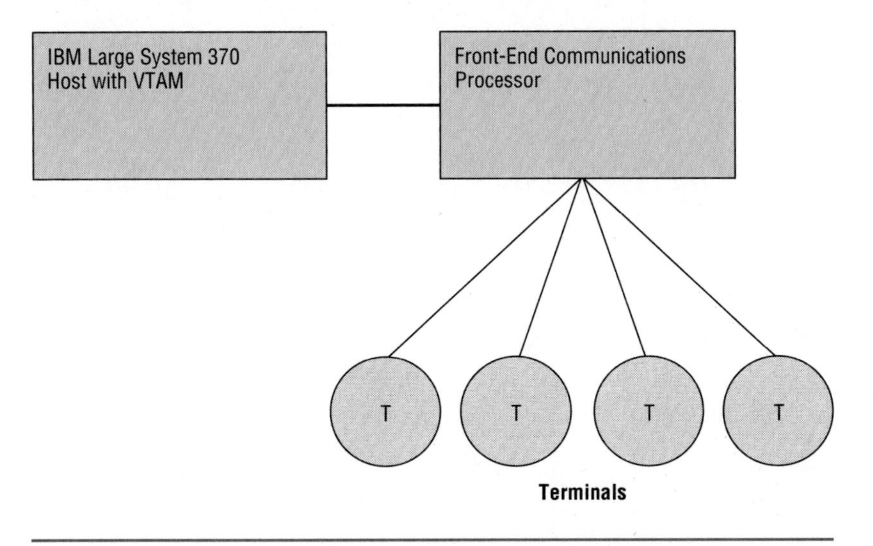

single data center location, as new processing applications are added. Under these circumstances, a star topology may be used. The multidrop topology may best serve to connect sites that are strung out across the country, creating, for example, a New York–Los Angeles network with a few drops between in the Midwest. The tree topology might be used when subprocessing areas correspond to different parts of the business, but this creates a need for coordination between the different business units. Exhibit 5.2 diagrams these topologies.

The fundamental building block of the IBM SNA network is the backbone network. This backbone network forms the infrastructure that IBM designed over time to accommodate many different services, such as X.25 packet switching (common in Europe) and the connection of sub-networks. The basic SNA networking function is designed to allow any terminal to connect to any application throughout the system, regardless of the geographical distance between the data centers. There are, however, many variations on this theme, and the complexities and varieties of IBM systems dictate that not all devices will link to all applications. **Total connectivity** is a stated goal for IBM to work toward, not a reality available to all users.

Exhibit 5.2

IBM Network Topologies

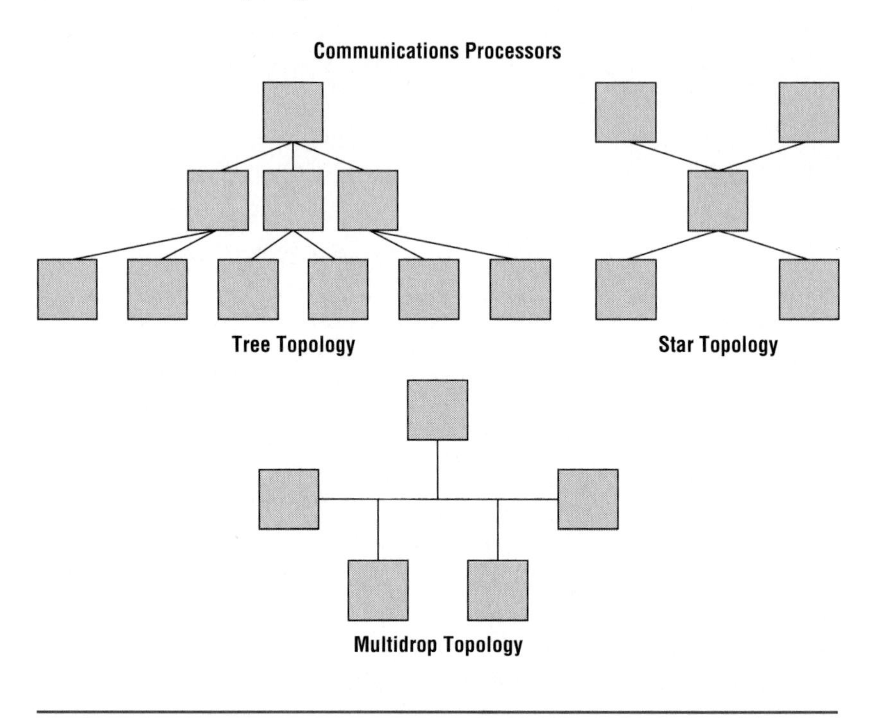

As the rise of minicomputers started to challenge the authority and control of the core mainframe systems, SNA adjusted to allow the minicomputers to join into the networks to share data and applications. During the early and middle 1980s, IBM faced severe criticism from users for its proliferation of different standards and different computer architectures. Systems 36, 38, 88, and others did not share data well with each other, while at the same time the user was receiving pressure to build integrated corporatewide information systems. These two tendencies were clearly at odds with one another.

IBM's later response was to introduce **Systems Application Architecture (SAA)**, which had as its goal the linking together of all systems at all levels. An important part of this was networking with minicomputers. Digital Equipment had challenged IBM hard in this area, since DEC's architecture, many argued, made it easier to link machines together. IBM's SNA was changed to accommodate minicomputers running from the network control processor. In addition, IBM changed its design somewhat to allow minicomputers to link to one another without going through the VTAM system on the mainframe in a system like that in Exhibit 5.3. This was a considerable departure from IBM's traditional strategy of controlling all network functions centrally through the mainframe.

As personal computers were introduced, the central MIS function again faced connection problems. How could one connect personal computers together to share printers and files? How could one integrate personal computers into the minicomputer environment? How could one make personal computers work with the mainframe?

As a first step, IBM introduced a personal computer that emulated the 3270 terminal. Thus, in some modes it would appear to the mainframe to be a 3270 terminal. This machine did not sell very well, since it provided little integration of the processing power on the personal computer and the processing power of the mainframe.

Another step towards integration was taken when IBM introduced its version of the local area network, based on the token ring system. This allowed the personal computer to share resources across the token ring network. For example, they could share files, printers or modems, and even applications (if the applications allowed it).

By the end of the 1980s, IBM was linking the token ring network into the main SNA network through the network control processor. The token ring could also link to minicomputers, including the new AS/400 system. The IBM solution was not the cheapest on the market, but planners could argue that at least the solutions being offered fit into a coordinated, long-term architectural strategy.

What this meant for the user and the planner depended upon their working environments. A new organization could consider a variety of very powerful smaller solutions based on personal computer workstations or minicomputer systems. However, for very large installations, the backbone systems of the major corporations, the planners found they were locked into a long-term developmental strategy in which gradually more and more disparate elements of information technology were linked together into the core system. The SNA networking strategies made that possible by linking the minicomputers and the workstations through the token ring local area network.

Exhibit 5.3

IBM Systems Network Architecture

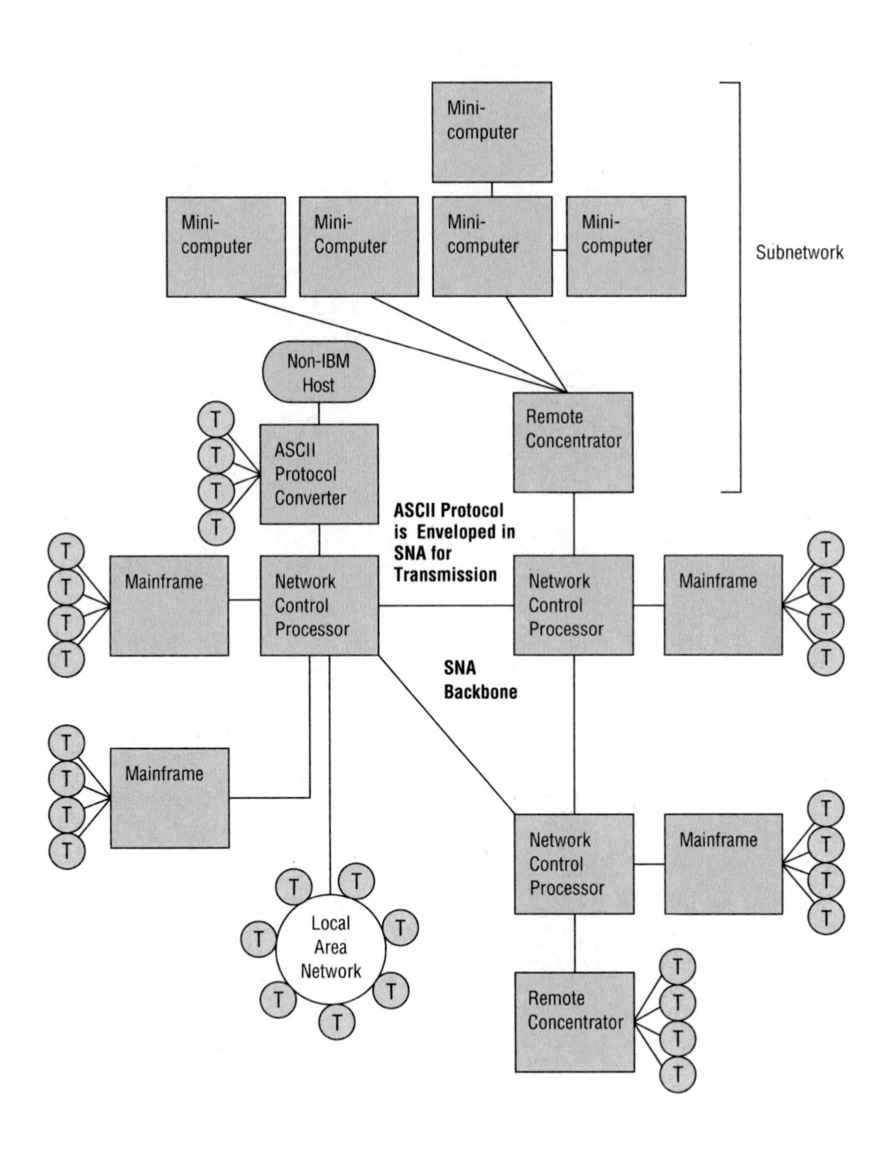

Exhibit 5.4

PBXs in SNA Network

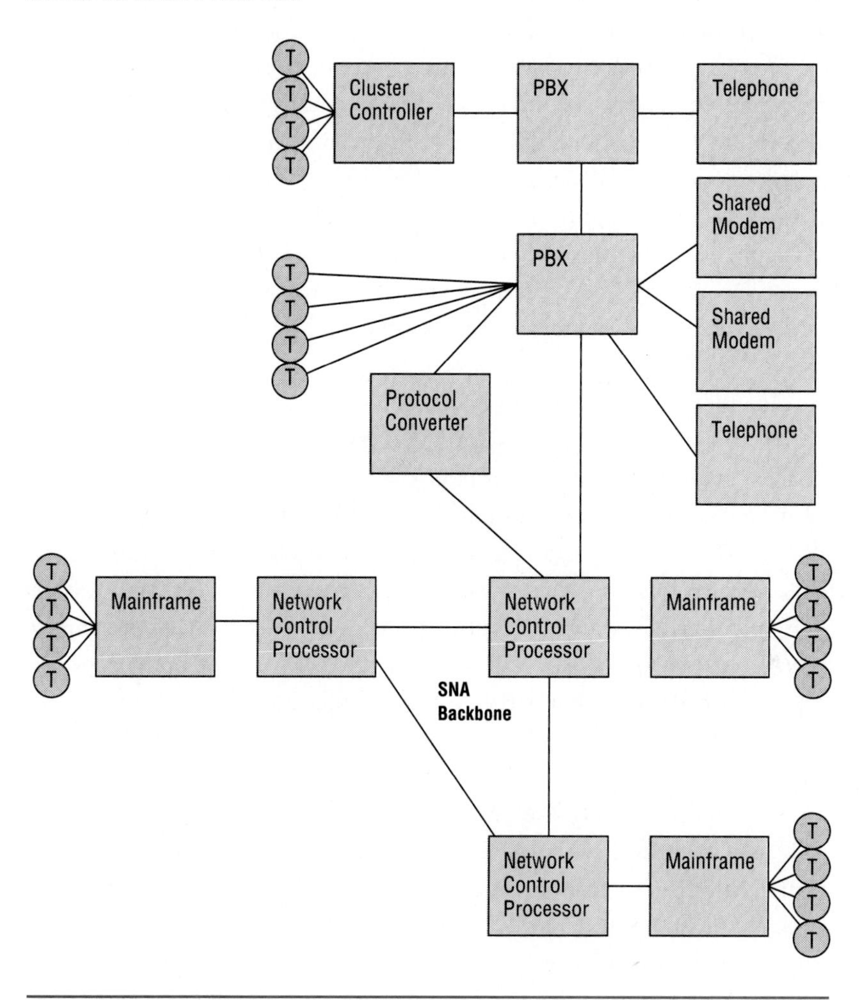

The SNA network strategy has grown to accept and handle other types of data communications, as illustrated in Exhibit 5.4. Using a PBX and a protocol converter, the SNA network is able to handle transmission of ASCII type information, as well. ASCII devices are very cheap, and almost any personal computer or terminal is capable of working in the ASCII format. Since many applications do not require much more advanced terminal capabilities and presentation (i.e., graphics) than is offered by ASCII, this alternative has proved to be popular. The protocol converter takes the ASCII encoded information and changes it to IBM code. Using this system, it is possible for ASCII terminals to access the IBM applications and operating systems that expect IBM equipment at the other end of the line. This alternative is, of course, much easier than rewriting the applications to work with the ASCII equipment. The PBX can also handle the pass-through of information from a cluster controller to the mainframe network. The cluster controller is linked to many terminals and devices to give them access to the mainframe network and its resources.

In many areas of the world, particularly outside of the United States, the most prevalent form of data communication is through public packet switched networks. In order to foster internationalization of data communications standards, many countries have adopted the ITU's CCITT X.25 standard for packet switching. For example, an organization building a network in France must almost certainly use the X.25 standard for data communications. It may also be very difficult to procure a private leased line. Instead, the PTT might encourage the organization, through tariffs and other measures, to use the public packet switched network. These circumstances require a different type of interface.

The SNA scheme also supports the integration of the traditional SNA environment with the CCITT X.25 specification for data communications in a system like that in Exhibit 5.5. It does this by a process of enveloping in which X.25 data are sent over the SNA backbone network "wrapped" in SNA code so it appears to be a standard SNA data stream. As the data reach the X.25 interface, they are unwrapped and sent through the packet network. If a set of standard IBM terminals waits at the other end of the X.25 network, the cluster controller changes the data back into standard IBM format. To the terminal, it appears that it is hooked up to the backbone SNA network directly.

IBM's architecture also supports complete integration of voice and data over the same backbone network (Exhibit 5.6 gives an example) without disturbing the SNA data communications channels. The key to

Exhibit 5.5

SNA Network Interface with Packet Switched Network

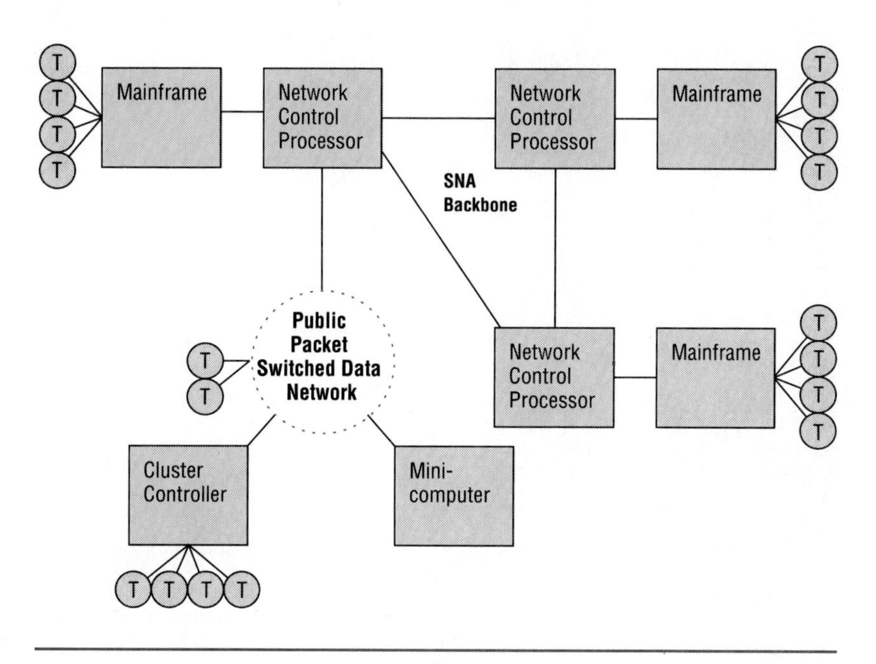

this is the placement of a bandwidth manager between the front-end communications processor and the backbone network. The bandwidth manager divides the large-capacity circuit of the backbone network into data and voice channels. The backbone network can consist of many T-1 channels, and therefore very large amounts of data and voice can be transmitted.

Since the front-end telecommunications processor is still linked into the mainframe, the bandwidth manager appears to be a separate network and the information passes through as though there were no interference. The switching of telephone circuits takes place through the PBX, which can dispatch each call in the correct direction, either internally to another branch office within the organization or to the outside. For calls that are either forwarded or placed internally, the organization is able to enjoy both security and savings from having its

Exhibit 5.6

SNA Network to Integrate Voice and Data

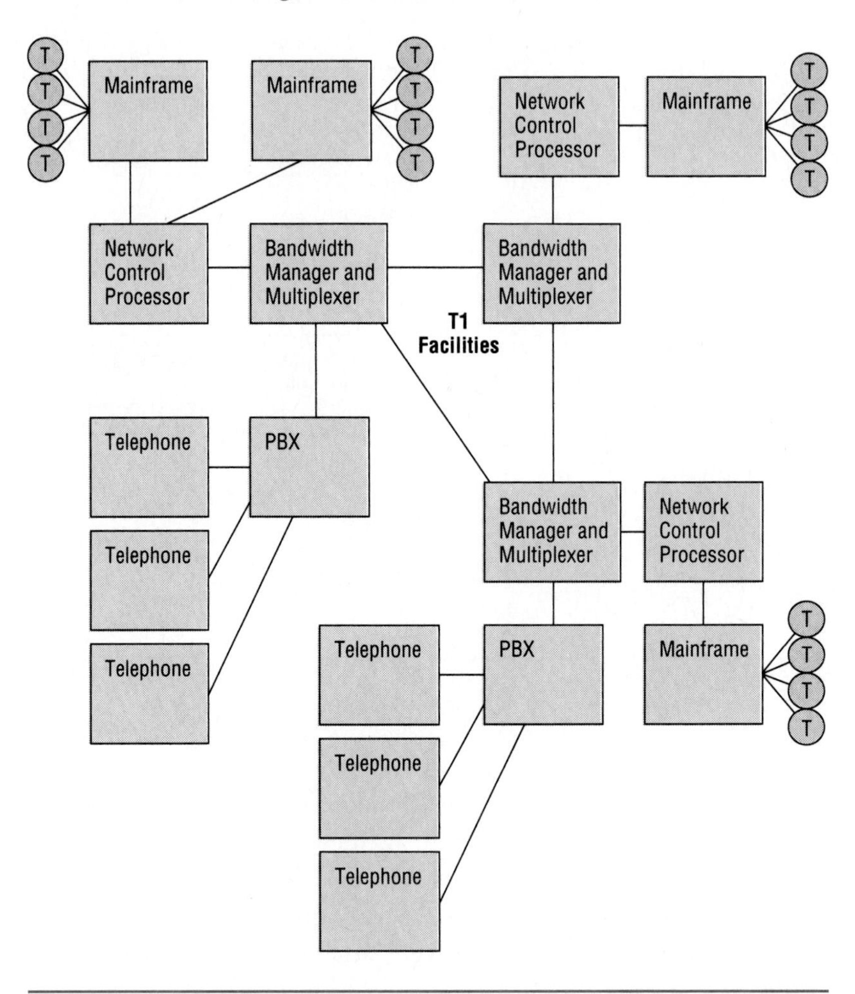

own network, thereby avoiding the public network with its taxes, charges, and occasional unreliability.

The Role of the Bandwidth Multiplexer

The high-capacity telecommunications channel carries data from information systems using the SNA protocols. However, voice traffic uses other channels over the same link. The bandwidth manager apportions

Exhibit 5.7

Bandwidth Manager's Role

The bandwidth manager apportions the telecommunications channel between voice and data traffic, depending on need. During nonoffice hours, it alters circuit allocation to accommodate more data transmissions.

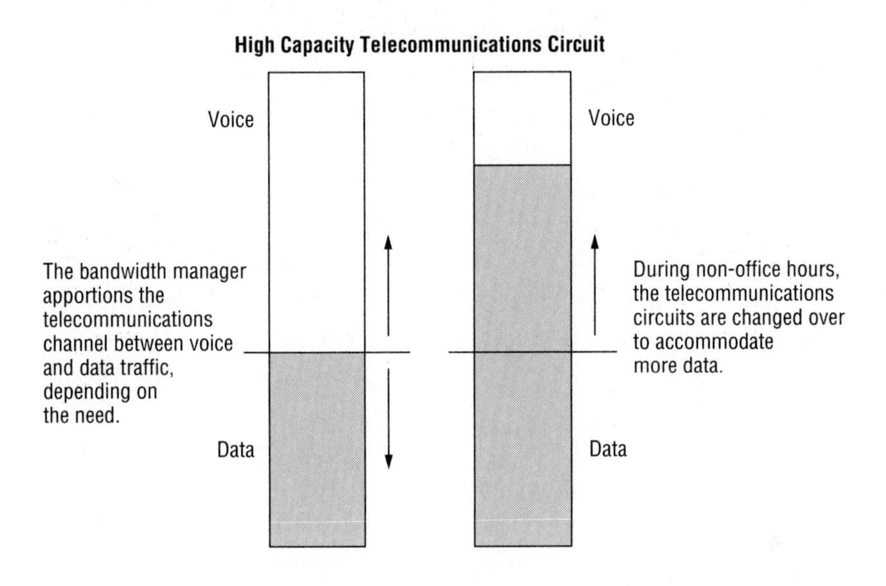

the circuits between voice and data applications depending on applications demand, as Exhibit 5.7 illustrates. For example, during off hours when voice demand drops, the network can be changed to handle more data traffic which may be flowing through the system during the off hours. This can represent a major savings, and allow maintenance operations, such as large batch jobs, to be telecommunicated during off hours. The bandwidth manager is also useful for temporary high-capacity applications. For example, video teleconferences can be transmitted temporarily on the backbone linkages. After the teleconferencing session is completed, the bandwidth can be reassigned to data and voice traffic.

With the introduction of the powerful 390 series, IBM took a large step in supplying integration across many sizes and types of machines, which prior to this had culminated in the 3090-600 series.

DEC's Enterprise Networking

As of the late 1980s, the Digital Equipment Corporation (DEC) had begun to capitalize on its ability to fully link systems at the departmental, campus, and remote levels. DEC introduced a data communications system in 1975 called **DECnet**. Remember that IBM's SNA was introduced at about the same time. Through the late 1970s and 1980s, however, a great deal of pressure began to appear at the international level, particularly in Western Europe, for standardization of computer networks. Users wanted ways of linking systems of very different vendors, and there were also concerns that fostering proprietary protocols and systems could give an unfair monopolistic advantage to computer companies that had commanding market shares. This criticism was targeted, of course, principally at IBM, which had and still has a dominant share of the information systems market throughout the world, except in Japan where it faces a restrictive industrial policy and, in some cases, superior technology.

At the same time users were complaining about IBM's systems being incompatible with one another, DEC was able to demonstrate a full line of upward compatible systems that communicated well with one another. For a DEC customer, *upward compatibility* meant that they could buy a small machine, and then as their company and its information processing requirements grew, they could add more DEC equipment or larger machines without having to rewrite or convert software applications and data. This was a considerable advantage to the customer, since conversion of applications and databases takes a very long time and is expensive and prone to errors. DEC's strategy was very heavily telecommunications oriented because, although it did not make the very large machines which were made by IBM, it was able to deliver considerable processing power through linking together machines in its networks. In addition, DEC promulgated an open architecture strategy which was intended to fully support the efforts of the International Standards Organization, particularly as regards the Open Systems Interconnection (OSI) reference model.

In order to accomplish this degree of interconnection, DEC based its networking strategy on the Ethernet local area network (LAN) standard, which was traditionally 10 mbps. Within an individual LAN, many devices could be connected and given network identifications.

For example, one LAN could have several VAXs with a server connected to many terminals and printers. Within that LAN (typically encompassing a department), users could access applications on the VAXs, which were also connected.

In order to accomplish its goals of interconnection, DEC made it possible to connect non-DEC mainframe computers, as long as they used the OSI reference model. In engineering environments, supercomputers such as the Cray XMP could be linked into the network. DEC provided an SNA gateway to connect the LAN to the IBM world for access to both the IBM mainframes and backbone SNA networks found in most very large organizations. Personal computers, such as those made by IBM, Compaq, and others, could also be connected to DEC local area networks for accessing applications. Also VAX stations could be added for high-speed processing.

When the departmental local area network strategy was understood, the user could then begin to link together various local area networks through DEC-supplied equipment called **LAN bridges.** If two local area networks were connected together through a bridge, then users on either of the local area networks could communicate with one another as though they were operating on the same local area network. A typical example would be a company with large local area networks on each floor of a headquarters building. In order to link the different local area networks, a LAN cable would be run through conduits from floor to floor. Each department would link into the floor-to-floor network through a LAN bridge. The effect of this was to allow workstations and terminals on, say, floor 1 to communicate with applications and data residing on a VAX on floor 3.

At the next level, DEC made it possible to link together different local area networks between widely distant geographic locations, such as from New York to Hong Kong, London to New York, Chicago to Atlanta, etc. Since there were several major media over which the information could be transmitted, DEC provided several different interfaces:

- For using public packet switched networks with the X.25 standard for data transmission (such as Tymnet and Telenet in the United States; Nordic Net, EDS, RETD, TRANSPAC, EDW, or EPSS in Europe; or INS in Japan), a gateway was provided. Any LAN could be connected to any other LAN in virtually any country where public packet switched networks were available.

- For using a high-speed digital network (such as an undersea fiber optic link), usually provided by an International Record Carrier, DEC provided a TransLAN bridge which could link local area networks over great distances to make them appear to be almost the same network.
- Satellite or leased lines could also connect routing equipment (DECnet Routers) which could switch pieces of information between different LANs. (The router ensures, for example, that information within a LAN stays there, and it passes on information destined to other locations through the network.)
- High-speed fiber optic links could be used to link together bridges across distances of around 3 kilometers or less, for example, from one building to another in a large industrial complex.
- High-speed microwave line-of-site links could link together LANs by using a special bridge, for example, to connect two sites up to about 5 miles (7 to 10 kilometers) apart.
- Using an **SNA gateway**, information from one LAN could be connected to another across a backbone IBM SNA network. This was a very useful feature for organizations that already had in place large IBM SNA networks, allowing them to add network flexibility without having to buy new circuits and set up parallel networks.

Once DEC came to a reasonable resolution of the networking problems between the many different LANs that would occur in a large private network in a business, it began to implement what it called "Phase V networking." **Phase V networking** was designed by DEC to be fully compatible with the International Standards Organization's **Open Systems Interconnection (OSI) reference model.** The OSI model provided for extended addressing of computers and other resources attached to the telecommunications network (including LANs, of course). This strategy was particularly important for applications such as Electronic Document Interchange, which involved linking networks between different organizations, or the interchange of information from one network to another.

The OSI model gives every computer and related resource a unique name, in much the same way that each telephone in the world has a unique number. When placing a telephone call from one country to another, for example, from New York to Rio de Janeiro, the call is being placed from one network to another. In the United States several private telephone companies exist, and placing calls also involves moving from one network to another. The area code system ensures that even if the local telephone number is the same in one city as another, the extended number will be different.

OSI uses a similar hierarchical system to ensure that each device connected to a network has a unique identification. Within the DEC Phase V system, this means that information and data can be sent from a private network to specific computing resources located on other networks. When transmission of documents includes images as well as data, or fully formatted and color documents, it will truly represent a great revolution in global telecommunications. Communication will become so fast, one wonders what will become of paper and that venerable institution, the Post Office!

Network management in the DEC system is both centralized and decentralized, depending on the size and complexity of the network. (See the example in Exhibit 5.8.) It should be clear to the reader that if a network has tens of thousands of nodes and points, the idea of total centralized control must be illusory. What type of staff would be needed, and how would network control actually take place?

The DEC approach to network control involves decentralization of certain key functions. This decentralization typically takes place as a shadow of the underlying local area network configuration and topology. Within the area of each local area network, a network group (as few as a single person) is in charge of network maintenance and control for that group only. This relieves the central network control staff from having to constantly manage and fix portions of the network which are most accessible from the local level.

The centralized network control function is typically put in charge of the overall network, and in particular the linkages between different geographic regions and centers. Central control is responsible for ensuring that the network node names and equipment names are consistent and logical throughout the network. It also acts as an electronic clearinghouse for control messages, complaints, and hot line activities.

Exhibit 5.8

DECnet Alternatives

DEC creates many alternatives for connecting distant local area networks.

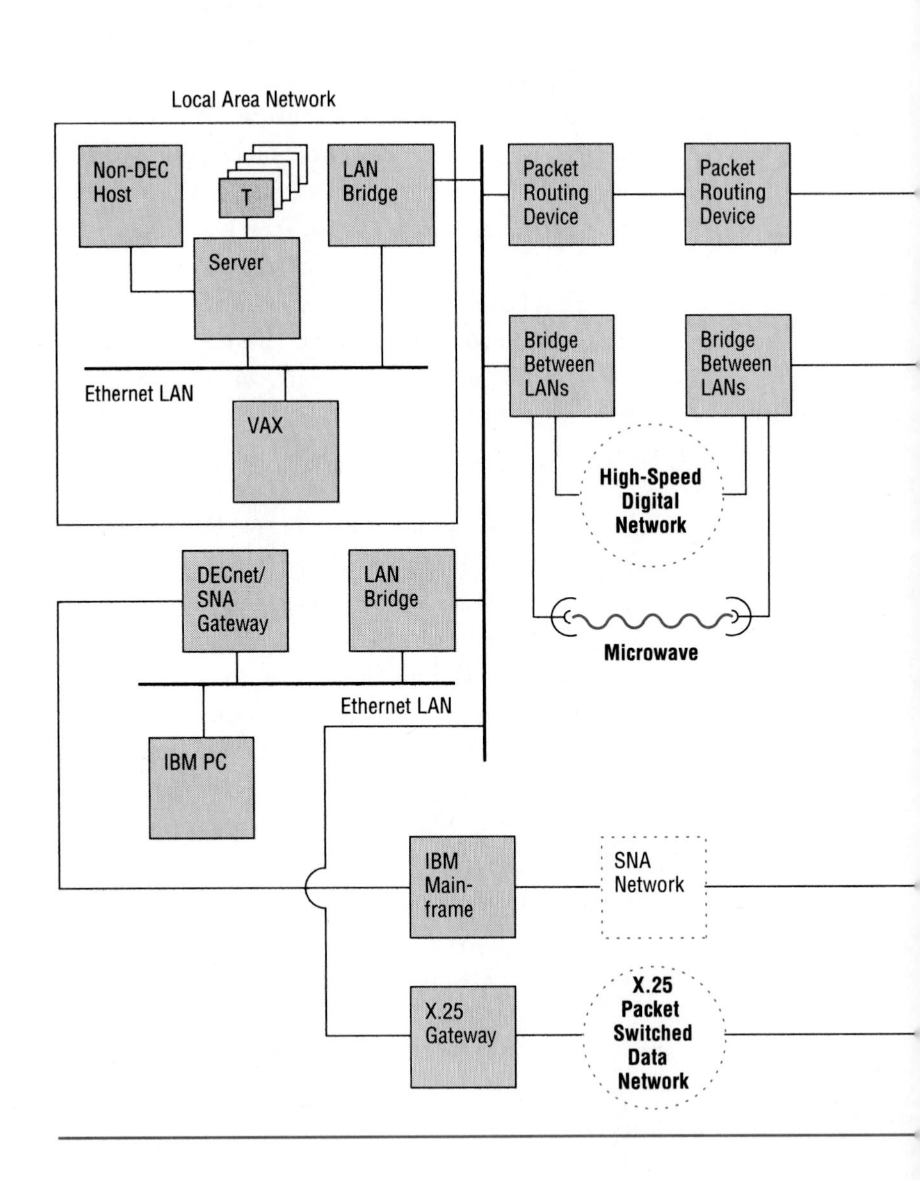

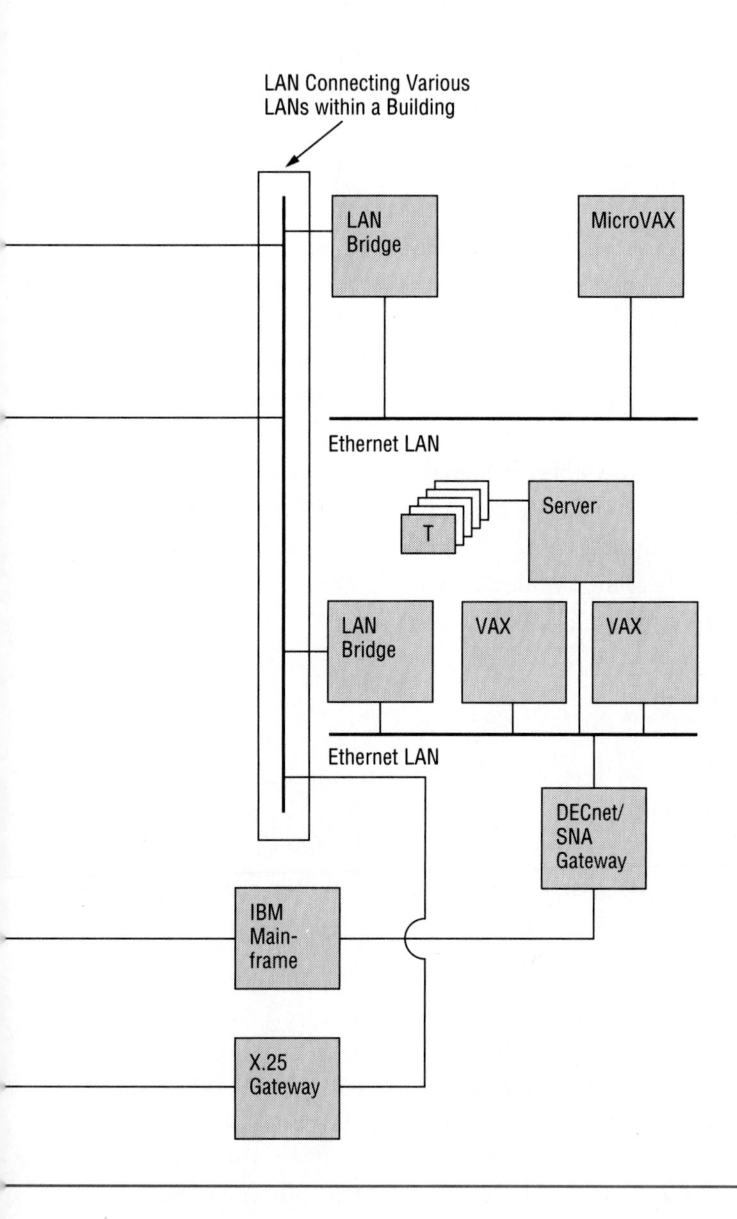
LAN Connecting Various
LANs within a Building
LAN
Bridge
MicroVAX
Ethernet LAN
T
Server
LAN
Bridge
VAX
VAX
Ethernet LAN
DECnet/
SNA
Gateway
IBM
Main-
frame
X.25
Gateway

It is very important to the DEC environment that the equipment from different vendors can be integrated into the network. Persons using a workstation or terminal at any location should logically be able to connect to major computing resources wherever they are located. For example, in an engineering environment, laboratories and design centers may be located in many different locations, but the organization as a whole may have only one supercomputer for running complex simulations. Using the network, everyone is able to contact the supercomputer to do their work. In addition, the ability to integrate many different vendor environments gives DEC a great edge as a provider of a communications medium between different applications environments. For example, in Europe, where many businesses have adopted information technology of local origin for national industrial policy reasons, the ability of the DEC system to interconnect helps them preserve their investment in this installed base.

The topologies available through the DEC approach to networking vary greatly. One key management challenge is always to determine the correct topology by matching it against the business' structure and goals. Stars, meshes, hierarchies, multidrop configurations, and more are supported by the DEC network approach.

Supercomputer Networks

The critical factor in understanding supercomputer networks is the transaction volume of the supercomputer. It is so great, it must be shared by many users simultaneously. In addition, the types of applications typically processed in the supercomputer environment are graphics and computation intensive. This has several implications for the network.

Determinants of Network Design

First, the supercomputer network is always centralized. This is because supercomputer resources are very expensive and relatively rare. In comparison with other types of information systems, only a handful of supercomputers are installed in North America and even fewer outside of the United States, since the supercomputer is an extremely powerful device that can handle a great deal of processing.

The number of users is great because the amount of processing available on the machine is very great. A supercomputer can handle a great number of transactions at one time. Therefore many users get involved. In addition, supercomputer resources are expensive and are therefore typically shared by different institutions. For example, 10 or more universities and their various departments may share a single supercomputer.

Users are highly decentralized. Typically the supercomputer performs a specialized type of processing which only a few types of persons use, and these persons are most often geographically widely spread. This puts a premium on the extensibility of the network.

The bandwidth required for quick response time is higher than for typical applications because many supercomputer applications involve a great amount of computation and graphics processing, for example, three-dimensional color modeling of organic compounds, or calculation and display of the stress points along an aircraft wing at every speed and every angle of turn possible. The movement of the great amount of information required to calculate, plot, and draw this type of graphic is great, so for the researcher to get a reasonably fast response time from the supercomputer, the telecommunications channels must be able to move these data quickly to the workstation after they are processed. Even after the data are received, the workstation still must be able to generate the appropriate graphics. So even with a very fast telecommunications system, it is difficult to maintain good subsecond response times.

Network Design Solutions

The result of these factors is that supercomputer networks almost always involve linking together other computers to serve as permanent data stores for the data used in the supercomputer calculations. These are called "slave" computers. It is less common for workstations to be able to handle the data volumes required for a major supercomputer calculation. These resources are shared. The workstation handles the user interface and generates the graphics that are needed. If it is a plain terminal, then the slave mainframe generates the graphics displayed at the terminal. The data stores used for the calculation, such as all the characteristics of an aircraft wing, or all the data on a topological formation, are kept on the Direct Access Storage Devices (DASD) of the slave host computer. This type of arrangement makes it possible to share the processing jobs between several layers of the information processing network.

The linkage between the supercomputer and the slave host computers is typically a very high-speed channel or local area network. Although low speeds of local area networks, such as the Ethernet standard of 10 mbps, are generally adequate for sharing printers, terminals, and workstation resources, they are not fast enough to handle the connections between a supercomputer and slave host computer system. For this type of connection, a high-speed channel typically links the supercomputer with other host computers over a local area network. As of the late 1980s, the supercomputer channel connection to other high-speed devices reached speeds as high as 100 mbps over a fiber optic linkage, but for a very limited distance (i.e., the distance between two machines in the same facility). Many other connections typical of supercomputers as of the late 1980s were in the 3 to 10 mbps range, relatively slow by today's standards.

The linkages between slave host computers and their users and dependent devices are typically based on standard telecommunications networking, rather than on a special protocol or network design specification. "Standard" means the off-the-shelf solutions provided by the vendor. There are great advantages in this, because the many users who do not access the supercomputer can link into their own host system and use the applications most suited to their work or research. There are additional advantages.

The most important advantage may be that the architecture available for extension of the network is supplied by the standard vendor, and not by the supercomputer vendor. Since most vendors, particularly major vendors such as IBM and DEC, have worked out many sophisticated techniques for networking, the logical task of extending the supercomputer network is greatly simplified. Additionally, the burden of managing the network is shifted away from the supercomputer to the slave host computers, leaving the supercomputer to do what it does best: perform high-speed calculations on complex problems.

To support sophisticated workstations that require a lot of processing, a local area network bridge may be used as an intermediary to the supercomputer instead of another slave host system. In this type of arrangement, the very high-speed local area network attached to the supercomputer is bridged to local area networks in other locations. The sophisticated workstation becomes the primary interface to the supercomputer. In the late 1980s, a great many sophisticated workstations emerged in the field, based primarily on 32-bit microprocessors, specialized dedicated graphics processors, and high-resolution screens. Most ran one version or another of the Unix operating system, which allowed multitasking. Building bridges between local area networks allowed the transmission of high- speed telecommunications between the workstation and the super- computer host.

Higher and higher bandwidth capacities over telecommunications channels will make much higher speeds available to link workstations directly to supercomputers. This will increase the ability of the research scientist to quickly do highly complex calculations with high-resolution color graphics interfaces. In the creation of sophisticated products such as aircraft or industrial designs, this capability will drastically reduce the costs of the design and testing phase of a development project. Exhibit 5.9 shows an example of a comprehensive supercomputer telecommunications system.

One example of the types of telecommunications solutions available from a supercomputer vendor is provided by Cray Research, one of the great pioneers in the development of supercomputing. By the late 1980s, Cray had clearly recognized the critical importance of efficient networking between its machines and those of other vendors. Its **Front End Interface (FEI)** provided a high-speed connection between the Cray machine and IBM, DEC, CDC, UNISYS, and Honeywell systems. Its speed was up to around 3 megabytes per second. In addition, Cray provided a **HYPERchannel** adapter which built a high-speed local area network onto the front of the Cray. Cray provided a high-speed fiber optic link for its FEI, a 100-megabyte-per-second channel to connect Cray supercomputers together, and a DEC VAX supercomputer gateway. Cray also promised to help with custom solutions to customer problems interfacing to other types of equipment.

Like so many other technologies, leadership in supercomputing has shifted to Japan, and it is questionable how long pioneering companies such as Cray will be able to survive.

Exhibit 5.9

Supercomputer Networks

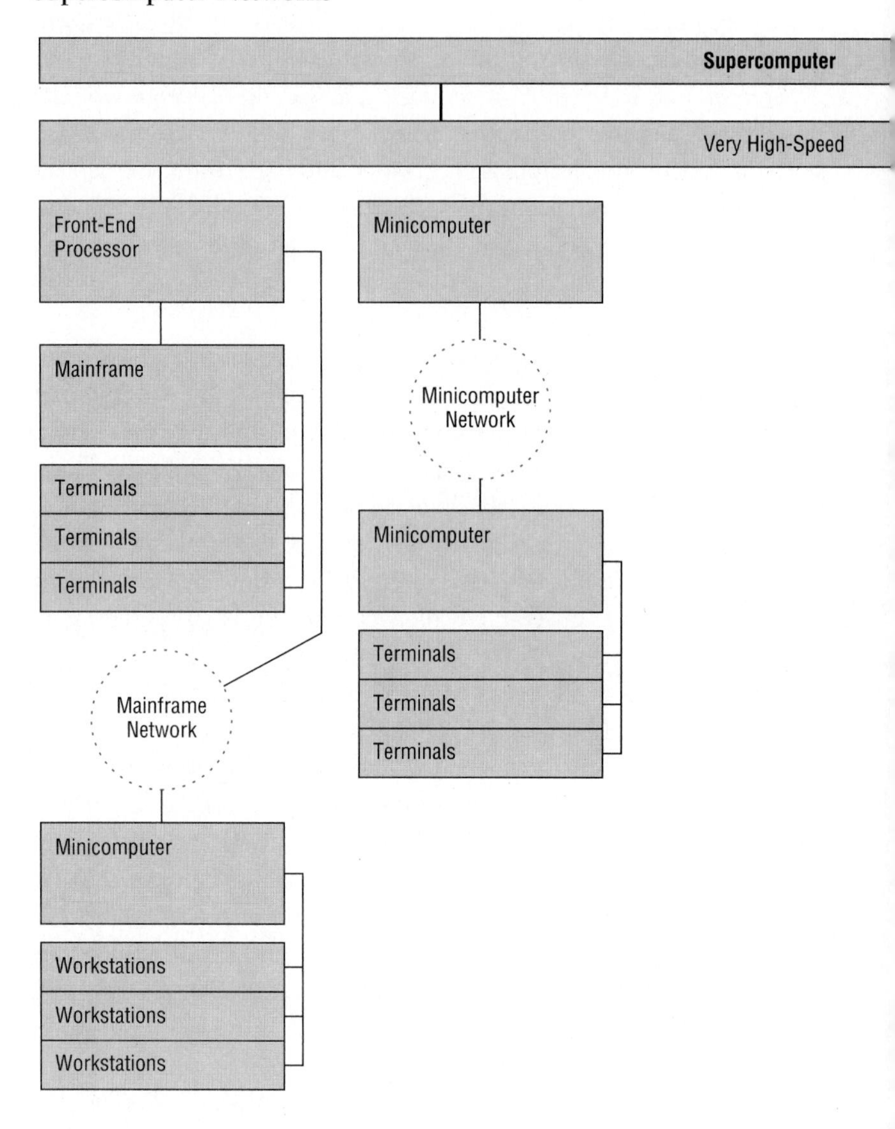

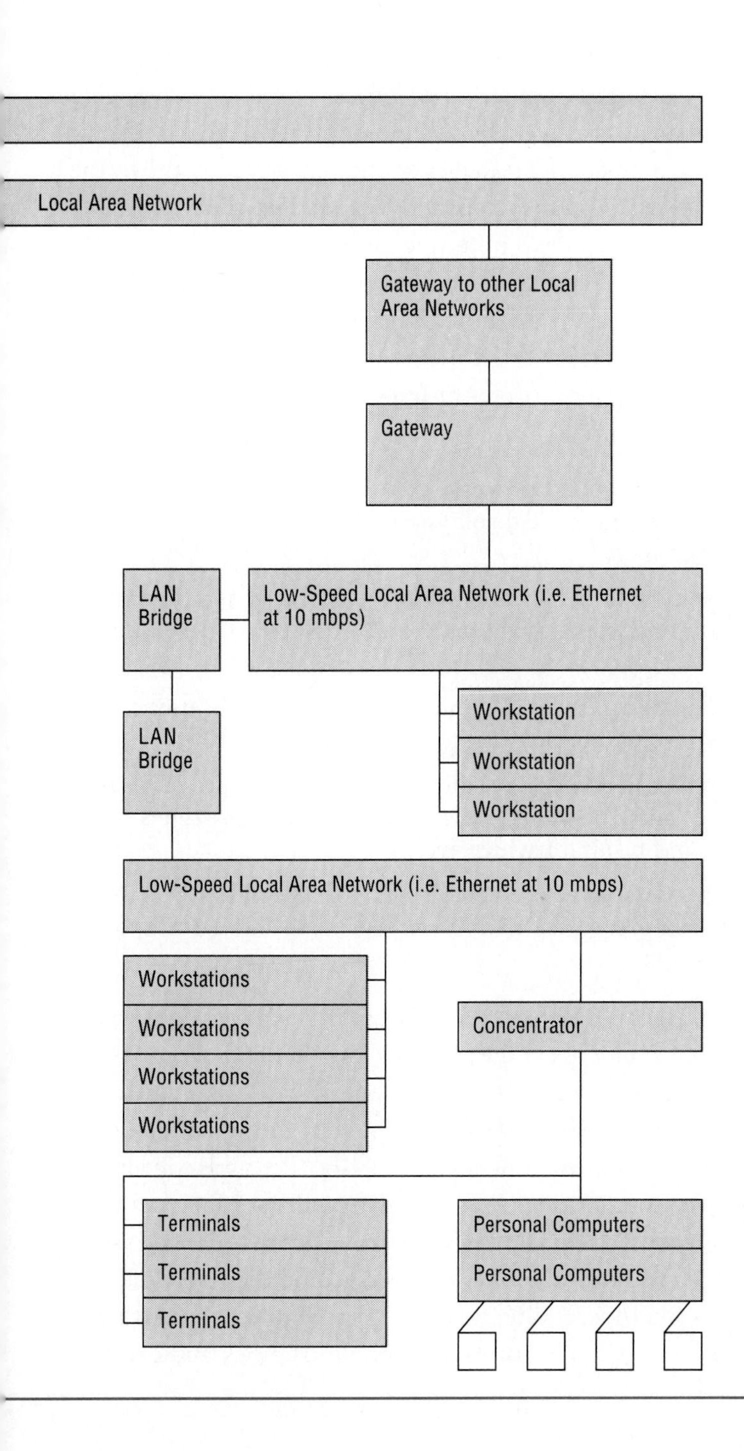

Local Area Network
Gateway to other Local Area Networks
Gateway
LAN Bridge
Low-Speed Local Area Network (i.e. Ethernet at 10 mbps)
Workstation
Workstation
Workstation
LAN Bridge
Low-Speed Local Area Network (i.e. Ethernet at 10 mbps)
Workstations
Workstations
Workstations
Workstations
Concentrator
Terminals
Terminals
Terminals
Personal Computers
Personal Computers

Note on Office Automation Strategies

The traditional role of office automation—discrete functional systems used by a focused target population of office workers—is changing rapidly. This change is driven by technology, the overall industry structure, and the competitive strategies of different vendors, and by the information workers who are constantly expanding their activities. Exhibit 5.10 summarizes some influences and outcomes of change.

Office automation has also continued to change because of better price performance in hardware, more efficient and easier-to-use software, and a rapidly improving user interface which has moved toward icons. Some of the major forces driving this change include the following:

- Hardware such as microprocessors, semiconductors, and optical and magnetic memory and displays are all giving rapidly improving price performance.
- Software is moving from the third to the fourth generation and is increasingly consistent at the user and application development levels.
- Advanced architectures are finding commercial applications, particularly for networked processors.
- These increases in power have allowed continual improvements in human interfaces, including icons, subsecond response times, voice interfaces, and resultant wider access.
- Thus, OA users are receiving systems that contain ever more functionality, are easier to use, and provide more transparent interconnections.

Simultaneously, the office automation industry in converging toward consistent standards at the physical, data interchange, and user application levels. For example, telecommunications standards are rapidly progressing toward ISDN, which will allow simultaneous transmission of mixed-media information. Also, LANs are converging on clear industry standards such as Ethernet and the Manufacturing Automation Protocol. Application standards such as IBM's System Application Architecture, Windows, Postscript, and others are moving toward broader interoperability. It should also be noted that artificial intelligence will be prevalent for executive workstations and decision

Exhibit 5.10

Office Automation Influences and Results

Drivers	Results
Evolution of product technology	*More capable, more usable products*
• Hardware advances	
• Innovative product architectures	
• Improving user interfaces	
Emergence of industry standards	*Greater interconnectivity, greater uniformity*
• Physical interconnection	
• Data interchange	
• User interfaces	
Refocusing of user buying habits	*Broader scope and larger scale of applications*
• Corporatewide solutions	Broader set of users considering purchases
• Multiapplication systems	

support systems soon, as major vendors move toward a consistent user interface for all applications, regardless of the underlying architecture or system. These developments will be promoted by movement toward a standardized application programmer interface which will further strengthen the development of environment-independent applications.

Although office automation has traditionally been focused almost solely at the departmental level within large business organizations, technologies, architectures, and standards promise both to deepen the penetration of office automation to the individual and to bridge the chasm between departmental and corporatewide information processing.

- Departmental information and functions are being pushed by enabling technologies closer to the end-user level as terminals gradually become intelligent workstations.
- Increases in hardware and system computing power are providing ever more powerful resources to manage the problems of network communications and interoperability.
- Intersystem interfaces and languages are helping departmental systems act as bridges and gateways to corporate data processing systems.

- As a result, work organization that has been focused primarily at the departmental level will become more closely integrated with corporatewide information systems. The nature of office work will become more interdependent.

Thus, office automation in the future will encompass corporate data systems; a broadly distributed telecommunications network; voice processing; complex document systems that integrate image, data, and even voice messages; and advanced messaging systems—all processed on powerful intelligent workstations. These trends will have several important effects:

- The vast amount of information locked in corporate data systems will be accessible to end users on a transparent demand basis.
- A broadly distributed network architecture will ultimately link all corporate systems together into a coherent environment with flexible rules for utilization of information technology.
- Voice capabilities will be integrated into many workstations, probably carried through integrated networks with data and video.
- Complex document systems will evolve to become more closely integrated with overall corporate information systems.
- Messaging systems will be universally available and include combined voice/document capabilities.
- Intelligent workstations with 32-bit microprocessors will form the basis for high demand on systems and a more highly active end-user environment.

In summary, the variety of improving technologies and more communicative architecture will enable the role of office automation to broaden its focus from the individual departmental level to the corporatewide level. In a loose sense, the heretofore deep chasm that has separated the mutually exclusive worlds of departmental office automation systems and the traditional centralized data processing function is being bridged with these enabling technologies. For vendors, this will continue to provide a wealth of opportunity provided that they can keep pace with the fundamentals underlying this evolution.

Office automation vendors continue to face a wealth of opportunities and challenges, provided they can keep pace with technology and demands for interoperability and ease of use. Office automation

Exhibit 5.11

Three Dimensions of Vendor Strategy

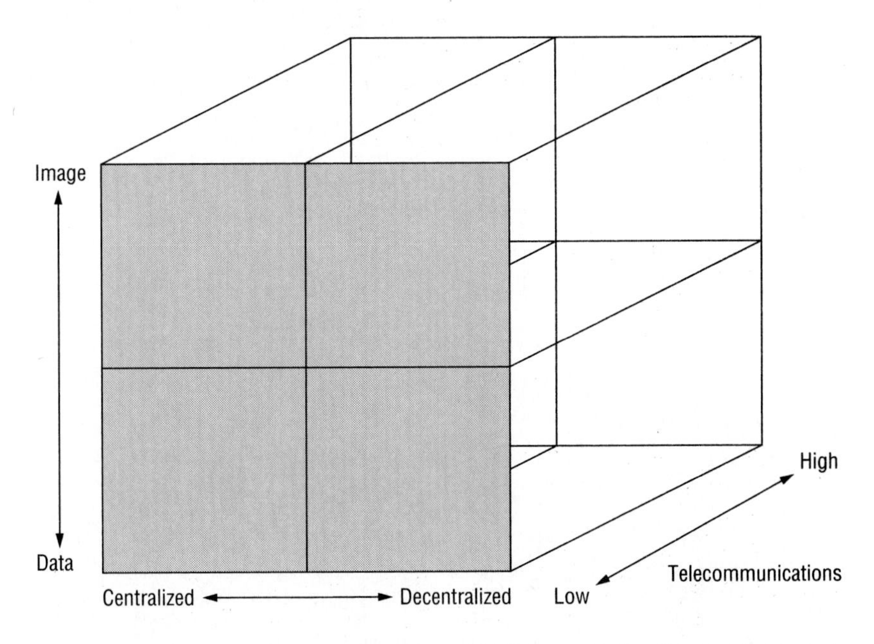

technologies and the associated vendor strategies can be sorted along three dimensions, as in Exhibit 5.11: centralized versus decentralized data processing, high versus low telecommunications intensity, and the degree to which images play a crucial role in the work flow.

It appears that the strategy for office automation pursued by IBM is intended to build from its strong presence in data processing. One indication of this might be the new products to make workstation-like word processing available through mainframe computers. In addition, IBM is making sophisticated word processing available through its minicomputer environments. As pointed out, IBM has a generally centralized strategy, although much of its architecture gradually began to change by the beginning of the 1990s. Except for the smallest fractions of information, IBM systems are data- rather than image-oriented and their office automation applications do not support a high intensity of telecommunications activities because they are located

Exhibit 5.12

IBM's Telecommunications Strategy

IBM's strategy focuses on data-intensive, centralized architectures.

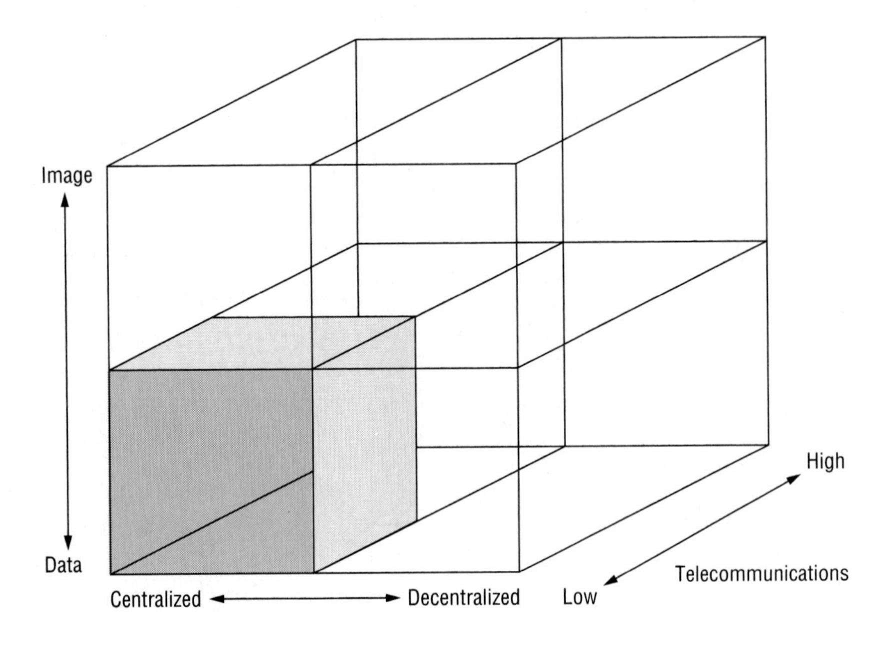

predominantly in a single location. Exhibit 5.12 illustrates IBM's strategy.

The minicomputer-based office automation strategies pursued by DEC and Wang (before it started to collapse financially in the late 1980s) involved a much higher degree of telecommunications intensity because of the distributed nature of the computer architectures. (See Exhibit 5.13.) However, for the most part, the data were handled and stored primarily in data, rather than image, form.

In contrast to these strategies, a company such as Ricoh, a large Japanese maker of image-related technologies such as plain paper copiers, facsimile machines, and other similar devices, is pursuing a highly image-oriented strategy which is both telecommunications intensive and highly decentralized. (See Exhibit 5.14.)

Exhibit 5.13

DEC's and Wang's Telecommunications Strategy

DEC and Wang follow similar decentralized, telecommunications-intensive strategies based on minicomputers.

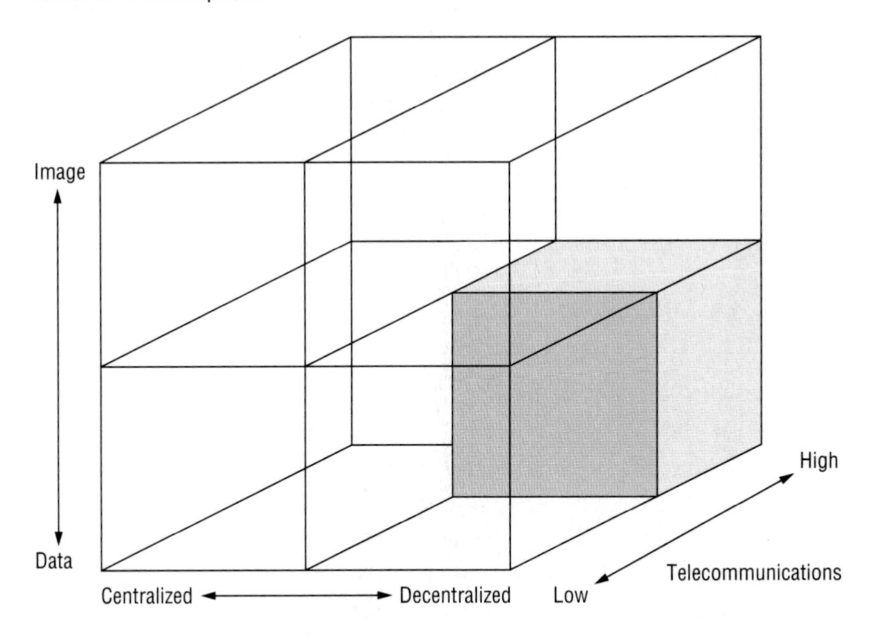

Exhibit 5.14

Ricoh's Office Automation Strategy

Ricoh's image-oriented OA strategy is highly dependent upon telecommunications.

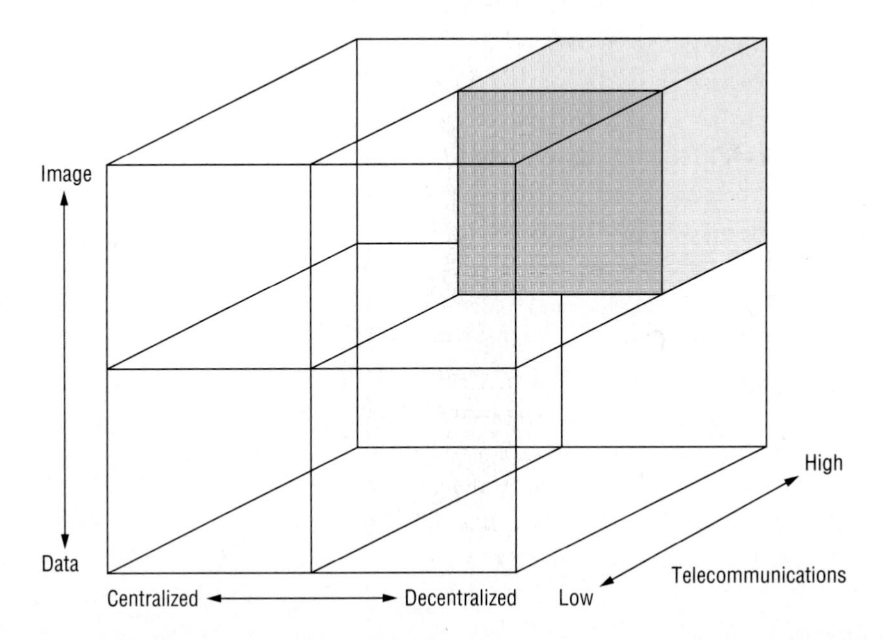

Summary

The largest installed base of information technology in the world has been provided by the International Business Machines Corporation. This central fact weighs heavily in all business strategy. IBM has gradually built up the ability of its equipment to provide more and more complex telecommunications networking. IBM's Systems Network Architecture (SNA) is, as a result, the most widely used system for telecommunications. SNA has gone through generation after generation of improvement as different layers of technology have been added. The Virtual Telecommunications Access Method (VTAM) helped separate the telecommunication functions away from the mainframe toward a programmable front-end controller. IBM faced severe criticism in the 1980s over its inability to provide a consistent application environment from one machine platform to another. As a result, it introduced Systems Application Architecture which will gradually fold all of its machines into a common system.

The key to the SNA system is the SNA backbone network which provides the essential infrastructure for the system. These backbones are built by linking together network control processors or, in more advanced versions, by linking the NCPs to bandwidth managers. It is no longer the case that an IBM network is automatically centralized.

The peculiar design of DEC computers has allowed a completely different approach to networking. DEC networks in a sense have no centers. Using a series of local area networks, the DEC machines allow users to string a great number of computing and peripheral resources together. For large distributed networks, LAN bridges are used to link together different LANs. Also, SNA gateways enable users to have access to IBM systems. The result is an extremely flexible system of networking which many believe is superior to any other solution.

The solution to networking of Cray Research supercomputers relies on a completely centralized architecture that links in "lesser" machines such as mainframes, minicomputers, or super workstations across a very high-speed local area network called a HYPERchannel. At the end of the 1980s, Cray was providing a Front End Interface which was linked to a fiber optic 100-megabyte-per-second channel.

Questions

5.1 Why is the IBM approach typically centralized and the DEC approach typically decentralized?

5.2 What is the role of the front-end processor?

5.3 What is a communications processor?

5.4 What is a LAN bridge?

5.5 What is a gateway?

5.6 Explain the role of a bandwidth manager in an SNA backbone network.

5.7 Can more than one communications protocol be sent over the same channel at the same time? Explain.

5.8 What are the differences between VTAM, SNA, and SAA?

5.9 How has IBM responded to the rise in the importance of the minicomputer during the 1980s?

5.10 Can non-IBM terminals using file protocols such as ASCII work well in the SNA environment? Explain.

5.11 What does a network control processor do?

5.12 What does a DECnet/SNA gateway do? Why would it ever be used?

5.13 a. Why is a supercomputer network always centralized?
b. Why does a supercomputer use a very high-speed local area network to make linkages with other information systems?

5.14 What type of businesses would be better served by the IBM solution to networking problems? What type would do better with the DEC solution? Does it really make any difference?

5.15 Review the dominant trends in office automation. What are some of the differences in the OA strategies of IBM and DEC?

Chapter 6

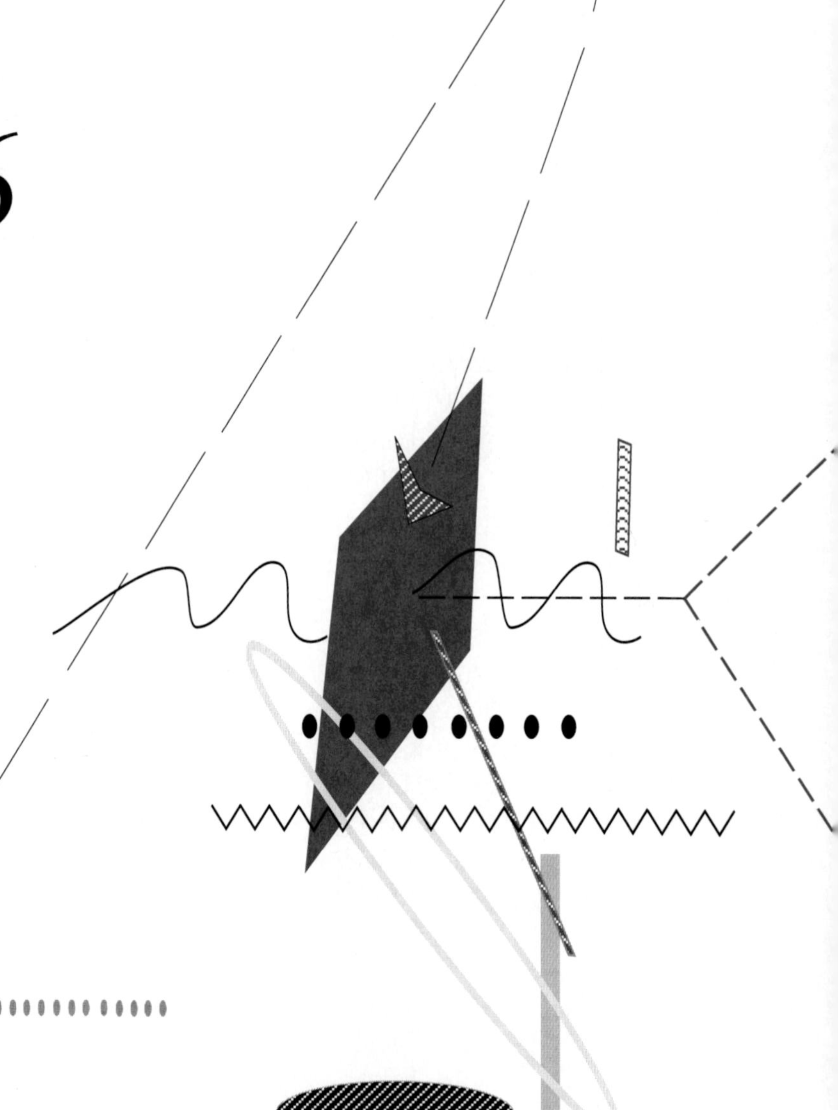

Network Control and Management

The control, management, and administration of highly complex networks is primarily concerned with keeping the network up and running at all times. Network performance is critical to almost all organizations, and yet like all technologies, telecommunications networks occasionally experience difficulty and even fail. The network control and management function must always be prepared for this event. In increasing numbers of organizations, network control is a 24-hour-per-day job. Network management includes managing changes in configuration, correction of faults as they occur, constant analysis of performance, careful accounting of costs, and careful administration of access and security to keep away hackers. Negotiation of contracts should ensure that after-sales services; moves, additions, and changes (MACs); and spare parts costs are agreed upon in advance. Network control is supported by highly automated interfaces which provide windows through which managers can see the network. Bypassing existing local telephone companies can cut network operating costs for long distance traffic.

Network Control

Network control is composed of many functions which should interrelate with one another. In simple networks, the network control and management function may be accomplished by the MIS group without much specialized training. However, in very large networks, particularly those that have a great degree of distributed processing at many geographic locations, it is necessary to have an independent management function staffed by highly trained specialists, to maintain certain critical functions on a 24-hour basis.

At any time during the day or night, a connection can be broken somewhere. As the size of the network increases, even occasional failures in the network quickly add up to a constant series of alerts to the central network control point. In addition, one must consider changes. As networks grow, new terminals and end nodes must be brought into the network. New lines with new characteristics must also be added. Each of these additions requires an updating of critical information regarding routing of information and identification of nodes. Again, as the size of the network increases, change and turnover go on constantly throughout the network as new and old nodes and other pieces of equipment become alternatively connected, disconnected, and reconnected.

In most networks, timing is critical. If a part of the network goes down, it must be put back into operational order as soon as possible. Imagine a financial trading network, a transaction network, a network involved in coordinating manufacturing between several large plants. What would be the cost in lost business and production of a network outage?

Several major functions of network management have been shown to be critical. (See Exhibit 6.1.) These provide a convenient classification of the many functions and actions which must be constantly accomplished by the network control group. In the course of keeping the network operating efficiently, these needs arise inevitably.

Each of the major vendors has a particular approach to network management. For example, IBM introduced a product called **NetView** which was responsible for facilitating complete management and problem solving for a large SNA-type network, typically of IBM computers. Companies such as NET Technologies have special

Exhibit 6.1

Components of the Network Management Task

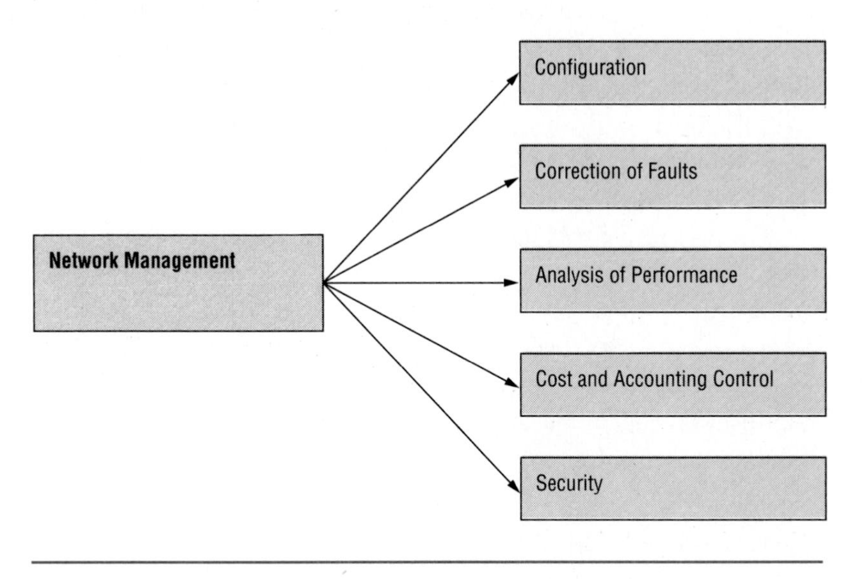

high-resolution workstations that present vivid graphics of the network to help the network manager isolate problems and initiate changes in configurations. Digital Equipment's network management approach is highly decentralized and relies on a great deal of automated updating of various directories and switching paths along the network. AT&T has developed a complex network management system that relies on AT&T equipment located throughout the network to diagnose problems very quickly and to make remote changes and modifications to the network. All of the systems have the ability to see into the network, many times down to the smallest level—that of an individual data device.

Functions of Network Management

Configuration Control

Management of the configuration of a network involves several key functions. As the network is constantly changing, it is important to keep track systematically of the hardware and software characteristics and locations for each portion of the network. All of the pieces of equipment should be registered and accounted for. In the case of a very large network, the rate of change on a day-to-day basis may be very great. Under those circumstances, each day will see many pieces of equipment either added to the network, taken off the network, or moved from one location to another on the network. The network control function must be able to log each of these changes and know at any point in time the exact physical and logical configuration of the network being managed.

The management of the required directories also comes under the category of configuration management. Whenever something goes wrong with the network, it is important to be able to immediately contact users and other personnel throughout the organization, regardless of how remote they are. Directories of personnel, their computer identification codes, and other relevant information are important to maintain this function.

Correction of Faults

One of the most dynamic and critical functions of network management is the correction of faults. When problems occur in the network they must be dealt with immediately. If one circuit stops functioning or deteriorates, back-up circuits must be brought into play. How long should it take to bring in the new circuits? In the 1970s and early 1980s, time required for restoration of service or activation of a new backup circuit was measured in hours or minutes. However, with fast microprocessor logic, the amount of time spent in bringing in additional circuit capacity has been reduced to minutes, seconds, and even milliseconds.

However, before a new circuit is brought into play to replace a failing one, the exact point of failure in the network must be identified. This may be difficult, as Exhibit 6.2 illustrates. How can you find the point of failure in a network of several hundred or thousand circuits? Is it the line, a modem, a terminal, a packet switch, a protocol converter? What is the point of failure? The function of identifying the exact point of failure of a network is called **isolation**.

Alarms may be generated by the software interface designed to monitor the network. After the network control operator sees the alarms, then all types of linkages between the nodes and data

Exhibit 6.2

Causes of Performance Degradation

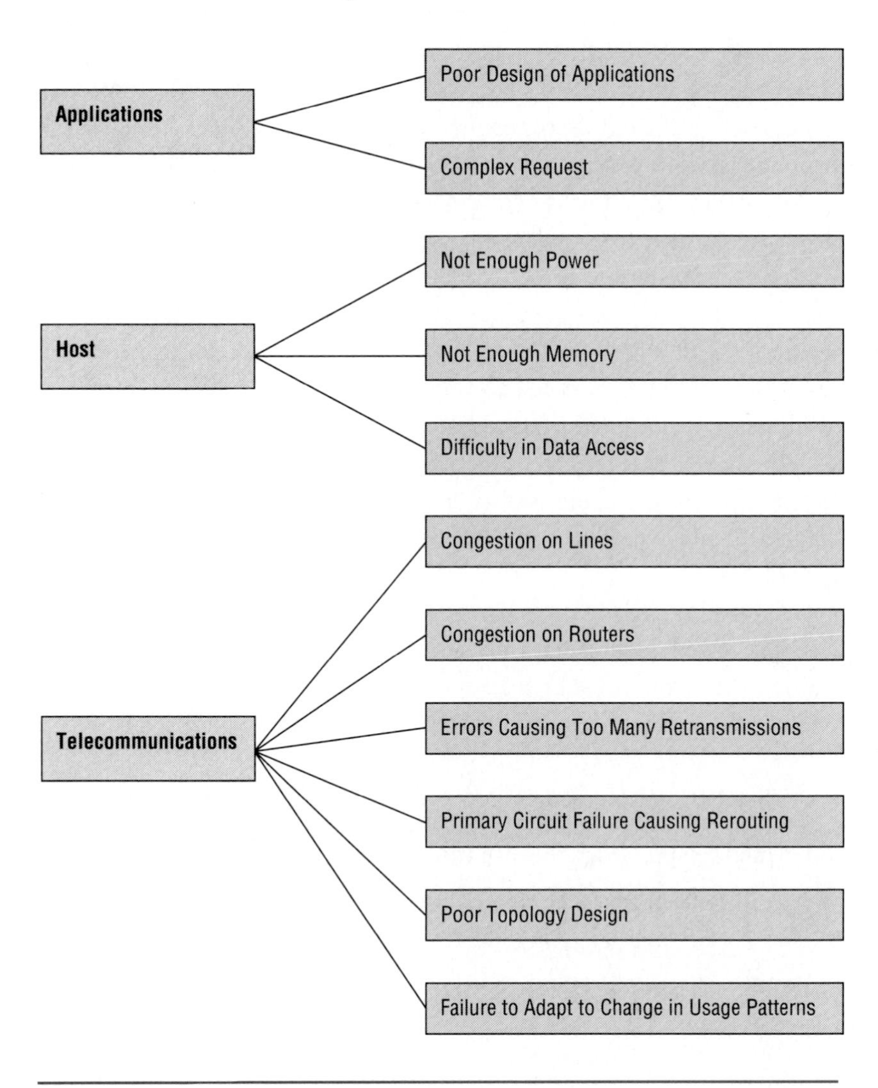

Source: Adapted from "A Common Sense Guide to Network Management," Digital Equipment Corporation, 1989.

communications equipment must be brought into play to isolate the problem. In many systems, such as that provided through AT&T, the central network control equipment is linked to all other nodes of a communications network through special channels set aside to communicate and carry information between different nodes in the network. These systems are designed to have simple built-in procedures to locate and isolate faults in the network.

After the fault is located, there are several possible responses depending on the nature of the problem. It may be possible to have the problem fixed remotely by resetting or readjusting some of the communications devices. In other circumstances, however, it is necessary to dispatch personnel to correct a specific fault. The entire trouble-tracking procedure is based on a system of **trouble tickets**. In the old days, the trouble ticket was actually a report of network failure written down on a small slip of paper. As the problem was located, the ticket was passed to the correct personnel for initiation of repair. With the development of automated systems and the widespread use of electronic mail between network managers and personnel, the ticket is no longer composed of actual paper, but instead consists of electronic mail messages passed around. It is still called a "ticket," however.

Finally, as each of the problems occurs, triggers an alarm, initiates a search for problem isolation, then results in repair and restoration of service, the tickets must be filed and analyzed as historical records. This enables the network management to understand any chronic problem that may not be apparent in the short run.

The task for management is to concentrate on the following factors:

- Quick response to any situation
- Drilling and practice for problem resolution
- Creation of a strong corporate culture of service and the importance of a good reputation

Analysis of Performance

When the network is functioning without fault, it still requires a great deal of network management. **Performance management** is the process of constantly analyzing the efficiency of the network. What type of response time is being delivered to users? For example, in an X.25 packet switched network, the response time can deteriorate seriously depending on the load of the network. During peak hours there may many times be great problems in this regard. This is why organizations that are highly dependent on fast response do not generally prefer to use public packet switched networks: they are unable to use network control techniques to minimize the response time throughout the system.

What is the quality of the service being offered to the users? Are they constantly facing a large number of resets and disturbances to their communications? Is the throughput of the system great enough? If the network is large and complex with multiple possibilities for sending data, is the load of telecommunications traffic being spread around in the best possible way to achieve the greatest efficiency? Are users able to set up their calls and make connections quickly enough without trying several times or suffering other headaches? These are some of the questions that must be asked during the process of managing the overall performance of the network.

Cost and Accounting Control

The financial aspects of network control can be highly complex, depending, of course, on the size of the network. Exhibit 6.3 shows a model of this function. For example, each of the leased lines must be paid for, usually on a monthly basis. For a network that extends to several regions in the United States and overseas to countries in Europe and Asia, an organization might be receiving monthly bills from several different telephone operating companies in the United States, from a long distance carrier moving data and voice from one state to another, from a satellite carrier if it has a private satellite linkage, and from several PTTs in countries of Europe and the Far East.

Although it is the job of the accounting function to stay on top of paying these bills as they come in, how can you verify that they are correct? With lines being set up, taken down, and rearranged on a regular basis, the network control financial operations group must have an internal system to verify that the charges are correct. Also, the group must keep track of all additional charges the network may incur, such as those for extra services, maintenance, and changes in services ordered. Some telecommunications service vendors rarely make a mistake in billing. Others, however, may make frequent mistakes. Depending on the size of the network, a single department or group may

Exhibit 6.3

Network Financial Management

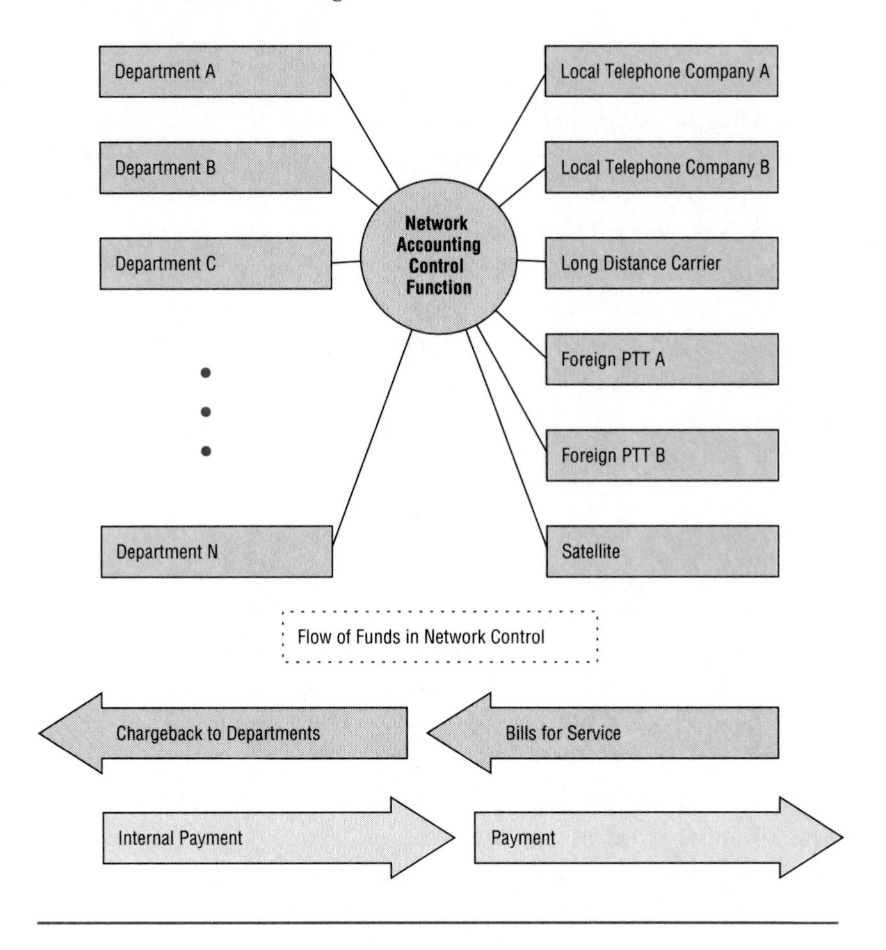

be charged solely with verifying the charges from telecommunications vendors.

In addition to the various telecommunications service charges from vendors, charges for equipment, maintenance, and other telecommunications goods and services may be incurred. Control over the incoming inventory of equipment, tracking its location, and registration of warranties, etc., all may come under the authority of the network control financial operations group.

In addition, there is the question of collection for the charges. Various departments within the organization use different amounts of telecommunications services. Depending on the philosophy of information technology utilization within the organization, an internal **chargeback** system may be initiated. The network control group must be able to substantiate the internal chargebacks it passes on to the various departments so as to be able to collect the internal funds from the departments using the telecommunications system.

Finally, the financial management group must have an overall view of the accounting and expenditure pattern of the telecommunications function to aid in planning expansion and continued operation of the network. As the expenditures are incurred from year to year, the financial management group can serve as a point of continuity. It can assist in the preparation of the budgets for hardware, software, telecommunications service resources, and retention of trained personnel. This critical management oversight function is a key input for the overall telecommunications plan.

Security

Another critical element in network management and control is providing security. How can the organization prevent unauthorized users from entering the network and sabotaging systems, stealing information, or otherwise disrupting the operations of the organization? Even if someone has illegally or illicitly entered the network, how can this be detected? What levels of security are required for entering into the various telecommunications systems and subsystems? Who is responsible for administering them?

The security function of the network management team is critical. In organizations with very large telecommunications networks, different levels of security may apply to different groups in the organization. The network may be divided into logical parts to which only certain groups have access. All of this is controlled and maintained through the network control and management function.

User Interface Technologies for the Network Manager

Over the years, the technologies available to network control specialists have changed and improved. The advent of color high-resolution monitors and multitasking or windowing for personal computers and workstations by the late 1980s greatly improved the user interface for network control. For example, color pictures and logical representations of the network appear on the computer screen. When a circuit has a problem, the representation of that part of the telecommunications linkage may turn a color, such as red, and an alarm might sound to alert the operator of a possible problem. Many systems also have a yellow condition in which a troubled circuit is identified and isolated before it completely fails. A yellow condition occurs when the number of errors on the link increase beyond a set level, indicating a gradual decay in quality for the circuit. When this is the case, a high-quality automatic network monitor will notice the change in condition and issue a provisional warning of a possible deterioration in the quality of a portion of the network. In addition, for highly complex network situations, graphic workstations typically give the operator the ability to zoom in on greater details of the network. This flexibility of a visually oriented user interface to network management based on real alerts and feedback generated by the network aids network management a great deal.

The multitasking network control monitor and operator consoles also must have the ability to cause changes to the network. The operators must be able not only to view changes in conditions, i.e., deterioration or failure of circuits, but they must also be able to initiate repairs. This is done remotely if at all possible, since almost any other alternative involves dispatching a human operator and waiting for the repair or change to come into effect. The commands to isolate faults and perform loopback tests (checking each loop of a circuit one at a time) should be executable from the network management control interface. Another important function of network control interfaces is the support of electronic mail to facilitate communications between different network control personnel who, in the case of very large networks, may be located in many different geographic locations, spanning time zones and continents.

Exhibit 6.4

Topics for Network Planning

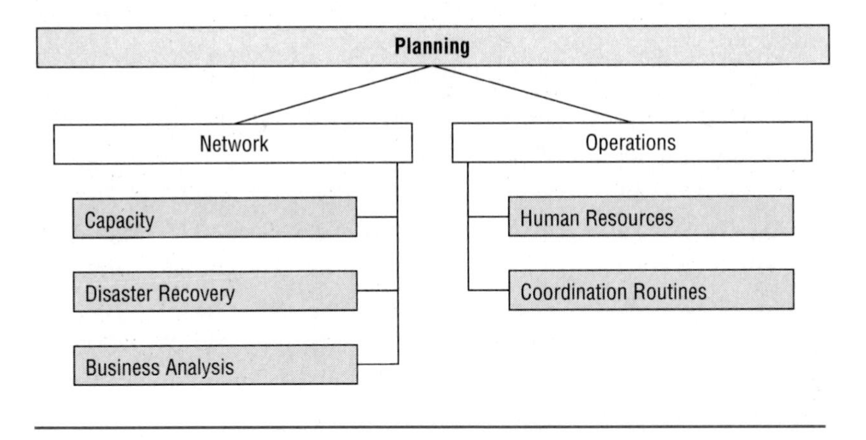

Planning

Another critical dimension of network control involves planning. This type of activity can be classified into several different focuses, outlined in Exhibit 6.4, each of which may require different types of experience, skills, data, and insight on the part of the telecommunications management.

First, the telecommunications manager is concerned with the human system, as it were, that supports the telecommunications operation. What types of skills are needed, and how is the team organized so as to get the most efficiency? How many people are needed, and what should they be paid? What activities can be undertaken to overcome the industrywide trend (at least in the United States) toward more and more turnover of qualified personnel? What type of people are to be promoted to management positions and what proven skills and qualifications should they display?

Additionally, the question arises of how exactly the telecommunications function should be organized. Management must specify standard operating procedures and actions to take in the event of situations that may arise from time to time. How should information flow between different subgroups in the telecommunications area? What is the correct operating procedure for each employee under various circumstances? These are some of the operations questions that must be answered by telecommunications management.

In addition to the planning and organization of human activity in well-organized operational guidelines for the network management personnel, another side of the activity focuses on the network itself. Determination of capacity and growth of the network is a major undertaking that involves a very technical look at the data traffic with estimates of the volume and bandwidths needed in the network. Where should extra lines be brought into the network? What are the best options in case the business grows?

Planning and control also involve a study of disaster recovery and back up procedures and options. What should happen in the network control center in case a fire, a flood, or even information terrorism destroys a telecommunications linkage? Who is supposed to do what? What are the correct procedures? How many times have they been rehearsed? What is to be reported to the rest of the organization? Who is in charge of documenting the correct procedures and when was the last time the disaster recovery plan was reviewed and evaluated? What types of disaster might most likely be encountered? These are some of the typical questions that might be asked.

In addition, planning involves thought at the highest level about the overall relationship between the telecommunications and information structure of the organization and its overall business mission, vision, and strategy. Taking the long-term view of network development and business strategy, the successful telecommunications planner is able to analyze the business strategy of the organization and then create or induce a parallel telecommunications strategy. This type of skill is of the highest order and should be thought of as a qualification of the very top levels of the telecommunications operation.

Programming

Network control also involves writing control programs and user interfaces that are easy to use. Also, each network administrative system requires a significant amount of maintenance and programming to customize these features. This must be accomplished by the telecommunications group.

Bypass

Bypass is the technique of setting up a telecommunications system that does not connect to the local telephone company. It is the building of an entirely private network without connection to the telephone company or PTT. Bypass is not legal in all countries, but in the United States it is possible because the law prevents the telecommunications giants from controlling the options available to businesses for building private networks.

There are two basic types of bypass, each of which has several variations:

- In the United States, traditional bypass involves linking into a wide area network without any contact to the local telephone company.
- In Europe, bypass involves setting up a network that does not come into contact with the PTT. For example, a bypass network from Paris to New York might involve placing a satellite receiving and sending antenna on the roof of the Paris headquarters and sending information back and forth without connecting to the local PTT. As of the late 1980s, this type of bypass was not allowed in Europe, although pressures were mounting for liberalization and even privatization of many PTT activities.

The entire issue of bypassing the telephone company arose first in the United States in the mid 1980s. Once it became clear to businesses that they did not have to use the same long distance carrier anymore, a great scramble began for cheaper means of telecommunicating information. Organizations were able to choose between different long distance carriers, but no deregulation allowed the creation of competition within the metropolitan areas. In Manhattan, there was not going to be an alternative to New York Telephone. (There had been talk at one time of allowing the cable television system to offer voice service, but this quickly died away after the cable companies realized the complexities of switching and the generally high level of service expected, quite different from the sloppy service levels they were providing for cable television.) At the same time, the freeing of competition between long distance carriers and the break up of the Bell System, which split apart the local telephone companies and the long distance companies, revealed the problem of cross-subsidization.

Cross-subsidization had been designed into the Bell System in its quest for universal service. The system's financial billing and accounts system had been designed so that long distance services subsidized local telephone service. This was because local telephone service is inherently more complex and costly to provide than long distance service on a per-call basis. With the break up, cross-subsidization was stopped, resulting in a sharp rise in local telephone rates. Businesses had to pay these rates. For a long distance call, it quickly became clear that the greatest cost of the call was not the long distance portion, say from New York to Los Angeles, but rather the local call from the business office to the local telephone office to connect with the transcontinental circuit provided by the long distance carrier.

As a result of these two pressures—rising local telephone rates and increasing flexibility and options—businesses began to search for ways to connect directly from their headquarters to the offices of their long distance carriers. If this were done, the company could avoid paying the local telephone charges for a long distance call.

Two key technologies were used in initial attempts at bypass: line-of-site microwave and direct satellite. (See Exhibit 6.5.)

- Line-of-site microwave systems involved setting up a microwave transmitter/receiver, typically on the roof of a headquarters facility, and then transmitting to the nearest microwave port facility of a long distance carrier.
- Setting up a satellite dish on the roof allowed firms to connect directly on an uplink to a satellite and thence down to another site of the company. This option completely bypasses even the long distance carrier, unless the satellite transponder space is being leased from a wholesaler or service provider.

In most countries of the world apart from the United States, the concept of bypass was treated with undisguised hostility by the PTTs, which saw as their mission the maintenance of complete dominance and control over all telecommunications in their respective nations. In many cases, these powerful PTTs were strongly linked with labor organizations, which in turn had political power to get their message across to any government bureaucrats or politicians who might be interested in liberalization.

Exhibit 6.5

Satellite Bypass Systems

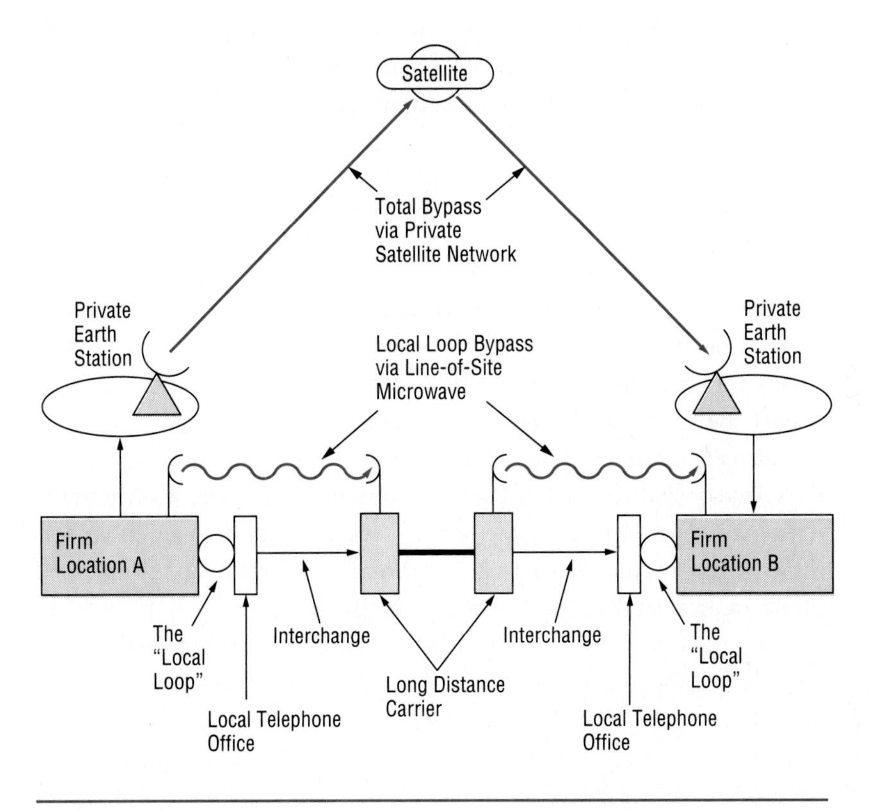

Because of this draconian monopoly, many highly efficient and technologically creative and innovative options were completely cut off from the realm of possibility. In practice, the business wishing to build its own direct satellite broadcasting and receiving system, even if the transponder space were available, would find that it was impossible to get a license to install and operate the ground station. This turgid rigidity still persists in most countries of the world, and is the subject of intense international negotiations relating to trade in services.

As a final note, the degree of bypass expected by the early projections never materialized. The types of systems available for bypass were never quite able to equal the superior service available through the local telephone companies, who in turn continued to improve their flexibility and options (at least in the United States). In addition, in some international environments, businesses did gain special exceptions for their telecommunications needs.

Negotiating Contracts and Understanding Telecommunications Law

Managers of the telecommunications function and financial teams with whom they may be working have a difficult task when making major telecommunications decisions. To decide on the best technical strategy is only the first step. Then comes implementation.

Implementation involves the negotiation of detailed contracts with the selected vendor to define the check points of the installation, and the timing and cost of each phase of the installation or change, as well as the on-going maintenance and operational support of the system. One of the greatest challenges is to negotiate the after-service characteristics of the relationship with the vendor. Some of the key provisions that have caused the greatest problems in the past involve **moves, additions, and changes (MACs).** Unless great care is taken in the negotiation phase, MACS and other services offered by the vendor may prove to be prohibitively expensive. In fact, it is many times the strategy of the vendor to sell the original service cheaply and then plan on making large profits to make up the difference after the sale is completed. MACs, repairs, maintenance, spare parts, etc., are some of the major profit generators used by many organizations to gouge the customer.

It is best when negotiating the contract to think about the entire life time of the equipment, including all the possible repairs and other ancillary costs the system will generate over that period, to the degree those costs can be estimated or expected. After this is known, it becomes much easier to build a comprehensive cost model of the true long-term effects of the purchase and the relationship with the vendor. Thinking about the purchase as a marriage between your organization and the vendor is a better analogy. Years later you will either benefit or suffer from the decision to choose one particular vendor over another.

Some of the critical details to negotiate include the cost and performance schedule for the following items:

- Moves, additions, and changes
- Spare parts
- Labor for repairs or modification of the system
- Guarantees of responsiveness for service calls
- Delivery of any enhancements or extra capabilities promised during the sales negotiations

Summary

Network management and control involve many interrelated functions. Management of the configuration involves connecting and disconnecting all of the different devices on the network as needs constantly change. For large networks, this is a very complex function. Correction of faults involves taking quick action when network performance starts to degrade. Analysis of performance must be done on a constant basis to see ahead of time if a problem is going to occur. Cost accounting and control are necessary to keep track of the considerable costs involved in network operations. Each local telephone company, long distance service provider, international record carrier, etc., must be paid, and the bills must be correct. The accounting information is critical for price/performance analysis. Network management must also carefully control the security of the network, which includes keeping records of use and user passwords.

These activities are supported by a series of tools that give network operators a view of the network and advise them of problems. In addition, another critical skill is managing the overall structure of the telecommunications function. Negotiation of contracts must make provision for exact pricing specifications for moves, additions, and changes (MACs), after-sales service, and cost of spare parts. Bypassing the local telephone circuit, called the *local loop*, can cut costs, but it is not allowed in most countries of the world.

Questions

6.1 Explain the details with which each of these functions is concerned:
   a. Configuration
   b. Correction of faults
   c. Analysis of performance
   d. Cost and accounting control and chargebacks
   e. Security

6.2 Explain the tasks involved in the different aspects of planning:
   a. Capacity
   b. Disaster recovery
   c. Business analysis
   d. Human resources
   e. Coordination routines

6.3 What are the critical details to consider in negotiating any telecommunications contract? What is the role of after-sales service?

6.4 What are MACs and why are they important?

6.5 What is the local loop?

6.6 a. What is bypass and why is it important?
   b. Why is bypass impossible in most countries of the world?
   c. Discuss various bypass technologies (satellite, line-of-sight infrared, microwave) and the advantages and disadvantages of each.

Chapter 7

Managing Global Telecommunications

The international telecommunications environment is governed by a group of institutions centered around the International Telecommunications Union. Business organizations can influence the development of both regulations and standards through International Telecommunications User Groups and through the International Chamber of Commerce. Many practical problems must be managed internationally including regulations on transfer of data across borders, strict requirements regarding privacy of personal data and information, and national requirements that all data processing take place within national borders, even when this is not the economically or architecturally optimal solution. In the international environment, many factors influence the effectiveness of operations. The critical fact is that it is impossible to operate a telecommunications system effectively without taking into consideration variations between countries and regions of the world.

Within the developed countries, these variations may not amount to much. However, in developing countries, firms may experience severe problems. This chapter highlights Brazil as an example of a telecommunications problem area and discusses variations between other nations. Generally, telecommunications across international borders has been the target of many efforts at regulatory reform, but the problem is still far from resolved after more than 15 years of international negotiations.

Coping with International Telecommunications Regulation

The regulatory control of international telecommunications is determined by many institutional forces interacting both within and across international borders. Business is forced to play a part in deliberations at each level. In the United States, because of the highly legalistic and adversarial nature of dispute settlement, the courts and their interaction with the Congress and the Federal Communications Commission have determined the type of environment available to business. At the international level, both international institutions and bilateral treaties determine the environment. Business must address these forces to have its interests represented.

International organizations such as the International Telecommunications Union, with its various CCITT committees and working groups (see Exhibit 7.1), constitutes a large arena in which business must make its voice heard. Within the context of bilateral treaties, the business interests in each nation state must work through their own governments to ensure their priorities are made known in the agreements.

Standards and the ITU, CCITT, and CCIR

A key international forum for business users of telecommunications systems is the system of **International Telecommunications Users Groups (INTUG)**, which operates under the auspices of the **International Chamber of Commerce (ICC)** in Paris. It is the function of the ICC to make the views of international business known to the right international organizations. The ICC receives recommendations and agendas from each of the Chambers of Commerce located within the member states. Within the member states, hierarchies of various local chambers act as grass roots organizations. The national chambers also make representations to various national governments to influence the negotiation process at the bilateral and even multilateral levels.

Within the United States, a vast **federal advisory committee system** ensures that the views and positions of the private sector are taken into consideration in the formulation of public policy. These lobbying efforts are necessary because, for the most part, the most

Exhibit 7.1

International Telecommunications Union: International Telegraph and Telephone Consultative Committee (CCITT)

Study Group Structure

I.	Telegraph operation and quality of service
II.	Telephone operation and quality of service
III.	General tariff principles
IV.	Transmission maintenance of international lines, circuits, and chains of circuits, maintenance of automatic and semiautomatic networks
V.	Protection against dangers and disturbances of electromagnetic origin
VI.	Protection and specifications of cable sheaths and poles
VII.	New networks for data transmission
VIII.	Telegraph and terminal equipment, local connecting lines
IX.	Telegraph transmission quality, specification of equipment, and rules for the maintenance of telegraph channels
X.	Telegraph switching
XI.	Telephone switching and signaling
XII.	Telephone transmission performance and local telephone networks
XIII.	Facsimile telegraph transmission and equipment
XIV.	Transmission systems
XV.	Telephone circuits
XVI.	Data transmission
XVII.	Digital networks

knowledgeable persons regarding international and national telecommunications issues are unlikely to be found in the government. This is particularly important for the assessment of effects, cost and otherwise, on business operations of various proposed or adopted public policies regarding telecommunications. Within the United States, the advisory committee system acts through the Federal Communications Commission, the United States Trade Representative, the Department of Commerce, the U.S. Congress, the Department of State, and many other parts of both the executive and legislative branches of government. In addition, a large amount of litigation is being processed through the legal system. All these government agencies provide structures for private sector access to the government policymaking process.

Depending on the size and scope of business operations of an organization, different levels of access are appropriate. At the international level, only the very largest telecommunications users appear to play significant roles in the policymaking process. These interest group activities typically are organized through the government affairs office of a business. They must be supported by a combination of business knowledge and awareness of the cost and technical issues surrounding telecommunications. In many cases, the interests of a business are relatively easy to identify; other times, contradictions may occur. For example, a company like IBM might support international standardization when proprietary network topologies better serve its self-interest. In the long run, however, it is the role of business to help determine and guide policymaking at both the national and international levels.

Participation in Development of Standards

Standards development involves a complex set of activities ranging across a very broad line of equipment and specifications. In electronics alone, standards cover electrical power systems, gyros and accelerometers, control systems, radar systems, antennas and propagation equipment, broadcast technology, televisions, circuits, consumer electronics, other electronic devices, software engineering technical standards, etc. In almost every aspect of electronic equipment, standards allow different vendors, suppliers, and component makers to produce interchangeable equipment.

In the area of telecommunications, agreement on and publication of standards helps software designers understand the infrastructure their software is expected to work through. There are several major local area network standards, as illustrated in Exhibit 7.2.

Local area networks are only one of many focuses of standards in the telecommunications area. Even the standards for local area networks must be accepted in the international environment to become true standards. The **International Standards Organization (ISO)** is responsible for coordinating the creation of standards internationally. It works through a system of careful consultation and management of

Exhibit 7.2

Family of Standards for Local Area Networks

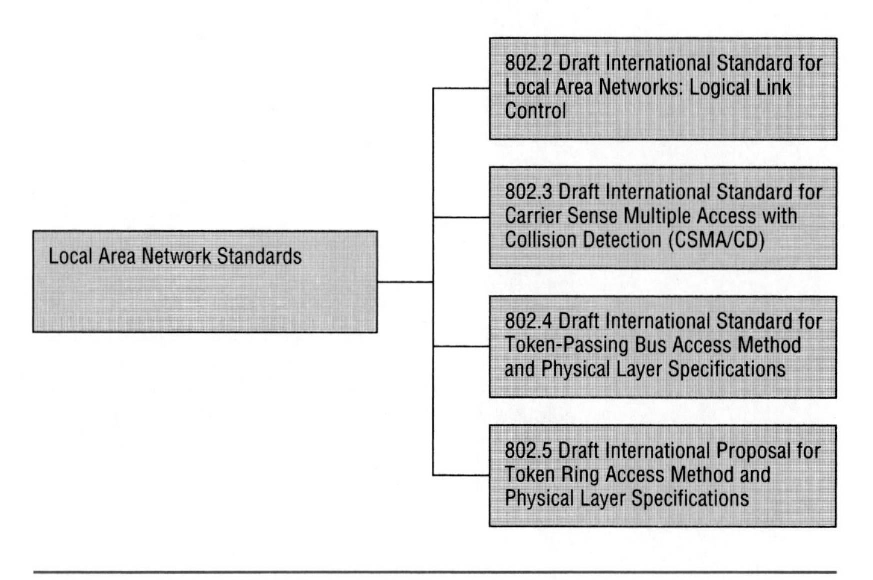

working groups responsible for investigating various standards under consideration. These activities come under the auspices of different national organizations. The **American National Standards Institute** is at the top of the hierarchy in the United States and reports to the ISO. Other national standards organizations are responsible for various standards activities within their respective countries or geographic areas.

The various national standards institutes and organizations coordinate through the International Standards Organization, which is organized as diagrammed in Exhibit 7.3. The result of this complex system of coordination is that it takes several years for an international standard to emerge. From a practical point of view, the business organization may be forced to cope with incompletely accepted international standards. Business organizations can play an important part in helping to influence the development of standards by participating in the various working groups that work to create the prototype standards. Exhibit 7.4 lists these groups for the CCIR. The key factor is to ensure

Exhibit 7.3

International Standards Organization Hierarchy

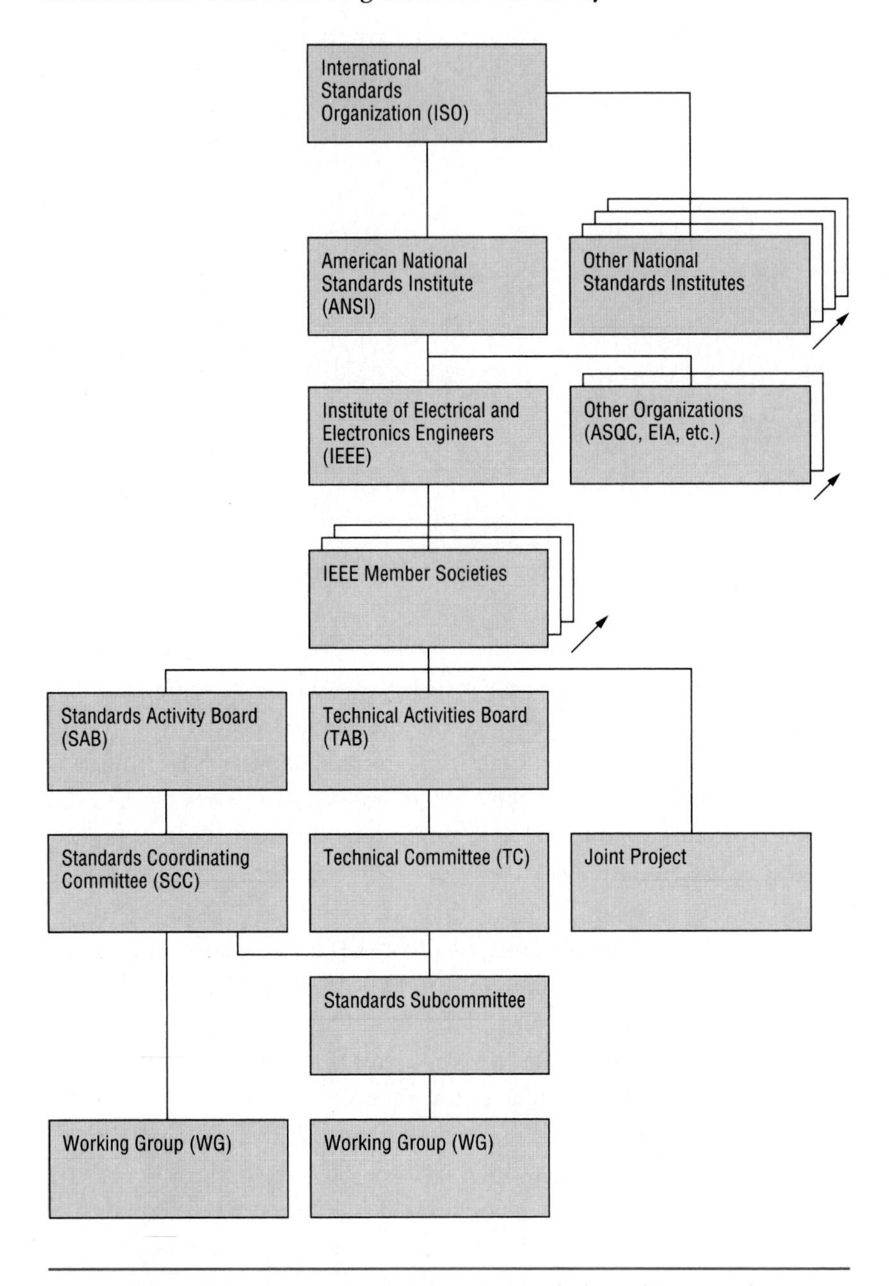

Source: Adapted from pp. 8 and 10 of "A Guide to IEEE Standards Development," February 1989, published by IEEE.

Exhibit 7.4

International Telecommunications Union: International Radio Consultative Committee (CCIR)

Study Group Structure

1. Spectrum utilization and monitoring
2. Space research and radio astronomy
3. Fixed service (point-to-point) at frequencies below about 30 MHz
4. Fixed service (point-to-point) using communications satellites
5. Propagation in nonionized media
6. Propagation in ionized media
7. Standard frequencies and time signals
8. Mobile services (as for maritime, aeronautical, and land mobile communications)
9. Fixed service using radio relay systems
10. Broadcasting service (sound)
11. Broadcasting service (television)

CMV. CCIR/CCITT Joint Study Group—Vocabulary

CMTT. CCIR/CCITT Joint Study Group—Transmission of sound and television signals over long distances

that current business operations are not harmed by the development of a standard.

Managing International Operations and Barriers to Data Flows

International operations of multinational corporations have special challenges and opportunities. The primary change in expanding internally involves the different assumptions concerning availability of technology. Each country and region of the world has variations in the telecommunications infrastructure, and these variations range from highly transparent to highly bureaucratic and unwieldy. There appears to be no direct relationship between the transparency of telecommunications in a country and its level of economic development. For example, Japan and France are far more developed than Hong Kong, and yet the former British colony's telecommunications infrastructure is more transparent to the external multinational corporation.

The planning, installation, and operation of international telecommunications and information technology systems relies on adapting to the peculiar nature of the environments found in different countries. There are several principal variations of the barriers that may impede the creation of efficient international networks. Transborder data flow regulations may force an organization to adopt a technologically suboptimal configuration. Differences in standards, interfaces, cost relationships, and tariffs may also heavily influence the possibilities found by the multinational. Ultimately, these regulations and controls over the ability to erect a telecommunications infrastructure may exert pressure upon the design of the information system, influencing such important questions as the location of data and the location of applications and hardware.

Transborder Data Flows

Transborder data flows take place whenever data or information moves across international borders. Technically, any movement of information and data, regardless of the medium and the form it takes, constitutes transborder data flow. However, in most circles the term refers to the telecommunication of data and information. Since any multinational corporation relies heavily on flows of information internationally for its operations, any system of regulatory control that hampers, retards, or raises the cost of transborder data flows has a direct effect on the ultimate cost position and competitive possibility of the corporation.

The debate on transborder data flow began in the mid 1970s in Vienna, Austria when various European governments threatened to license or regulate international flows of data. In different circles, and influenced by different ideologies, the question of transborder data flow flowered into a broad complex of issues. For example, Marxists argued that the distribution of computing resources around the world was unequal and favored developed countries, and that international telecommunications systems only served to make this type of unequal informational structure stronger, to the detriment of developing countries. Business interests in the western developed nations argued that in order to conduct commerce in the most efficient way, it was necessary to have completely free exchange of information internationally.

Much of the debate turned on a conflict between the United States and European nations. The Europeans argued that controls on transborder data flow were necessary to control the negative social effects of the computer. U.S. interests argued that just as with manual information and data, companies using more advanced forms of technology, i.e., telecommunications systems, should not be punished by restrictive regulations that would ultimately stifle innovation and hurt profits.

Several years of negotiations on this issue, held primarily through the **Organization for Economic Co-operation and Development**, led to adoption of the **Guidelines on Privacy and Transborder Data Flows of Personal Data.** These guidelines were aimed at providing protection to the individual, but came under criticism because of the administrative overhead they would require.

Privacy Regulations

The privacy concerns in international telecommunications arise from the experience with the Nazis during World War II, who used public databanks to hunt down Jews and others. The events in Europe at that time left a deep impression on the European psyche regarding the need for state protection of privacy and confidentiality of sensitive information on individuals. At the time the Guidelines were being negotiated, no existing international protection standardized the way different countries treated the confidentiality of computer-held data.

The short-term effect of privacy regulations was to make it very difficult to import and export information on individuals across international borders. This type of prohibition quickly resulted in conflicts: what about mailing lists, internal company records, and centralized personnel records?

In addition, the question of legal liability arose. Who was responsible and what were the penalties if damage resulted to an individual's interests from a breach of data processing rules? How were inspection and auditing to occur, and what would be the accepted procedures? These issues eventually led to strict regulations on the export of information on individuals out of host countries.

In-Country Data Processing Requirements

In-country data processing (ICDP) restrictions surfaced as a result of controls on the international telecommunication of individual name-linked data. Instead of being able to telecommunicate information at will, as determined by the efficiencies of system design and maximization of investment in information technology, it became necessary to build separate data processing facilities within countries.

For example, within a relatively small geographic area such as Europe, it would be very easy to base a large, centralized data processing environment in one center, such as Brussels. However, because of ICDP requirements, it was not possible to build large, centralized systems. Instead, separate data processing centers had to be built in each country. This tended to change the telecommunications requirements drastically. Rather than being able to set up large time-sharing systems, organizations moved toward more distributed systems in which different data processing centers were hooked together.

When designing an international telecommunications network, it is therefore critical to review both local legislation and practice in order to determine the type of information architecture to be required. After that is known, the details of the telecommunications systems can be worked out.

Leased versus Volume-Sensitive Pricing

One enduring conflict in building international telecommunications networks involves the availability and conditions for getting leased lines. In the United States, for example, leased lines are purchased (leased) from a carrier, traditionally AT&T, a local telephone company, a regional Bell operating company, etc. After the business leases the line, it is able to use whatever technology it likes to maximize its return on the investment. A firm typically will use advanced modems and multiplexers to cram as much information and data as possible down the leased line, at the highest speed possible. As a result, the number of bits per second increases dramatically.

Before the unification of telecommunications being brought about through the development of the EC in 1992, individual European countries had attempted to control this activity within their borders. Certain PTTs, particularly the West German Bundespost, threatened to change the way in which tariffs for leased lines were set. The proposed change would replace time pricing, i.e., set fees per month, per week, etc., with a system of volume-sensitive pricing. **Volume-sensitive pricing** means that no matter how fast the information and data travel through the line, the pricing reflects the amount of information that flows through the line, not how much time the line is used.

Businesses objected to this type of pricing arrangement because it negated the benefits of increasingly efficient types of multiplexing. If a business was going to be charged the same price for each piece of data being transmitted, completely divorced from the issue of how efficiently the data were being transmitted, then there was no incentive to use advanced technology. Matters became worse when it was hinted in the press that the change from time to volume-sensitive pricing was intended precisely to gradually force business users off private leased lines and onto the public data communications circuits.

The additional issues of not being able to build private networks and the lack of competition in international and national telecommunications systems complicated the debate. Gradually, the world has been moving toward privatization and the introduction of competition into telecommunications. However, the legacy of the natural monopoly theory of telecommunications lives on in many countries throughout the world.

In this type of business environment, telecommunications planners must realize and account for the fact that it is not possible to implement optimally efficient networks on an international scale. The reason is that policies in various environments will reduce the technological possibilities available.

Examples of National Variations

Hong Kong and Singapore

Hong Kong is remarkable for its entrepôt activities and the type of openness it maintains. As opposed to the types of restrictions placed against the telecommunications plans of multinational corporations operating in developing countries around the world, Hong Kong tries to maintain a very open system with as many linkages as reasonable to the rest of the world. A result may be the tremendous degree of openness evident in Hong Kong, and enjoyed by its business community. With Hong Kong's openness, it is quickly emerging as the de facto capital of much of the economic activity in China, but of course this is not the view of the relevant ministries in Beijing.

In addition to the very open telecommunications policy enjoyed by companies telecommunicating into and out of Hong Kong, Singapore, another of the Five Dragons of East Asia, has gone one step further. The policymakers in Singapore have realized that telecommunications, far from being something a country might wish to restrict to promote development, is something a country may instead wish to enhance as a means of attracting the foreign overseas operations of multinational corporations. This is a way to change a country's factor endowments.

In order to try to attract business away from Hong Kong, particularly in the financial services sector, Singapore has placed much emphasis on its telecommunications channels. It has:

- Worked to link Singapore with the rest of the world through undersea fiber optic cables
- Worked to lower the cost to the user of telecommunications services as much as possible
- Cooperated as fully as possible with the building of international satellite linkages

The result for international business operations is that they face little if any difficulty in setting up international computer and telecommunications operations. It is not too strong to argue that the freedom enjoyed by businesses in the telecommunications area is largely responsible for helping Hong Kong and Singapore develop into and remain major financial centers of the world.

Brazil

Brazil is an example of another type of developing country that has adopted a rather radical series of steps in its information technology and telecommunications policies. These policies prevent multinational corporations from freely setting up multinational information technology and telecommunications systems. Brazil does not allow the free transmission of data over its national borders, although it imposes little if any limitation on data transmission within Brazil.

This highly restrictive policy toward transnational computing systems poses a serious challenge to an international system of arrangements based on both free trade and freedom of information. Some of the best-known limitations posed by the Brazilian information policy include the following:

- Restrictions on importing computer and telecommunications equipment
- Strict requirements for application and approval by the **Secretaria Especial de Informatica** (SEI) for any information technology expenditures
- Prohibition against importation of any technology if a "substantially equivalent" Brazilian product is available
- Prohibition against access from within Brazil to external databases and remote computing services
- Strict licensing and lengthy approval processes for any type of international computing linkage

These types of problems faced by the multinational corporation doing business in Brazil may become a prototype of actions to be taken by other developing countries.

The Brazilian information policy has a long history. The goal of the policy is to achieve independence for Brazil for the manufacture and creation of information technology and telecommunications services. By protecting its internal market, Brazil is able to guarantee the capital accumulation necessary for investment of its own industries. Otherwise, it is doubtful the Brazilian government could afford the large amounts of funding that would be necessary to conduct the required research and bring the needed products to market.

The types of controls Brazil places on the building of international information telecommunication networks is indicative of the type of policy corporations may find in many different countries, including other countries within Latin America.

Korea

Korea is another one of the Five Dragons which have developed in East Asia in the post-war period. Korea has stunned the world by quick movement into shipbuilding and heavy construction, particularly in the years leading up to the oil crisis. Korea then moved rapidly into the information age with the rapid fabrication of semiconductor memory devices of 64K, 256K, then 1 megabyte and more.

Korea is another country that has protected its information technology and telecommunications infrastructure as it built its own system. Korea protected its internal telecommunications system by using administrative control and guidance as it opened up to the outside world. The effect of this on business was to prevent development of an efficient telecommunications system until the late 1980s and early 1990s. Korea's effort was boosted substantially by the massive laying of fiber optic cables for the 1988 Olympiad in Seoul. For a time immediately preceding the Olympic Games, Korea was installing more than 3 million lines per year.

As part of its technologically incrementalist infant industry protection industrial policy, Korea extended protection for local manufacturers at each stage of the product development and progression cycle. For example, in telecommunications, it protected markets for first 300 baud, then 1200, then 2400, then 4800 baud modems from imports until Korea was producing its own equipment. Once equipment was being produced in Korea, the market was for all practical purposes closed to the importation of foreign equipment.

The effect on businesses operating in Korea was to make it very difficult for a while to operate highly efficient international networks. Gradually in the late 1980s, as Korea brought more and more international record carriers into a gateway to Seoul, financial services organizations as well as others managed to interconnect into world-class systems. Combined with gradual financial liberalization, the revolution in telecommunications was engineered by the Korean authorities to develop the infrastructure in the quickest way.

The Korean case shows that strong prohibitions or other controls over the availability of telecommunications services may derive from a strong national reason, perhaps based on the need to rapidly develop the country's infrastructure. This was the case with Korea, and is certainly the case in many other countries. It is still an anathema to the Koreans to have foreign ownership or participation in the development of their own telecommunications system, notwithstanding technical collaboration with such giants as AT&T.

One could conclude that when developing strategies to operate sophisticated telecommunications in the developing world, regulatory and technical problems can be anticipated with near certainty. It is understandable in many cases because of the needs of elected representatives with their corresponding domestic political baggage. They must prevail in running the government and setting policy that will affect the telecommunications strategy of the multinational corporation.

Summary

In managing international telecommunications, firms can take advantage of an institutional machinery through which they can make their concerns known to the international community through such bodies as the International Telecommunications Union, the International Chamber of Commerce, and the International Telecommunications Users Groups. These organizations help to establish the technical standards through a variety of committees known as CCITT groups, each of which focuses on a single topic.

Setting up international telecommunications networks can raise some difficult problems for the manager. Regulations on transborder data flow, guarantees of privacy, and the occasional requirement for in-country data processing can all heavily influence the development and deployment of telecommunications systems. Occasionally, standards for interconnection may be used against the company wishing to set up shop, but this has become less of a problem with the introduction of new generations of technologies.

National variations are severe and in order to successfully plot a strategy, it is necessary to study each nation separately. In Brazil, for example, a relatively severe regime imposes tight controls; in an entrepôt such as Hong Kong, the government imposes very few controls over international telecommunications.

Questions

7.1
- a. How can a firm participate in setting international technical and regulatory standards for telecommunications?
- b. Discuss the roles of the International Telecommunications Union, the International Chamber of Commerce, and the International Telecommunications User Groups.

7.2 a. What does *CCITT* stand for?
b. How is it organized and what does it do?
c. What does *ISO* stand for?
d. How does the CCITT relate to the ISO?

7.3 What is the difference between leased versus volume-sensitive pricing and why does this matter?

7.4 Why has transborder data flow become an issue?

7.5 What types of regulations typically are imposed to protect the privacy of individual data?

7.6 How can standards be used as a nontariff barrier to trade?

7.7 Discuss the telecommunications environments in Hong Kong and Singapore. Why have they adopted the type of regulatory regime they have?

7.8 Why does Brazil prevent remote data processing?

7.9 What is the long-term prospect for a completely transparent global telecommunications network?

7.10 a. What types of skills do you think are necessary to manage international telecommunications?
b. How do they differ from the skills needed for managing networks that do not have international connections?

Part ■ *II*

Strategic Business Applications

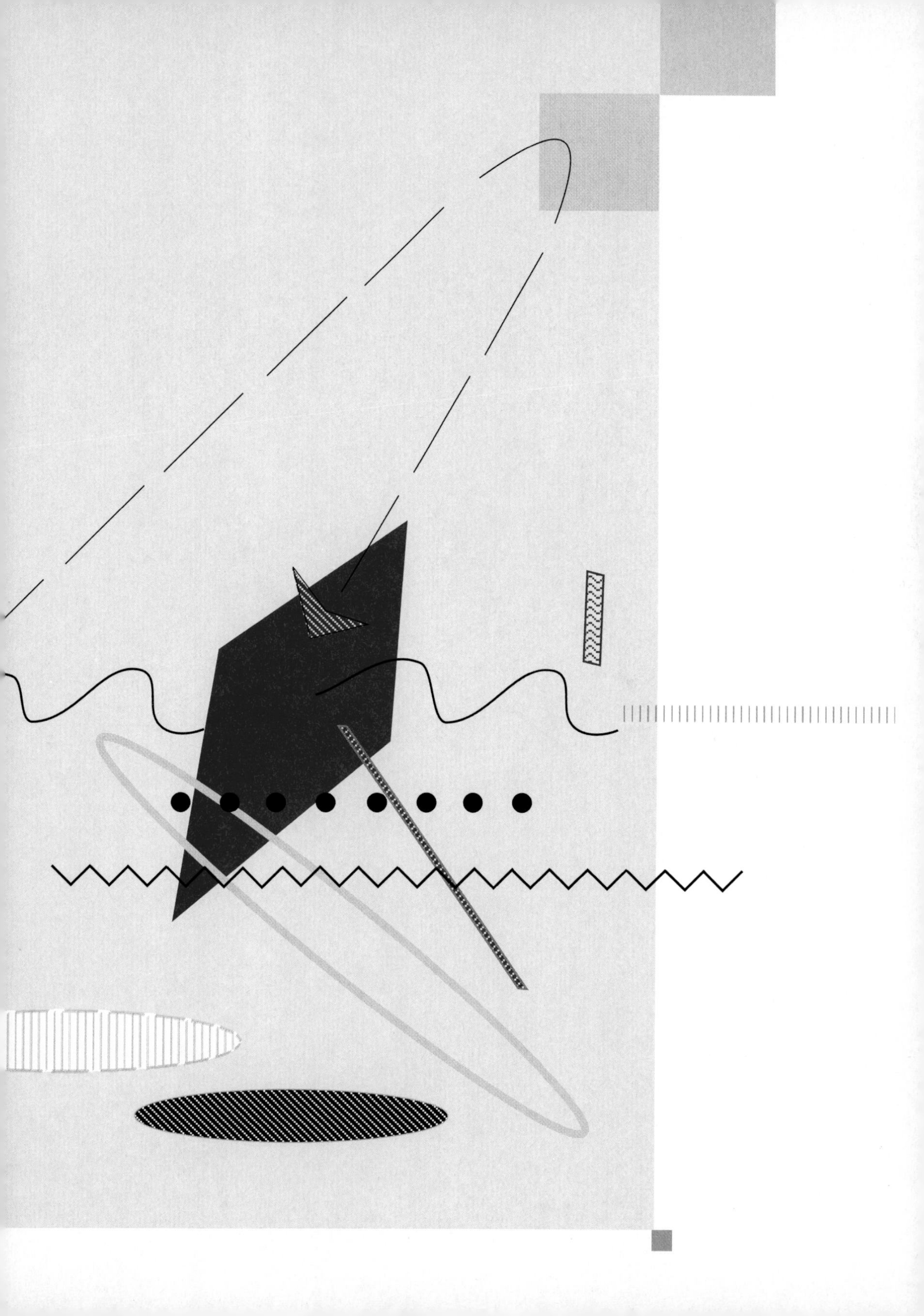

Chapter 8

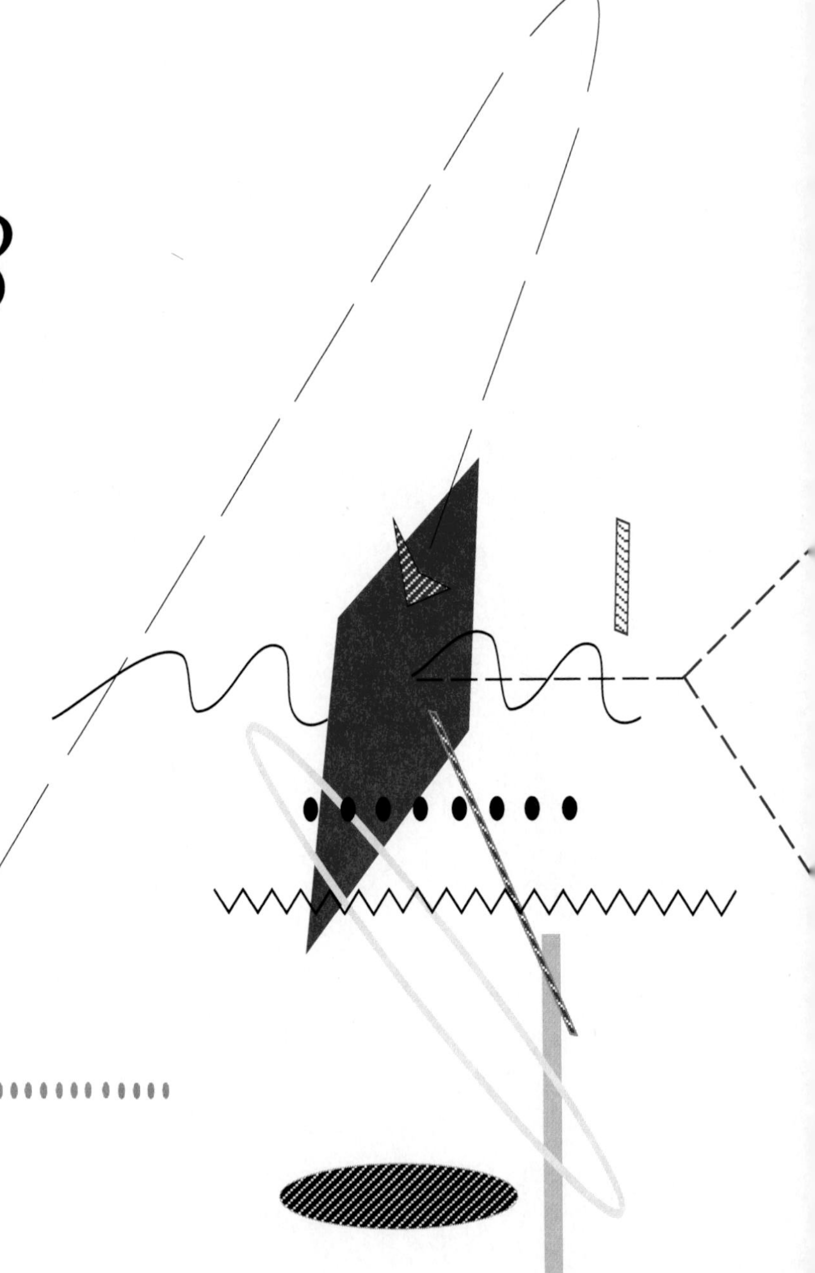

Telecommunications and Competitive Advantage in Business

What is the role of telecommunications in business today? How does telecommunications contribute to competitive advantage? What are the ways it is used? What have various businesses done to maximize their sometimes substantial investments in telecommunications?

This chapter introduces the seeds for the fundamental idea in this book: that telecommunications can give strategic advantage in business competition, as shown by many case studies. However, underlying the idea of using telecommunications for strategic advantage is a model that provides a useful framework for categorization of ideas.

This chapter presents the idea of strategic linkages and value added chains in organizational structures. It shows that much of the strategic advantage deriving from telecommunications comes from linkages both outside and inside the organization. Linkages outside the firm are considered particularly strategic. Value added chains can show that at every level of business operation there is a series or pattern of information processing and distribution that can be made more efficient through telecommunications.

The chapter closes with a review of several categories of strategic advantage deriving from telecommunications. This prepares the reader for study of the following chapters, which take a deeper look at how strategies are developed.

Strategic Linkages through Telecommunications

Telecommunications is a way of linking together both computers and people in a way that eliminates the physical distance between them. Telecommunications can also be used to destroy the physical distances between different businesses, for example, between a manufacturing firm and its subcontractors. This may be difficult to realize, because it is so much a part of our life. In order to model the efficiency and competitive effects of implementation of both internal and external telecommunications systems, we can develop a value added or process model of the firm, as in Exhibit 8.1.

In its simplest form, a firm is composed of many value added chains through which it takes in material, processes it, and then outputs it to the market. Some of the strategic thinking in the early 1980s, following Michael Porter, viewed the firm this way. It is a very simplified model, but it has accomplished the purpose of getting telecommunications managers to think outside of their boxes, i.e., to think in a larger context than the immediate confines of their jobs.

There are a few problems with using the simplified value added chain analysis:

- It is difficult to model a complex firm involved in many markets and many different products.
- It is difficult to distinguish between the sources of internal efficiency versus external efficiency.
- It is difficult to model cooperation between firms.
- It is difficult to model the activities of different firms involved at different stages of a total value added chain for a product.
- It does not apply very well to the world's most successful businesses, the Japanese *gurupu* type.

Another value added approach to understanding the strategic utilization of telecommunications technologies is to examine all of the firms involved in supplying a final product. (See Exhibit 8.2.) This could be termed a **complex chain analysis**. Although slightly different, this type of model can help explain the existence of interorganizational systems and how they can both cooperate and compete with one

Exhibit 8.1

Strategic Telecommunications Linkages: Value Added Model

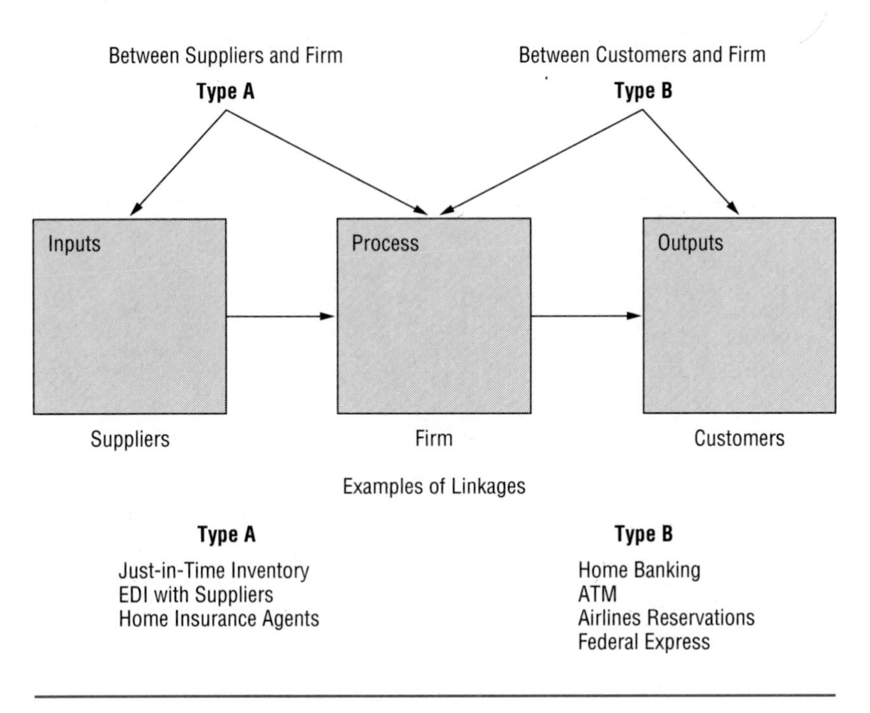

another. It also helps model the extremely successful Japanese *gurupu* organizations.

In this model, telecommunications linkages can connect the firm to both suppliers and customers, as well as to other organizations. Competitors would probably not be logically linked together in many countries. In the **Porter model**, the linkages with suppliers can reinforce the relationship between a firm and its suppliers.

In modeling the Japanese experience, the Porter model appears a little antiquated. There, linkages are to a great extent already taken for granted, since firms are linked together by the *gurupu*. This restrains a business from making strategic linkages to suppliers and customers.

In this more successful industrial model, the suppliers are not pressured and played off against one another. Instead, telecommunications can aid in making the linkages much more efficient and better for

Exhibit 8.2

Value Added Chain for Final Product

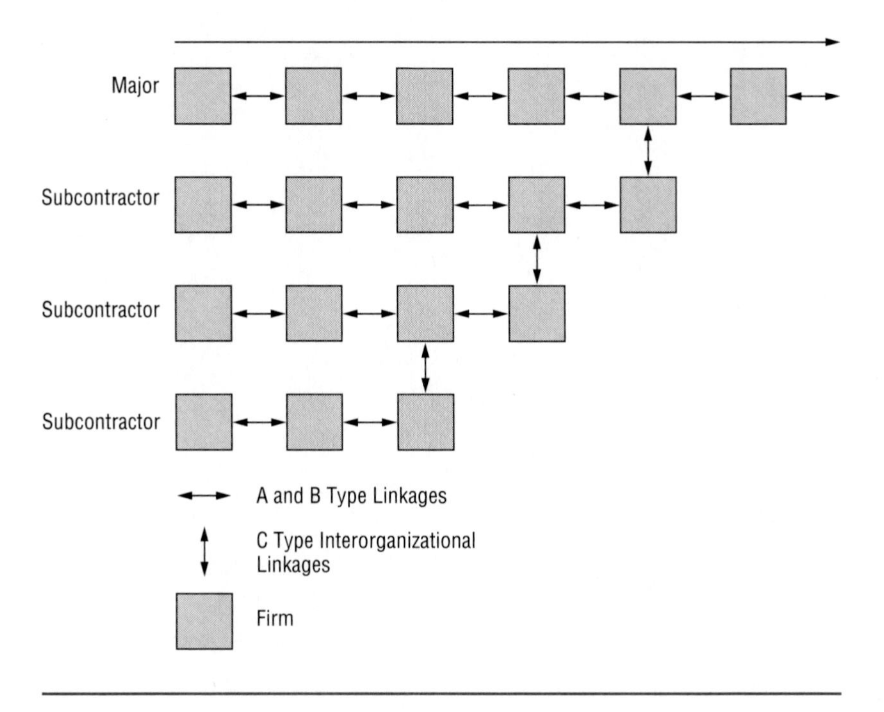

the entire *gurupu*. These types of linkages are a subset of what is called the *interorganizational system*, although it refers more to the information technology aspects, rather than the telecommunications aspects of such a system. As one of their functions, these interorganizational systems link together parallel companies in the exchange of information through telecommunications channels.

The production of each industrial product involves many steps and components. Each of these components or steps may be completed by a different business, as is the case with suppliers or subcontractors. In addition to the backward and forward linkages between firms involved in producing a specific product, parallel linkages join the principal product creation chain to other subcontractor chains that support the main chain at one place or another. To the extent that these types of telecommunications linkages are set up for the purpose of coordinating the manufacturing or a related process, they are of a slightly different class than the well-publicized strategic linkages.

Levels of Value Added Telecommunications Chains

In addition to studying the interorganizational linkages, and the linkages with suppliers and customers, study of strategic telecommunications systems involves a look at what happens within the firm. It is easy to see that telecommunications linkages with organizations or entities outside of the firm are strategic in nature, but if telecommunications within the firm also gives a strategic advantage, such as with internal efficiencies or cost reductions, then these linkages too are strategic in nature. Furthermore, just as the bulk of paper-based information flows within the organization, more than 90 percent of it within individual departments, so too are telecommunications linkages within the firm critical.

The value added chain, seen at the highest level, links together firms involved in working with a single product. However, within each of the firms involved, there is another level of analysis. Within each firm, the individual departments or sections of the corporation or *gurupu* must also be linked together. (See Exhibit 8.3.) At this next level of telecommunications analysis, each of the major sections of the firm processes and transmits information within individual departments.

At each level, telecommunications linkages can give strategic advantage to the firm, but it is generally conventional to think first about the strategic picture, then go into details on the lower levels of information flow and its corresponding telecommunications infrastructure.

The ideal telecommunications plan and infrastructure should be linked together so that each of its levels supports the information processing and telecommunications of data and information at the corresponding information system level. At each level of the value added chain, different efficiencies can be gained. This raises several problems for the planner, however:

Exhibit 8.3

Levels of Analysis for Telecommunications Linkages

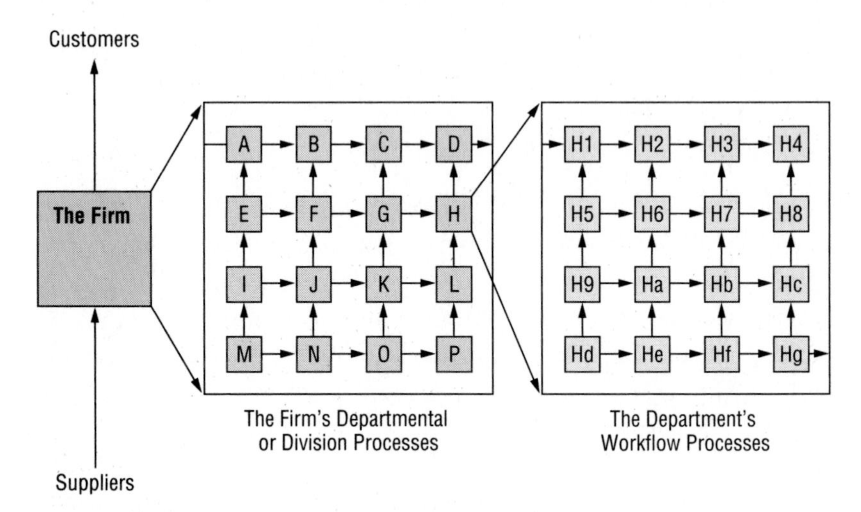

Type A and B linkages with suppliers and customers Type C linkages with other organizations within *gurupu*	Linkages between divisions or departments within the firm	Linkages within work groups or departments
Examples		
Centralized time-sharing Electronic Document Interchange system Centralized mainframe with remote (nonfirm) access	Centralized mainframe system Distributed minicomputer system Large interdepartmental batch systems	Local area network linking workstations Some wide area networks linking workstations in different locations involved in the same work

- Which level of value added analysis offers the greatest potential for payoff from telecommunications?
- How can the benefits actually be measured?
- How can telecommunications linkages support information needs at different levels?
- How can the best opportunities be identified?
- What are the best technologies to employ at each level?

In practice the linkages are rarely so neat, and the structures are constantly in the process of change, all at different rates. This is an important point to which the student should pay special attention. Ideas and concepts presented in textbooks are rarely perfectly implemented. The world is not perfect. Ideal plans or methodologies and their clever answers may be little more than snake oil disguised as elixir. Students should always keep in the backs of their minds the limits of knowledge, and the frailty of generalizations.

Types of Strategic Advantage

As the 1980s passed, several families of strategic advantage through telecommunications emerged. (See Exhibit 8.4.) Some of the gains came from systems that were completely internal in nature, not involving contact with other organizations, customers, or any outside entities. These internal systems were primarily concerned with the internal command and control of the business. They generated a number of telecommunications problems:

- How could telecommunications make it easier for top management to control day-to-day operations?
- How could telecommunications shorten the intervals between various critical steps in the business process?
- How could telecommunications radically reduce the costs associated with business operations?
- How could telecommunications help the firm manage inventory to gain a strategic market advantage?

Other gains came from systems that used telecommunications to build a better interface with the customer. One of the management rages during the late 1980s was customer service. Companies began to realize that satisfying the customer well was the key to building a long-term relationship. Some analysts argued that one of the keys was providing customers what they wanted in the shortest period of time. Others argued that the key was to provide excellent, personalized, high-tech, high-touch customer service. Others simply pointed out that telecommunications could be used to deliver services. Some of the questions raised included:

Exhibit 8.4

Categories of Strategic Advantages from Telecommunications

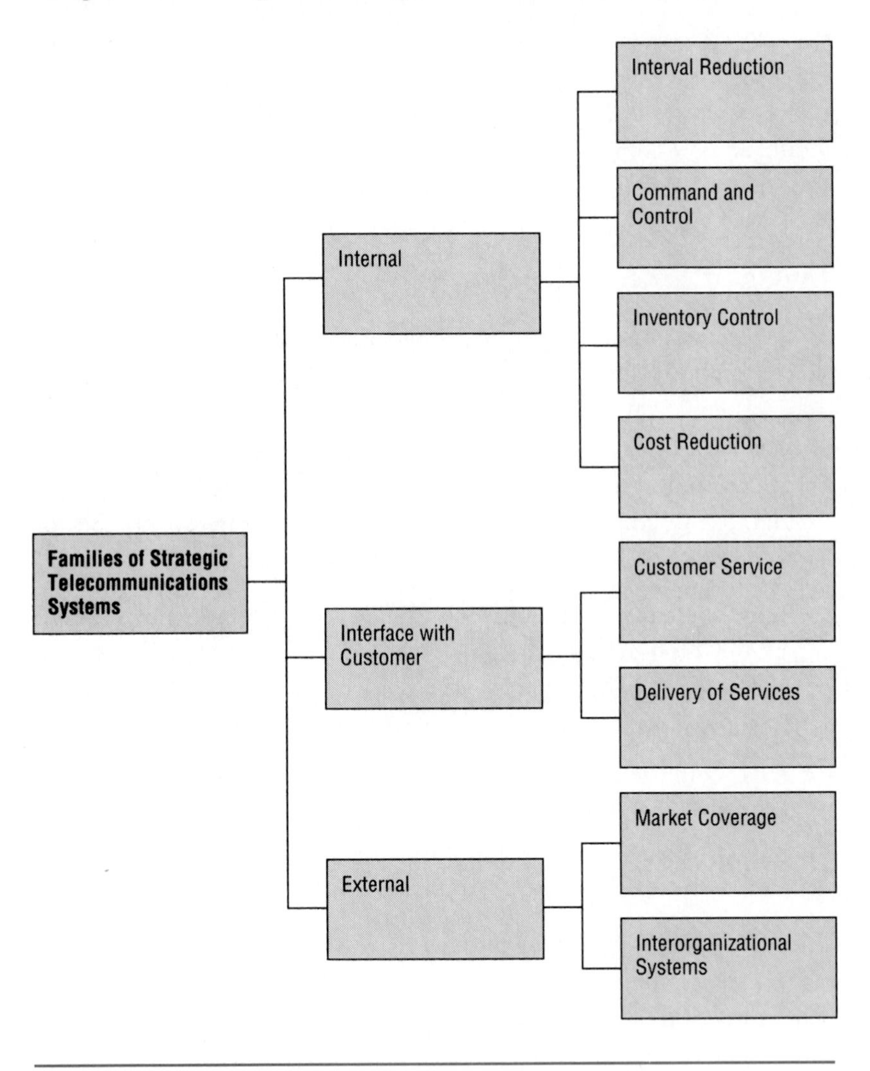

- How could telecommunications improve customer service?
- How could telecommunications help a firm deliver services to customers in a way that would yield a competitive advantage?

In addition, the late 1980s brought much discussion regarding the linking together of the business with entities on the outside. This area of discussion received the greatest amount of press attention, and also saw some of the greatest developments. Telecommunications linkages with external entities could provide greatly expanded market coverage. It was also possible to use telecommunications to process orders and generally automate the vast array of standard, paper-based transactions that had been burying business under piles of paper for years. **Electronic Document Interchange**, the movement of standardized, paper-based instruments through electronic telecommunications channels, helped greatly in this effort. Some of the questions raised by this type of telecommunications linkage included:

- How could one use telecommunications to greatly expand the market coverage of the business without incurring prohibitive extra costs?
- How could telecommunications automate the cumbersome paper-based interfaces with external entities?

All of these developments in the late 1980s began to show business that telecommunications had finally moved out of the back room into the forefront of corporate thinking about the overall strategic development of the business. Long seen as merely an overhead item, a cost that was little more than a mandatory drain on the business profits, telecommunications had shown that proper use of its technologies could provide great strategic advantage. Top management began to think in terms of internal and external linkages. They began to wonder how information technology and telecommunications linkages for moving information and data—including data about customer wants and desires—could be used to improve the business.

By the 1990s, telecommunications had finally regained the critical position it had held approximately 150 years before.

Summary

Telecommunications is one of the key driving forces in modern business. It is a conduit through which information in a firm flows. Virtually all data and all voice communications travel at one point or another through a telecommunications system.

However, besides studying the use of telecommunications within an organization, observers of business strategy have looked outside the firm. In particular, telecommunications linkages between a company and its customers and suppliers have been the center of a great deal of thinking about how telecommunications can provide competitive advantage.

The use of value added chain analysis is key to looking at the impact of telecommunications both within the organization and without it, and as a combined unit taking all flows of information, voice, and data into consideration. Telecommunications can, therefore, provide both efficiencies and advantages in all of the dimensions considered: internal, interdepartmental, between firms, and between the firm and its customer base.

Questions

8.1 What is the difference between telecommunications inside an organization and outside an organization?

8.2 a. What is the basic concept of the value added chain?
b. How does it help us understand business strategy?

8.3 How can telecommunications help a firm obtain strategic competitive advantage?

Chapter 9

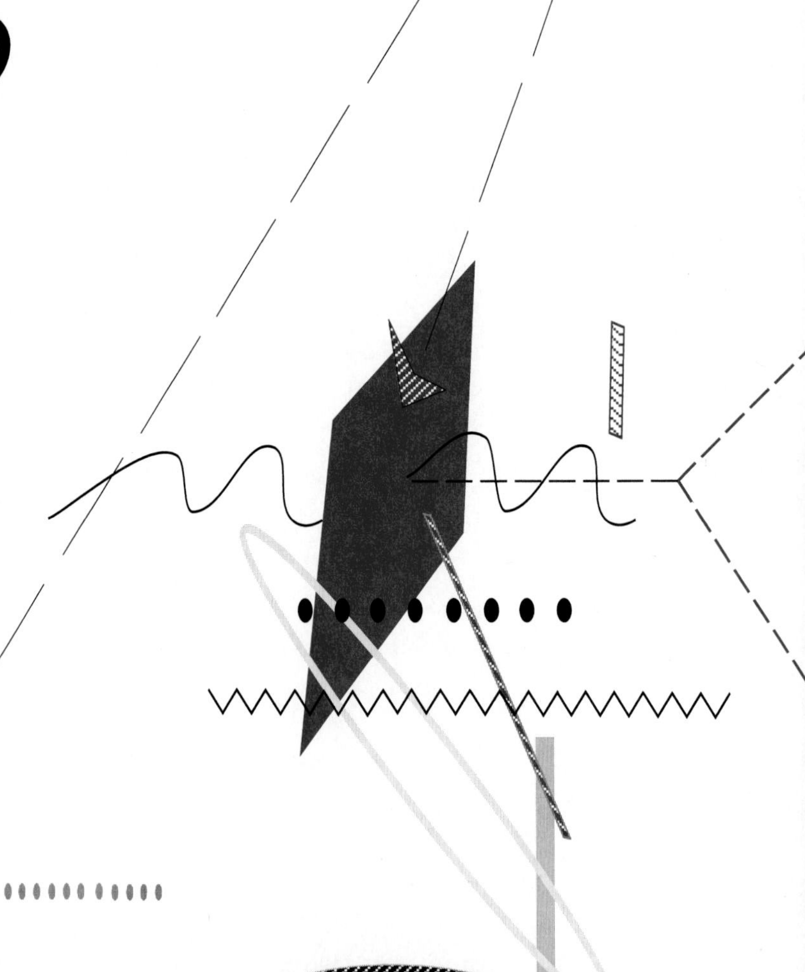

Rapid Market Penetration

How can telecommunications be used to greatly increase market coverage without a great increase in costs? What type of information must be moved from the center of the organization to the periphery in order to conduct business? What type of alternative infrastructure can telecommunications provide for delivery of services and information that in the past have been provided by other means?

This chapter discusses the idea of using telecommunications to aid very rapid market penetration. One of the key technologies making this possible is the communications satellite.

Telecommunications can be used to greatly increase market coverage by linking the corporate center with highly remote locations, where previously the central office had little if any presence. The primary underlying effect of this is to greatly increase the velocity of information flows between the center and the periphery of business activities. This increase in velocity is due to the abandonment of physical movement of information and data in favor of telecommunications.

The Electronic Newspaper

The developed world's largest circulation newspaper is *Yomiuri Shinbun* of Japan. Like other famous newspapers around the world, it has built up a system of delivering its editions to remote printing plants for publication. This electronic distribution of the newspaper allows it to capture much greater market share than would ever be the case if it had to depend on the physical distribution and logistics arrangements prevalent in traditional systems. Other famous newspapers have adopted the same approach, using technology like that illustrated in Exhibit 9.1

- The *Financial Times* of London uses satellite transmission to publish its paper throughout North America, Europe, and the Far East.
- The Gannett newspaper *USA Today* was one of the great innovators in the United States. *USA Today* is available in virtually every location in the United States every morning. This type of saturation coverage of an entire market the size of the United States could only be achieved with utilization of telecommunications technologies.
- *The New York Times* publishes through five printing plants located in different parts of the United States. It is able to produce a regular New York version as well as a West Coast version of its newspaper, with significant editorial differences in the front page and features sections of the editions.
- *The Wall Street Journal* is published through several plants in the United States in the same way as *USA Today,* and has thereby become a national newspaper.

In each of these cases, the key to competitive advantage lies in using telecommunications to break through physical distribution bottlenecks and gain much wider market penetration. In some cases, the financial structure of the newspaper publishing company changes as local publishing plants can take differing shares of ownership and risk in the publication of that newspaper in particular regions. In other cases, the use of telecommunications enables a great newspaper to penetrate major metropolitan centers that in the past depended on air delivery of the information for financial and other decision making.

The widespread use of telecommunications, in conjunction with other changes in technology including electronic composition with direct writer input of stories, has radically changed the economics and

Exhibit 9.1

International Electronic Newspaper Publishing Systems

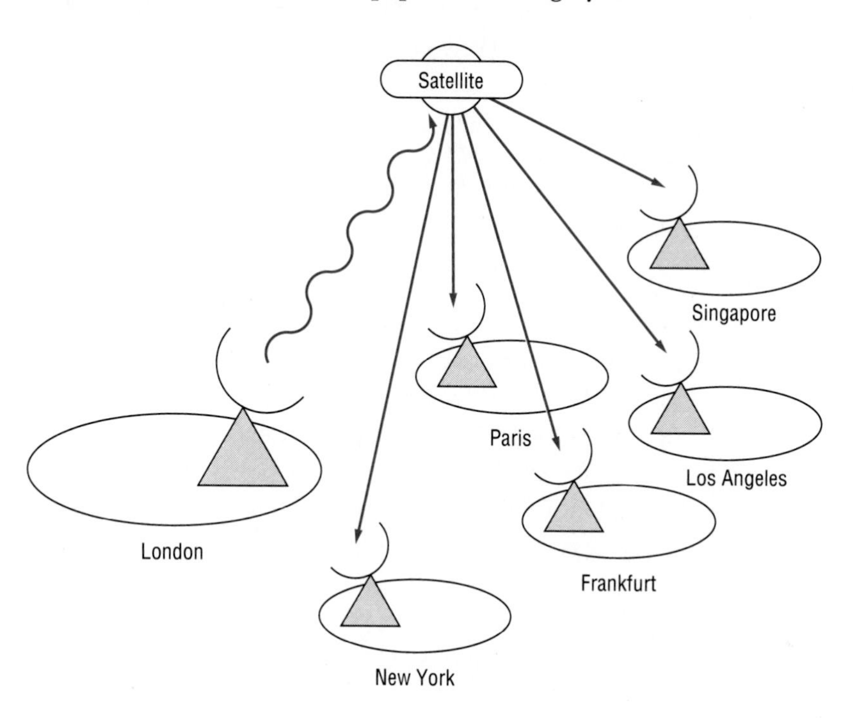

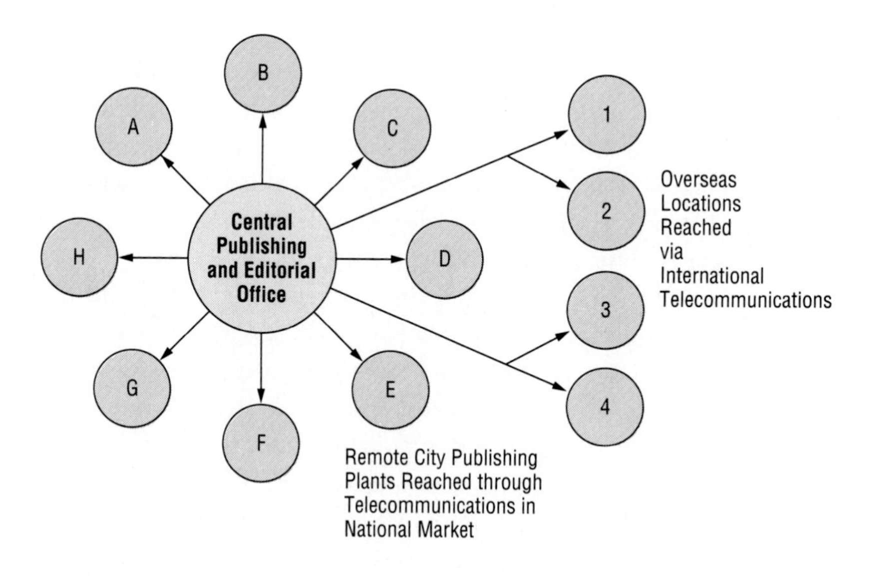

nature of the newspaper industry. In some cases it may eventually help to squeeze some newspapers out of local markets, since telecommunications changes the definition of a "local" market to include entire countries in some instances. The world is seeing the quick emergence of global newspapers.

The Electronic Delivery of Information

In the late 1940s and early 1950s, some of the world's futurists predicted that soon the world would see the elimination of newspapers and books, and their replacement with electronic systems that would enable the reader to view text on a screen, thereby eliminating the need for paper. They predicted the death of the newspaper as the world knew it.

In the late 1980s, 30 to 40 years later, this type of vision had finally come true, although in a slightly different form than was expected. The regular newspaper based on paper is still used as before for reading of the news. However, in research on a subject, electronic information technology plays a critical role.

The use of telecommunications networks to link together information providers and information users has developed rapidly since the early 1960s with the development of large-scale packet networks accessible through dial up to the local telephone company. During the early stages of the systems, users could access only titles and abstract information on a variety of publications and articles. Later, however, companies led by Mead Data started to provide full-text copies of articles. Finally, in the latest stage of the process coinciding with the development of the CD ROM, full scanned images of the relevant articles have been made available to the user.

The use of telecommunications technologies for widespread dissemination of information was also pioneered through the development of videotext technology in the early 1980s. **Videotext** was a way to send visual information down the telephone line into a television receiver or personal computer. The United Kingdom pioneered the effort. Eventually there were several standards, including a North American, a Japanese, and a French and German standard. This proliferation of presentation standards for videotext systems limited their development.

Also, other developments in technology, such as HDTV, and the failure of videotext delivery systems to reach low enough costs to convince consumers the service they bought was worth what they paid, meant that videotext ultimately failed to become a widespread phenomenon.

Again, as CD ROM publishing proliferated, videotext gradually fell by the wayside and failed to become a standalone, highly successful commercial venture.

This does not, however, invalidate the continued possibilities of using telecommunications to deliver information to customers and the public on a very large scale. Companies are finding in some cases that the information they have developed internally can be repackaged and sold to the public as a profitable operation in itself. In addition, companies are finding that making their services more visible to the customer through telecommunications enhances customer service, as Federal Express has shown.

Federal Express

Federal Express courier services has been in the forefront of utilization of telecommunications for competitive advantage. As it analyzed the needs of its customers and the way it was conducting its business, Federal Express developed the concept of linking its customers directly into its scheduling computer through the provision of terminals in the customers' offices. Using these Powership terminals, customers are able to log onto the system and ask Federal Express to pick up a package. This type of solution results in several advantages for Federal Express. Customers with Powership terminals are more likely to order their services through Federal Express, and less likely to change to a competitor. Also, on-line verification and entering of critical pick up and delivery information by customers themselves has a few important advantages:

- Blame for errors in shipping and address information is not placed on Federal Express, since it did not enter the information.
- The costs of filling the order are reduced because a Federal Express employee does not have to stand by on the telephone to take the order, thus eliminating the overhead that would involve.

- Billing and statement inquiries are available to the customer directly, again without the direct intervention of a Federal Express representative.

The system also results in better operational control of pick ups since it may even out the timing of the order process.

The Federal Express case is an example of a company's extra investment in telecommunications to link with the customer for provision of important information. Although there is a net cost of building the system, it produces a payback not only through better business and closer relationships with the customer, but also through the reduction of per-transaction costs of Federal Express customer service representatives.

Banking and ATM Networks

The revolution in **automated teller machine (ATM)** deployment has been a telecommunications-based expansion of market coverage. The first moves into electronic banking were made by Citibank of New York, the largest bank in the United States and the 15th largest bank in the world, in the early 1970s. Citibank needed a way to address the large retail consumer market, but without having to incur the very high costs of providing human tellers. Citibank recognized that automation could replace the tellers for various simple services: making deposits, getting information on accounts, and getting cash. Citibank pioneered the first ATM machines and then priced its services so that consumers would gradually move away from dependence on window tellers and become more dependent on the machines.

In addition to reducing its cost per transaction through its machines, Citibank was able to offer banking services 24 hours per day, a great improvement over the traditional banking hours, which were typically 10 a.m. to 3 p.m. only on weekdays. This additional feature of the service was a great advantage to consumers.

The Citibank machines were linked together with the central computing facilities through telecommunications linkages. When the customer inserted a magnetic card into the machine and requested a transaction, a telecommunications linkage with the host computer was established. When the machine dispensed money, the customer's account was immediately debited in the same amount.

For the machines located on the same premises as a Citibank branch office, the telecommunications linkages were in place already as part of the established system linking the central computing facility with the branch office terminals. Shortly after the introduction of the machines, Citibank began aggressively to build extra locations for the machines, some of them completely unstaffed. Each of the locations used a telecommunications linkage to ensure that as soon as the customer took money out of the machine, the account was immediately debited. Additionally, the computer kept a central record of the security passwords (known as personal identification numbers or PINs) which the customer used to validate the transactions on the ATM.

Citibank's strategy was to increase market penetration of the retail consumer sector of the financial services market and thereby achieve economies of scale for its transactions systems. Ultimately, it realized, the cost per transaction would be lower than its competitors'. Citibank has achieved this goal riding on the back of telecommunications toward radically increased penetration of the market. By using telecommunications, Citibank was able to extend the reach of its mainframes directly to the consumer.

Everyone appeared to benefit: Citibank was able to lower its cost per transaction by limiting the role of the human teller. At the same time it could take its teller population and train them for more advanced and personalized customer services and attention. For the consumer, the automated telecommunications-based system offered the new prospect of 24-hour-per-day access to funds.

Telemarketing

Telemarketing blends several types of telecommunications systems into an integrated marketing, order taking, and billing network. The types of telecommunications technologies include television, telephony, local area networks, packet switched networks, and dial up batch telecommunication of orders. These types of telemarketing systems are particularly popular in the United States.

The firm wishing to use this type of system first advertises on a national, regional, or local television system. This type of advertising is different from national campaigns in that it is typically carried on local or independent channels at a fraction of the cost of major, network advertising. During these advertisements, the consumer is given access to a toll-free telephone number and told which credit cards the company will accept.

A customer who calls the number typically reaches a large telemarketing center handling a variety of products and campaigns at the same time. Either through the opening conversation with the customer or through a record of the exact line on which the customer placed the call, the telemarketing operator is able to tell which group of products the caller wants. Using that information, the operator calls up the relevant information on the computer screen, supplied through a local area network from the mainframe computer of the telemarketing center. After answering questions from the customer with information provided on the screen, the operator can place the order.

Placing the order involves taking down the name and shipping address of the customer, as well as information on the form of payment, in particular the type and number of the credit card being used. Typically, the telemarketing center is connected directly through a packet switched network to the credit verification computers of the major credit card systems. When the information concerning the card number and the amount of the purchase is entered into the computer, the validity and credit limit of the card are immediately checked. This is not a lengthy process. Typically, the Visa network requires about 3 seconds of total time for verification.

After the purchase has been verified, the order is complete and the telemarketing center typically batch transmits the total orders for the day to the firm selling the product. From there, the orders are shipped, and payment is made from the credit card companies to the firm. The telemarketing firm takes a fee based on the amount of business generated.

Typically, the toll-free telephone centers have emerged as a separate business in themselves. This gives the seller of a small item the chance to make a one-time promotion over television, then have orders accepted for a small fee. This development has opened up vast regional and national markets to different goods, without the investment typically required to build national brand presence.

This type of integration in a marketing strategy of different media and types of telecommunications systems is an indicator of developments in the future. It is only a stage of evolution in the development of telecommunications systems that carry visual, voice, and data information by different media, i.e., the television, telephone, and packet switched networks, respectively. (See Exhibit 9.2.) In the future, all these types of information will be carried together in integrated networks.

Another important factor in telemarketing is its flexibility. It is possible to set up an instant network to market an item on a short-term basis. Furthermore, the addition of television allows quick changes in prices, depending on demand, which the system makes known immediately.

Toll-Free Numbers

Businesses can offer a variety of products and services through telephone sales. Telemarketing systems have been established to sell virtually every kind of product. In addition, service inquiries, warranty questions, emergency numbers, and more are performed through toll-free numbers that can serve to bond customers to the corporation. A toll-free number can help a business in several ways:

Order Taking • When customers know there is no charge for the call, they hesitate less in making the call. The fact that the order is instantaneous can take advantage of customer desire while it exists. It has been shown that the ability to satisfy this instant gratification increases orders.

Response to Advertising • Since there is no cost for the call to the customers, they can call even if they do not wish to buy anything, but only want further information. It is a good way to get names for mailing lists. Subsequent sales might result through direct mail advertising. Even when this is not the case, the response is a good way to get insight about customer demand for products and services, and can allow limited polling and survey work.

Traveling Employees • A toll-free number provides a generally less expensive method for employees to call back to the office when they are on the road. This is particularly helpful for sales personnel.

Exhibit 9.2

Telecommunications for Telemarketing

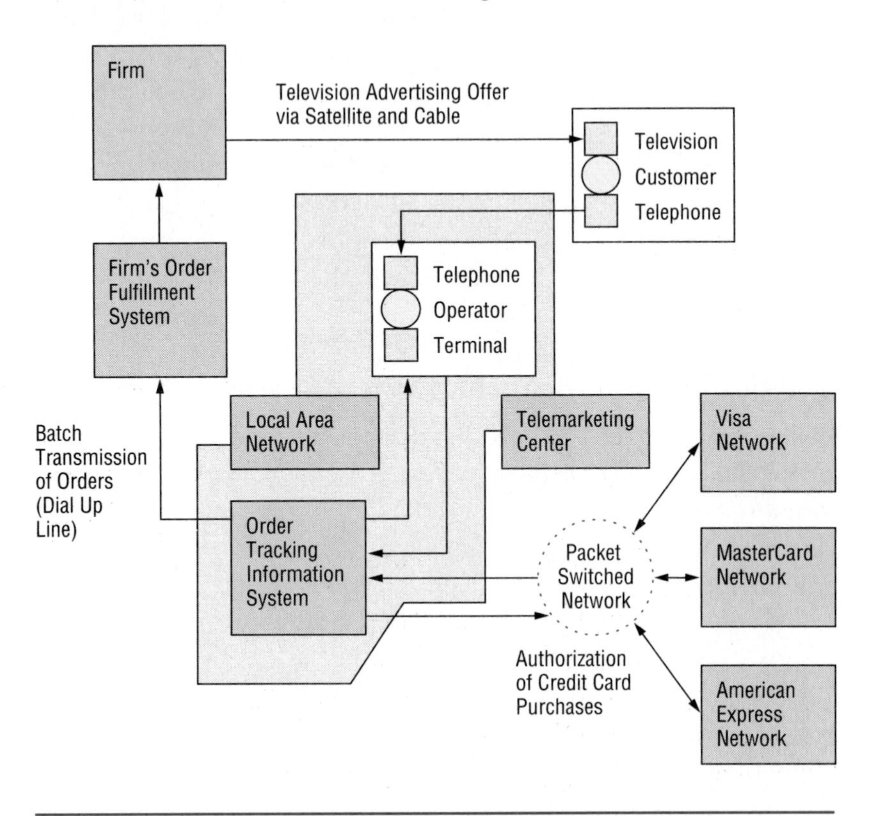

Customer Service • Frequently, major food companies place toll-free numbers on their products asking for suggestions, complaints, or other information. They get few calls per year, but the customer views this as evidence of confidence and honesty on the part of the selling company. In the software business, many companies set up toll-free numbers for customers who have purchased the product. In many cases this has resulted in hundreds of calls from customers who have not properly read their manuals asking very elementary questions. In some cases, the trend may be away from this type of service, or toward charging for this type of phone service. Still, it has helped many users in the past.

Instant Credibility for New Start-ups • Providing a toll-free number appears to provide instant credibility to the firm, helping customers to perceive it as being more of an institution. However, in many cases the answering service is subcontracted out to a professional service that handles calls for many different organizations.

Test Marketing of New Products • Products can be marketed in one part of the country for a time before committing to wide-scale production. The use of the toll-free number helps in obtaining very rapid responses about consumer preferences.

Telethons and Fund Raisers • In addition to business use, charity, nonprofit, sports, and other organizations promoting various social causes have benefitted by using toll-free service. Occasionally, business does a great deal in helping facilitate charitable works of this nature.

Thrifty Corporation

The Thrifty Corporation was a drugstore and sporting goods chain with approximately 1,200 stores in the United States. It was growing at more than 10 percent per year. In addition to building its own stores, it had acquired Big S and Gart Brothers. Its information network was based on a host IBM 3090 model 200 and an IBM 3083. At each of its stores, Thrifty had placed IBM Series 1 minicomputers. Its telecommunications system was based on dial up telephone lines.

As Thrifty grew at its rapid rate, it found physically increasing its telecommunications network to be costly and burdensome. Thrifty needed to support its growing telecommunications-intensive applications. The use of dial up became costly and administratively burdensome as a result of the rapid rate of growth. Thrifty was having difficulty handling the growing volume of business data.

Thrifty installed a VSAT system using Scientific Atlanta equipment supported by many advanced applications including:

- On-line credit card verification
- Pharmacy applications such as transmitting information on drug dispensing and pricing, medical insurance claims, and interactive transactions on patient benefit eligibility

- Electronic point-of-sale information to automatically record sales and inventory data on the host computer and to update pricing information
- Energy management systems at its stores
- Video teleconferencing for merchandising, employee training, and customer service applications

Thrifty's VSAT system is based on a third-generation Scientific Atlanta X.25 network with the capacity to provide video and data to any remote location in the network. Thrifty has found use of the VSAT system to be more reliable than its former terrestrial telephone network was.

Summary

Telecommunications can be used as a competitive weapon to gain very rapid market penetration. This has been shown in several ways: by the spread of the distributed newspaper, through the proliferation of automatic teller machines, and in other ways. The use of telecommunications allows the firm to distribute its products and services electronically instead of physically, thus saving a great deal of time and expenditure.

Questions

9.1 Discuss the use of satellites for gaining rapid market access using as an example distributed newspaper printing.

9.2 How did telecommunications support the rise of the automatic teller machine and what was the competitive effect?

9.3 a. How can very small aperture terminals be used to aid a retailer in growing rapidly?
b. How are they different from other types of telecommunications technologies?

Chapter 10

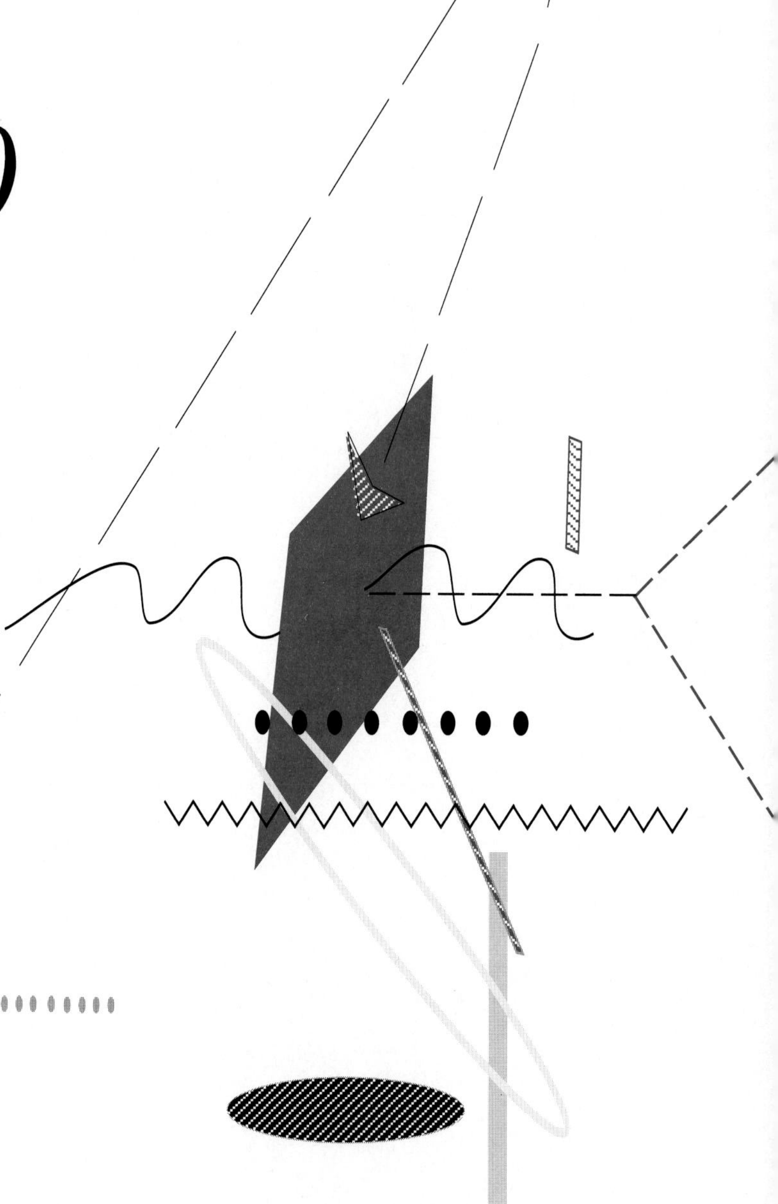

Slash Time from Your Activities

Telecommunications can also be used to slash the amount of time it takes an organization to respond to changing conditions. This is critical in providing customer service. Customers want what they want when they want it. They don't want to hear "later."

The effects of saving time can be very great. If, for example, inventory is involved, and if the organization is large, then even small savings in overall timing of movement of inventory can impact the bottom line of the corporation. Also, anyone who has ever waited in line too long, been put on hold until forced to hang up, or suffered any similar type of experience knows what it means to the customer for things to move rapidly.

The Time Factor

Telecommunications can be used within the firm to greatly increase efficiency in various functions. Efficiency can have a great many variations, but the concept of interval reduction is emerging as one of the most important for gaining competitive advantage. **Interval reduction** is the process of shortening the response time to customer demands. For example, how long does it take for the customer to take delivery on an automobile that has been custom ordered? How long does it take for a company to get back to the customer with a price quotation on a service it offers?

What makes these questions interesting is consumers' changing expectations, which push the successful firm to reduce the amount of time it takes to deliver satisfaction. In some cases, the response time to the customer is one of the most important criteria in determining satisfaction with a product being offered.

This concept of interval reduction was summed up by Stanley M. Davis in his book *Future Perfect* (Reading, MA: Addison-Wesley, 1989).

The larger rules operating here are:

- Consumers need products and services ANY TIME (i.e., in their time frame, not the providers').
- Producers who deliver their products and services in REAL-TIME, relative to their competitors, will have a decided advantage.
- Operating in real-time means no LAG-TIME between identification and fulfillment of the need.

The examples below show how various corporations have used telecommunications to accomplish these goals.

Toyota Provisioning

In the early 1980s, Toyota of Japan was approaching the limits of returns on investments in automation and manufacturing technology. **Just-in-time** inventory systems, which had been for the most part pioneered by Japanese corporations, combined with widespread use of subcontractors and manufacturing automation, had radically reduced the amount of time it took to produce an automobile. Human labor had been greatly reduced, and Toyota was clearly one of the most efficient automobile manufacturing corporations in the world.

The problem was that Toyota had essentially overinvested in manufacturing automation, and underinvested in back office provisioning systems. It had radically reduced the amount of time required to produce an automobile, but it had done little to reduce the amount of time needed to process the order, that is, little had been done to reduce the time between when an order was placed and when the customer actually received the automobile.

The result was that it was taking about 2 days to manufacture a car from beginning to end, and about 25 to 30 days to process the order. When all costs were factored in, Toyota came to the stunning conclusion that it was costing more to process the order on a car than to manufacture the car!

The engineers who had been working so hard over the years to reduce to a bare minimum the amount of time to manufacture the car became irritated at the "laziness" of the bureaucratic side of the company. Toyota started to reorganize its operations with a view to radically reducing the provisioning interval. The goal was to make it possible for a customer to walk into a car dealership in Osaka, order a car that did not exist (that is, that was not in inventory) and have that car delivered within 48 hours.

In order to do this, Toyota utilized international telecommunications networks to the fullest. Dealers and distribution centers were linked up in an ever-increasingly complex web of information systems. Orders came to be processed electronically through the telecommunications system. The headquarters in Toyota City became very quickly aware of precisely which orders should be filled, and were able to plan production accordingly, thus further reducing the waste and inefficiency of holding unsold inventory.

The Toyota telecommunications system links the entire electronic flow of information logically from the order point, either within Japan or overseas, ultimately to the production and manufacturing centers and their control and scheduling programs. It was a tremendous achievement for Toyota, and has given them a clear competitive advantage over other companies. That competitive advantage can be measured both by the rise in customer satisfaction at the faster response from Toyota and by the reduced costs of delivering a product from the factory to the customer.

Garment Industry Ordering with Portable Computers

The garment industry operates through different phases in its production cycle. Ideas for new products or designs are collected, then the goods are manufactured in a low-cost area of the world. Afterward, the complex mechanism of marketing and retail distribution begins. Most of the major clothing manufacturers have operations spread around the world. Ideas for designs for the next year's fashions come many times from Europe, typically Paris and Milan. The management decisions and headquarters control may be located in New York. Manufacturing, however, typically takes place in Hong Kong or another area of the Far East with cheap labor such as Seoul or Manila. The goods are then shipped back to the major markets of the world and distributed through retail outlets, many times at fabulous mark-ups!

One of the critical points in the process is the ordering season. This is when the designers have prototyped the clothes they are going to offer for the coming season. Buyers from all parts of the market flock to the major garment centers such as New York and place orders. After the orders are received, the company then knows how many items to have produced in the sweat shops of the Far East. It is during the ordering phase that the most critical decisions regarding pricing and quantities have to be made.

The problems of coordination and quotation are very critical during this ordering process. Much of the ordering takes place during conventions in which all of the different companies show their goods. The representatives of the companies take orders on the spot, most often in hotel suites located at the convention. The problem is that when this ordering takes place, the company representative must call the headquarters to get price quotations and delivery information. Delays or other problems with getting quick, reliable information may cause the potential buyer to lose interest. Competition is fierce. Also, should one offer be made, the buyer can compare quotes from other sellers and attempt to get the best deals. Several critical time-related factors come into play:

- What is the speed at which delivery at a specific price can be confirmed by the seller?
- How easy is it for the salesperson to negotiate changes in prices to meet the needs of the buyer?
- How can the seller become aware of the types of deals being offered by competitors, and how fast can he/she respond?
- How fast can the orders be transmitted to headquarters?

It is at this critical juncture that information technology and telecommunications can play an essential part in helping the seller gain competitive advantage. Companies have found they can resolve a few of these issues by linking their sales agents with portable personal computers and terminals, which in turn are linked together into an electronic mail network. The types of systems used depend on the sophistication of the mainframe or minicomputer at the corporate headquarters. If the system is sophisticated enough to host its own electronic mail system, then the salespeople dial up the communications controller or modem bank directly through the regular telephone circuits. (See Exhibit 10.1.) They then establish communication and send and receive information.

On the other hand, the firm may use a commercial electronic mail system with access points through a major packet switched network. The user dials a local telephone number from most locations, connects to the packet network, then establishes contact with the commercial electronic mail host computer. The difference is that users at both the remote sites (i.e., the sales conventions) and the headquarters dial up to the same commercial electronic mail computer instead of having their own in-house electronic mail system.

Electronic mail (many times abbreviated E-mail) enables the user to send and receive messages from many different locations and to and from many different people in an asynchronous mode. Asynchronicity in communication trades off the delay in receiving the message against the amount of time each individual saves in reliability and response time when mail is requested from the system. The sender has only to transmit a message once to be sure it is received. In addition, time is saved from

Exhibit 10.1

Dial up Remote Telecommunications System

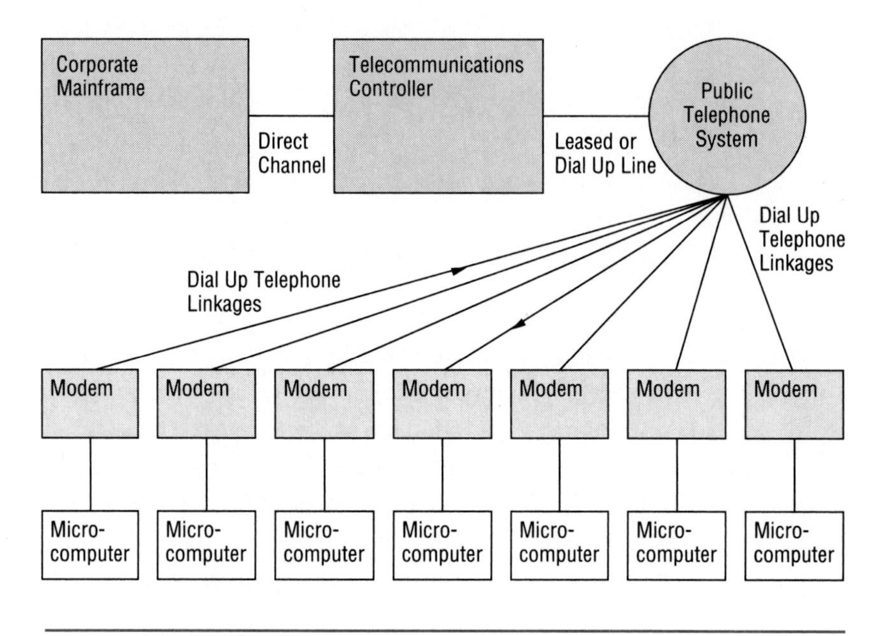

the point of view of the individual because all messages may be handled at once, instead of being a constant interruption during the working day.

Most of the universities in the world are linked together by a private network called Bitnet. An electronic mail message that was carried from the Decision Sciences Laboratory at the University of Arizona to the Information Systems Department at New York University's Stern School of Business is presented in Exhibit 10.2. The electronic mail message is held in an electronic "envelope" which identifies the sender, the recipient, and the route of the message through the system. The type of electronic mail system used on Bitnet was originally developed for the Department of Defense Advanced Research Projects Agency Network.

Getting back to the garment industry example, the benefits of using the electronic mail option carried through a telecommunications network can be very great. Orders are processed more quickly, the

Exhibit 10.2

Sample E-Mail Message

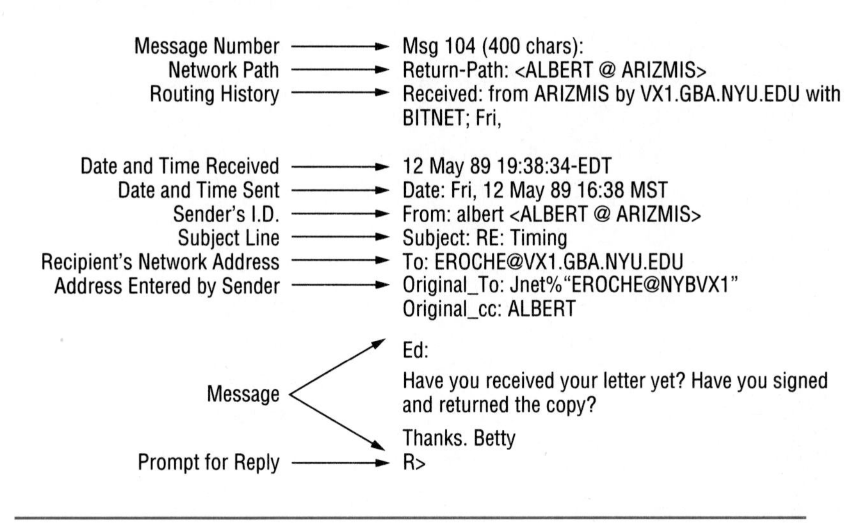

company finds itself able to respond to changing market conditions (i.e., competitive bids) faster than before, and the availability, price, and schedule of each product are immediately accessible to the salesperson. The salesperson is not tied up trying to conduct all of the business by telephone, freeing that person to cater to the customer more closely.

The implications of this can be vast for the entire sales and marketing industry. Linking salespersons through telecommunications to a source of information and deal-making resources enables them to transact business and make commitments and closures faster than their competitors (at least until the competition copies the technique and is able to offer the same type of service).

Insurance Claim Filing by Independent Agents

Many insurance companies rely on a vast group of independent agents to represent them in the field, penetrate local markets, and sell insurance. In addition to sales, local insurance agencies are responsible for filing claims on behalf of their local customers in the event of emergencies. As anyone who has waited for money to come from an insurance company knows, in the mid 1980s it appeared that insurance companies were quite leisurely about the speed with which they settled claims. Much of the communication between the insurance agents and their mother companies was transacted essentially on paper. Besides the paper trail, additional information might be transmitted or received from the local office through the telephone.

In addition to transmission of information based on the filing and settlement of claims, another important informational relationship existed. In selling insurance policies to customers, the agents had to make contact with the headquarters location. Selling insurance is a complex job. Many factors determine prices. For life insurance, the age, health, and previous medical history of the customers, as well as other factors, must be considered in pricing. Pricing also may be complicated by the variety of policies and services offered. Differences between localities and states also play important roles.

In order to properly price their policies, the agents frequently had to communicate with the headquarters of the insurance issuer. Some of the problems encountered in this arrangement centered around customer satisfaction and, ultimately, fairness:

- How to ensure that changes in conditions and prices made by headquarters are immediately reflected in the policies issued by agents
- How to eliminate the errors of the paper-based system, which called for writing out claims and policies, sending the paper to headquarters, getting it keypunched, then sending back a computer confirmation, which may or not have been entirely correct
- How to reduce the lag times in the process caused by the paper transmission of information, generally through the mail

One of the solutions adopted was to link the agents to the headquarters mainframe computer using dial up personal computers. Each personal computer was linked to a modem, which in turn could be used to dial up the headquarters' mainframe computer using regular telephone lines. This enabled the remote location of the agent to stay in constant contact with headquarters.

Exhibit 10.3

Economics of Dial up versus Leased Lines

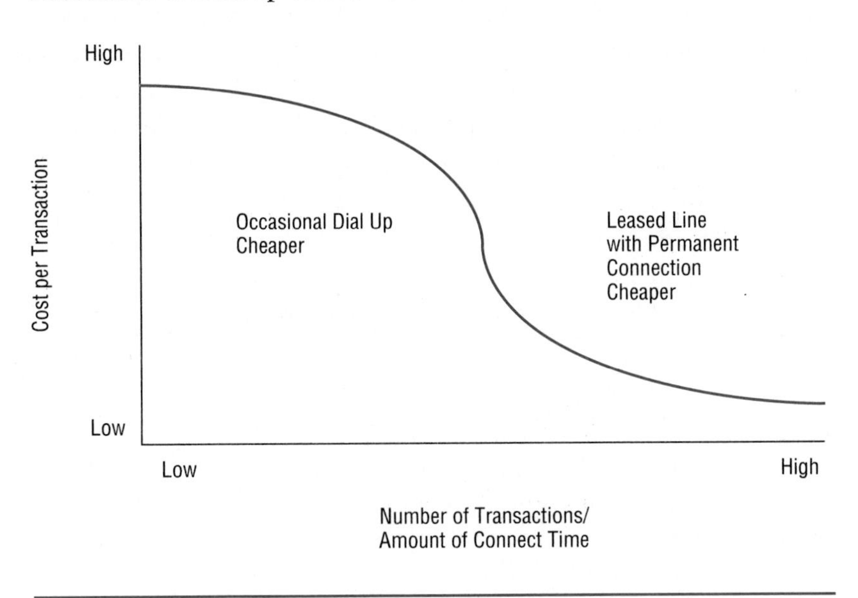

This link between the independent agents and remote locations enabled several critical operations to be accomplished:

- Downloading key pricing information from the mainframe computer
- Uploading claim information to the corporate mainframe without having to file paper forms, thus avoiding double entry of information and delays caused by use of paper and mail
- Uploading new policy information for a customer buying insurance
- Making inquiries to the corporate headquarters about a claim or adjustment that had been made

A question might arise as to why a fixed hard-wired terminal wasn't used. The answer lies in the economics of communications. (See Exhibit 10.3.) With a leased line connecting the remote terminal to the mainframe, there is a fixed cost. The line must be rented from the telephone company for a fixed monthly fee. In addition, the cost of the other equipment used, including modems, terminals, etc., increases the cost of the system. Given the fixed cost over time, it should be clear that the more transactions the system processes per month, the lower the cost per transaction. There is a point at which the cost per transaction becomes small enough to justify the expense of the leased line hookup between the remote terminal and the headquarters' mainframe computer. When the cost per transaction is very small, it adds little to the cost of processing each transaction, so the cost charged to the customer does not have to change much.

However, if the cost per transaction is very high because of low volume on the leased line, then it is enough to have a significant impact on the price the firm must charge the customer. Under those circumstances, the expense may either drive away customers, or become unrecoverable.

The alternative is to use the occasional dial up line. Using an occasional connection means that the computer-to-computer connection is established only when it is needed. For example, a remote location might establish a connection only once per week, or a few times per day. Under these circumstances, the cost of the dial up is only the price of a telephone call. As long as usage remains relatively low, the total cost of occasional contact is lower. It is only when the amount of dial up usage increases greatly that the leased line option begins to look attractive, as Exhibit 10.4 shows.

These economics are at the heart of the dial up options used by many insurance companies in the late 1980s. This type of economic tradeoff was typical before the massive introduction of the ISDN concept, which changed the economics of leased and dial up lines. ISDN eventually made much of this distinction obsolete, since data transfer capabilities came bundled with the regular telephone service. The insurance companies realized that they did not have the volume of transactions needed to support the cost of the hard-wired leased line option. They chose to use occasional dial up by the personal computer workstation using standard equipment and modems.

The implications of this approach go beyond the insurance example. Any large company or group of companies in a sector that uses independent dealers or agents has the option of using personal computers in an occasional dial up mode to maintain connections between the

Exhibit 10.4

Break-Even Point for Dial up and Leased Line Systems

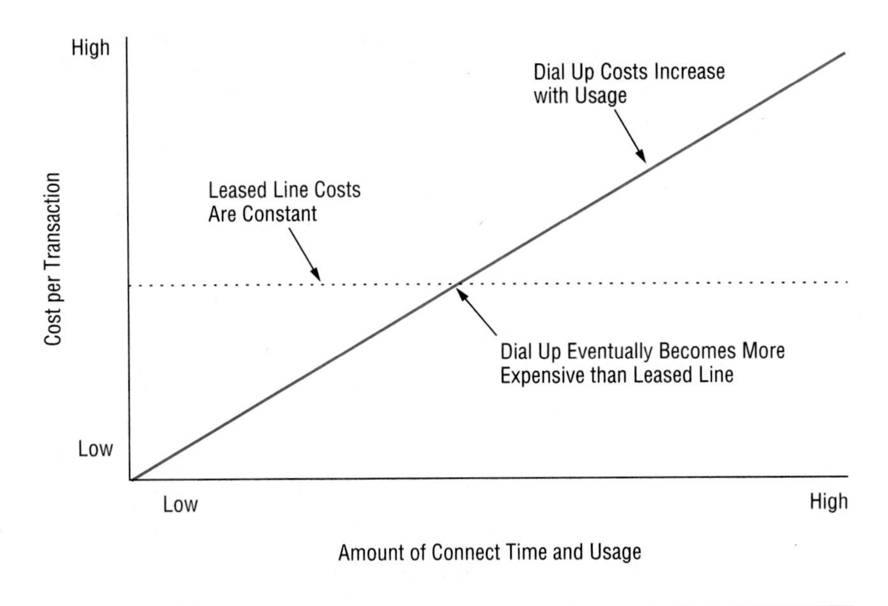

headquarters and the field. This type of operation can greatly reduce the interval for many internal transactions, which in the past have been heavily paper-based.

Woolworth's Automated Credit Card Authorization

The F. W. Woolworth Co. is a large, multinational retailer that owns and operates over 7,700 retail stores with $8 billion in sales in 1988. The company organizes itself into two major operating segments, general merchandise stores and specialty stores. A general merchandise store is a large (20,000 square feet) store selling a wide variety of low-margin inventory. By contrast, a specialty store is a much smaller (2,500 square feet) store that carries a very narrow line of high-margin inventory. Woolworth has many stores divided into these two segments, as shown in Exhibit 10.5.

Exhibit 10.5

Woolworth Operations

	Number of Stores	Sales (percentage of total)	Profits (percentage of total)
General Merchandise Segment			
Woolworth United States	1,100	25%	20%
Woolworth Canada and Germany	541	35	28
Specialty Segment			
Kinney Shoes, Footlocker	3,434	31	49
Various other specialty stores	2,664	9	3
Total	7,739	100%	100%

The F. W. Woolworth Co. began operating Woolworth stores in the United States 110 years ago. The U.S. Woolworth division has several striking characteristics, which are fundamental:

- The stereotypical Woolworth store is dated, both in physical form and inventory content. It is trying to survive in a very competitive market.
- Woolworth has a very strong corporate culture. In keeping with many retail establishments, there is a resistance to change and an in-built conservatism.
- The U.S. Woolworth division is often criticized for being a dinosaur—large, lumbering, and clumsy.

This is not the type of environment where one would expect to see remarkable innovations in telecommunications.

The senior management is acutely aware of its situation in the United States Woolworth division. Many believe that Woolworth's is not, and never will be, on the cutting edge of retailing. Such energy is directed towards the specialty operations. However, as long as the U.S. Woolworth division contributes positively to profits and cash flow, it will remain in operation.

It is not surprising that state-of-the-art information and telecommunications technologies do not thrive in this environment. By its very nature, information technology involves significant costs, far beyond those of hardware and software. Therefore, heavy investment in information technology is generally justified on the basis of future benefits. This is particularly difficult in the case of the U.S. Woolworth stores whose return on investment is only half that of the company's high-flying specialty store segment. During 1988, only 10 percent of the company's total capital expenditures were allocated to the U.S. Woolworth division, while this division contributed 20 percent to profits and 25 percent to sales. In spite of this, an opportunity was identified to use relatively low-cost information technology to improve the authorization and processing of customer credit card sales.

Authorization of credit card sales involves communication with the credit card companies (American Express, Visa, Mastercard, and Discover are accepted in U.S. Woolworth stores) to determine that the credit card has not been reported as lost or stolen. Under the manual system being used at Woolworth, checkout clerks consulted a book containing the account numbers of missing credit cards to determine whether a credit card was fraudulent. The manual authorization of an individual credit card sale took an average of 2 to 3 minutes, but could extend to 10 minutes.

Although Woolworth appeared to be saving money with this manual system, it also presented several problems. The information contained in the books was always out of date. By the time the books were prepared, printed, and delivered to the retailer, a fraudulent credit card could have been in circulation for 1 to 2 weeks. Also, the manual system lacked adequate internal controls. A manual look-up system relies heavily on the honesty of the salesperson, who can circumvent the look-up procedure without detection.

In addition to these problems, the manual system costs sales dollars because it takes too long. Checking a credit card number against those in the books is a time-consuming process. A customer dissatisfied with long checkout lines is prone to walk out of the store rather than wait in line. Such walkouts are particularly distressing to retailers who have gotten the customer so close to finalizing the sale. In addition, any sale for an amount over a set maximum would have to be manually phoned into the credit card company. This is especially time consuming and annoying for both the sales clerk and the customer.

These inefficiencies of the manual system discourage use of credit cards. The American consumer has become accustomed to shopping with credit cards, however, as demonstrated by the fact that credit card sales have been accounting for an increasing proportion of total sales. During the past five years, sales on credit have increased an average of 15 percent per year, while total sales have increased an average of 2 percent per year. The Woolworth management, therefore, realized that an inefficient authorization system was costing the company sales revenues by not encouraging the use of credit cards.

Once a transaction is completed, processing credit card sales involves sending the credit card media at the end of the day to the credit card companies for reimbursement. From the moment a credit card sale is generated, the retailer has an account receivable with the credit card company. From a cash management standpoint, it benefits the retailer to settle the receivable with the credit card company as soon as possible so as to have use of the funds.

Under the manual system, illustrated in Exhibit 10.6, the credit card media were manually accumulated at the end of the day, balanced with the registers, and physically mailed to Nabanco in Fort Lauderdale, Florida via the U.S. Postal Service. Nabanco, a credit card processing company, acted as an intermediary between retailers and the credit card companies. Upon receipt, Nabanco optically scanned the media and stored the information on magnetic tape.

This information was then electronically transmitted to the appropriate credit card company, which then released funds to the retailer. The manual processing system had several drawbacks for Woolworth:

- Manual processing increased the discount rate by 0.8 percent. It is inefficient and costly for Nabanco to manipulate and optically scan the physical media. This additional cost is passed on to the retailer through an increased discount rate. The discount rate is the percentage withheld by the credit card companies and Nabanco from each sale as compensation for their services.
- With manual processing, it took 9 to 10 days on average to receive the funds. Using the U.S. Postal Service to transmit the media to the credit card companies resulted in an increased float, which cost the retailer the opportunity of using the money. Float is the length of time from when the account receivable is created by the transaction and when it is ultimately settled.

Exhibit 10.6

Manual Credit Card Processing System

With manual processing of credit card sales, it can take 9 to 10 days to receive funds.

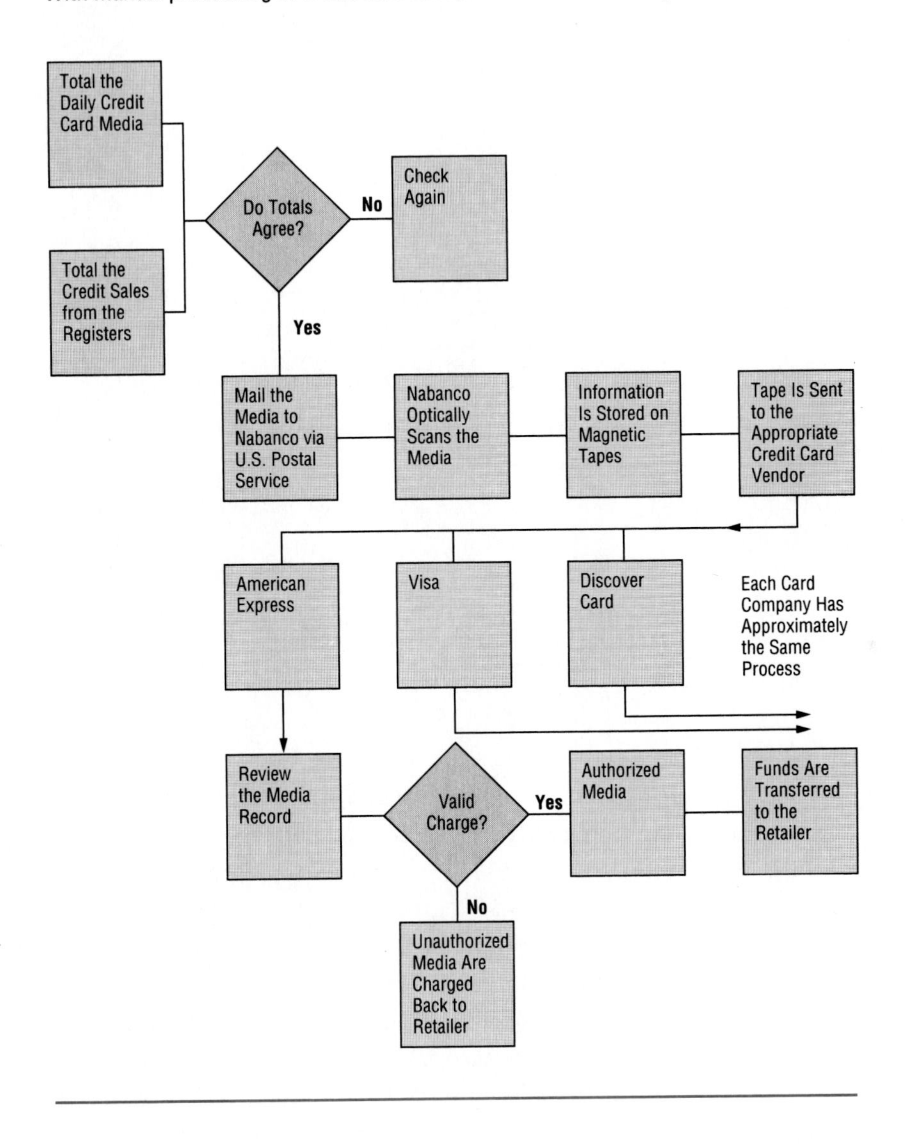

Exhibit 10.7

Processing Intermediary in Electronic System

Nabanco redirects the credit card authorization to the appropriate credit card company.

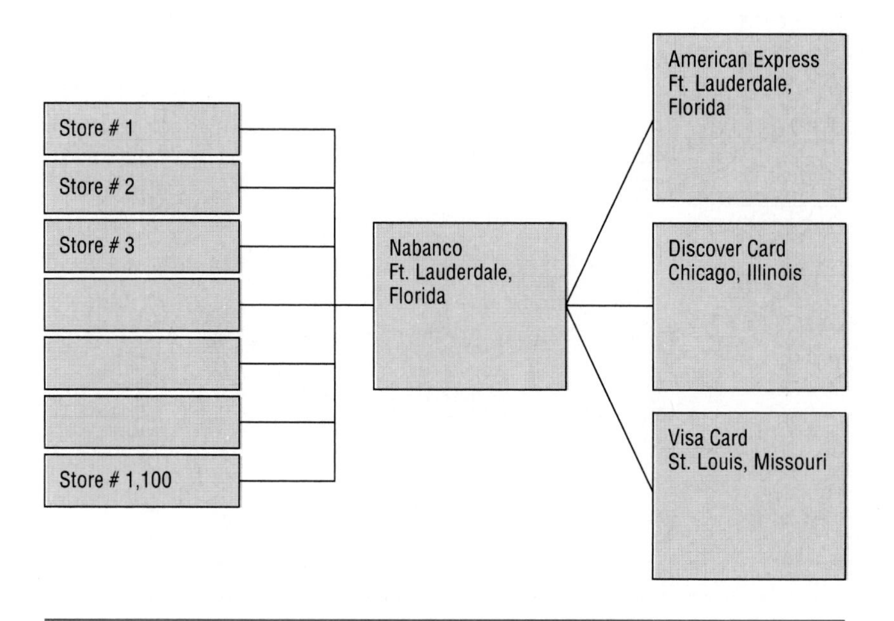

These problems were solved with the installation of two types of telecommunications technologies: Automated Bank Credit Authorization Terminals and Data Capture Terminals. When the system was up and running, approximately 5,500 authorization terminals and 1,100 data capture terminals telecommunicated between 1,100 Woolworth stores nationwide and Nabanco in Ft. Lauderdale. Nabanco redirected the transmission to the appropriate company. (See Exhibit 10.7.)

Bank Credit Authorization Terminals are small machines (approximately 4 inches by 6 inches) placed at the points of sale in the stores. The customer's bank credit card is inserted into the terminal and credit authorization is received automatically from the credit card companies. The entire procedure takes approximately 35 seconds and alleviates the need for manual look-ups and phone calls on credit card sales.

The system uses telephone lines already in existence at the stores. Since a large number of Woolworth stores are quite old, often as few as two phone wires were available for all the telecommunications needs of the store.

The system adopted by Woolworth utilizes Nabanco as a credit authorization host serving all of the accepted cards. The Nabanco host acts as a dispatcher, transmitting each credit card sale to its appropriate company for authorization. This allows the credit card authorizations to piggy back on one another until they get to Nabanco. This uses fewer telephone lines at the store.

Data Capture Terminals accumulate the data on each finalized credit card sale. These terminals are maintained in the back room office of the store. At the end of each day, all credit card information is electronically transmitted directly to the credit card company. This replaces the previous mailing of the media and reduces the float on that day's sales. Exhibit 10.8 illustrates the telecommunications links of the system.

The total cost of the new system was approximately $2.6 million with an annual operating cost of approximately $0.9 million. An average of five Authorization Terminals and one Data Capture Terminal are required for each of the 1,100 Woolworth stores. Each Authorization Terminal costs $155 and each Data Capture Terminal costs $788. Exhibit 10.9 summarizes the costs of the system.

This telecommunications technology system benefits all parties: the customer, the retailer, and the credit card companies. For U.S. Woolworth, annual net benefits conservatively approximate $0.6 million, as shown in Exhibit 10.10. A reduced discount rate results in a savings of $800,000 per year. This represents the reduction in the charges from the credit card companies. The credit card companies reduce the discount rate as an inducement to retailers to switch to the electronic look-up and processing system. The annual savings represent a 0.79 percent reduction in the discount rate on approximately $100 million of credit card sales.

Improved customer service results in over $300,000 in savings per year. This represents the value of reduced walkouts. Woolworth estimated that its average sale on credit totalled $24. If the new system reduced walkouts by one per store per month, a benefit of $316,000

Exhibit 10.8

Linkages in Electronic Credit Card Processing System

Sales are approved within 35 seconds. Funds are received within 2 days of the sale.

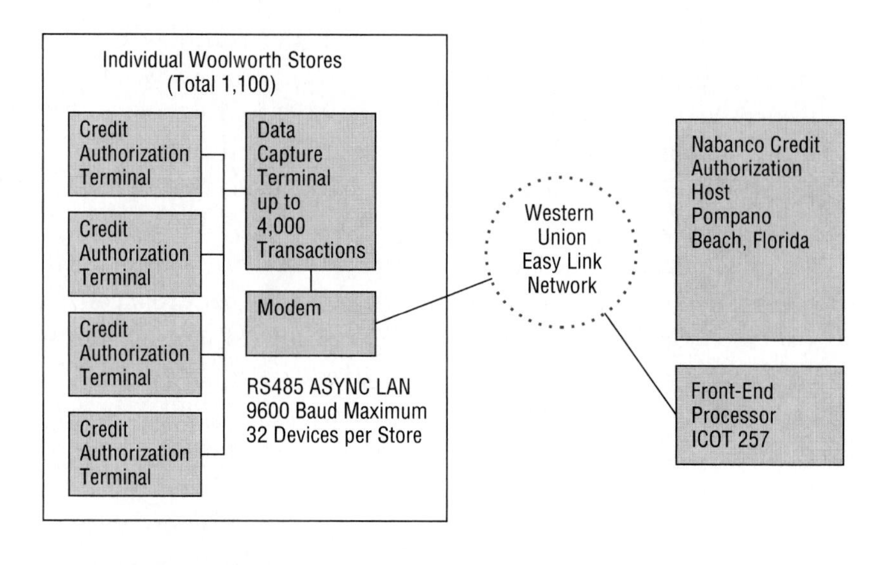

Exhibit 10.9

Woolworth's Credit Card Authorization System

Total cost: $2.6 million		
Authorization Terminals:	5 ↔ 1,100 ↔ $155	$ 852,500
Data Capture Terminals:	1 ↔ 1,100 ↔ $788	866,800
Wiring ($573/store):	1,100 ↔ $573	630,300
Installation and Training ($240/store):	1,100 ↔ $240	264,000
Total cost of equipment:		$2,613,600
Annual operating cost: $0.9 million		
5-year depreciation of equipment ($2,613,600 / 5):		522,720
Cost of approximately 120,000 local calls per month to local Western Union switches at 8 cents per call:		115,200
Local access lines (approximately 600 stores need separate business phone lines at an average of $35/month):		252,200
Total cost per year:		$ 890,120

Exhibit 10.10

Annual Net Benefits to Woolworth

	Annual Savings
Reduced discount rate	$ 792,620
Customer service improvement	316,800
Float improvement	154,384
Elimination of media postage	104,025
Chargeback savings	$79,500
Total annual savings	$1,447,329
Minus annual costs	890,120
Resulting net benefits	$ 557,409

would be realized. This is an extremely conservative estimation of the reduction in the number of walkouts per month.

Reduced float results in at least $150,000 savings per year. This represents the benefit of having the funds from credit card sales available 7 days earlier under the new electronic processing system. This benefit was calculated using an 8-percent borrowing rate on 7 days of credit card sales ($100 million annual credit sales divided by 360 times 7). As the cost of borrowing rose, this benefit would also increase.

Elimination of media postage results in a savings of $100,000 per year. This represents the benefit of not having to pay the U.S. Postal Service to physically transmit the media. Reduced chargebacks result in a savings of nearly $80,000 per year. This represents the benefit from a significant reduction in returned media and chargebacks for nonauthorized transactions.

While the benefits of this new system clearly outweighed the costs, there was some resistance from the U.S. Woolworth division management. They were not enthusiastic about the expenditure for this new system, nor about the thought of installing it in each of their 1,100 stores and training the employees to use it. As a compromise, it was agreed that the system would be tested in 14 stores. After the test proved successful and the employees easily learned how to use it, the new system was installed in all the U.S. Woolworth division stores before the 1988

Christmas shopping season began. The system was deemed a major success within 6 months of beginning operation.

Summary

Toyota showed that radically slashing the time interval for ordering an automobile can result in both higher profits and much better customer satisfaction. What could be better? Telecommunications can play a truly critical role.

In the internal ordering process, such as in the garment industry example, faster remote processing of information can result in significant competitive advantage. In insurance claims filing, telecommunications can also help greatly in reducing the time it takes to perform critical activities.

The Woolworth example shows clearly the critical relationship between interval reduction and customer behavior.

Questions

10.1 Discuss the concept of interval reduction and tell how telecommunications can bring it about.

10.2 How important do you think it really is to give customers quick service? Is it really worth the expense?

10.3 Discuss how linking up activities taking place in the field with activities at corporate headquarters can help improve efficiency and business operations.

Chapter 11

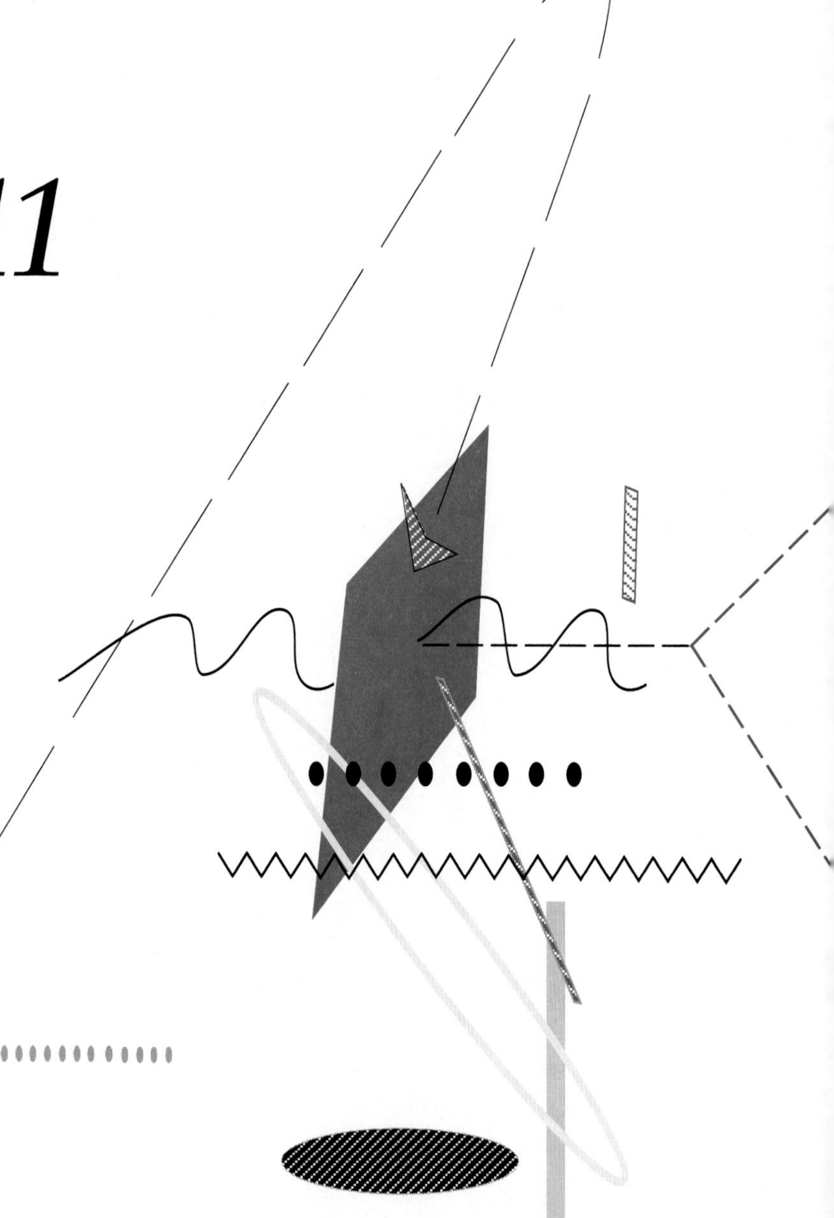

Give Superior Customer Service

Whether caused by the yuppie mentality on the consumer side or the increased emphasis on inventory management and just-in-time techniques for both manufacturing and sales stocking, in the 1990s, customer service is one of the critical factors in achieving success. Customers have very little loyalty compared to the past. At the same time, they have many more options available than in the past. As a result, it is important to give customers what they need when they need it.

The role of telecommunications in this job is to help coordinate geographically widespread activities, to distribute intelligence and sophistication to customer sites, and to increase the information velocity between the firm and its service or customer base.

The idea of "customer" should be extended to include any group receiving a firm's services or products. This chapter discusses several examples showing how telecommunications can be critical in getting the customer what is needed at the right time.

Chemical Bank's ServiceLine™ and ServiceXtra™

Chemical Banking Corporation is the name of a banking group that has existed in various forms since 1824. It is a global financial services institution, offering a full range of financing, risk management, and advisory services to corporate and institutional clients worldwide. It is the acknowledged leader in serving midsized and small businesses in the New York metropolitan area and in the southwestern United States. With assets of more than $6.7 billion in 1988, it was one of the largest bank holding companies in the United States, and one of the top 50 or so banks worldwide.

After an internal restructuring in 1987, the Consumer Banking Group (CBG) was split off from the retail banking and middle-market services division to form its own group. Its mandate was to focus on individual and small business customers and eventually to become the premier consumer banking institution in selected regions of the United States. The CBG desired to broaden and deepen customer relationships, increase loan origination, improve productivity, and differentiate the bank through superior customer service.

Unfortunately, Chemical Banking Corporation (ChemBank) faced a very competitive market in Manhattan and throughout its territory. Surveys commissioned to analyze how ChemBank's Consumer Banking Group compared with its competition in various areas indicated that it was edging away from its closest competitors (Chase Manhattan Bank and Manufacturers Hanover) but it still had a very long way to go to catch up to Citibank, the retail banking industry leader in the nine-county region around New York City. Although ChemBank was clearly on its way to firmly establishing its number two spot, its focus became catching up with Citibank.

Along with its commanding market share, Citibank had initiated a number of technological innovations that set new standards in retail banking. Automated teller machines (ATMs) had revolutionized individual banking by duplicating simple teller functions with a secure machine accessible 24 hours a day. Home banking was developed with the hope of capturing the young professional market. A centralized telephone facility channeled phone traffic of the branches. CitiTouch™

service enabled customers to access information about their accounts through touchtone telephones.

ChemBank's surveys showed that its own customer satisfaction was low. The surveys administered both by internal service quality administrators and by outside marketing research firms revealed that ChemBank customers had a considerable number of complaints. Some of the most common were long waits in lines in the branches, long lags between applying for loans and getting responses (turnaround time), and overcrowded branches.

As a result of this information, a focus on customer satisfaction became a management priority. In the beginning of 1988, customer satisfaction became the official top priority. It was realized that in an age of increasing sophistication in the retail banking industry when products are constantly changing, customers will choose the bank that offers the best service. Furthermore, customers will stay with the bank that has provided them with a good banking experience. By focusing on customer satisfaction and service, ChemBank was hoping to create a niche in the market.

The solution adopted by ChemBank was to set up ServiceLine to provide external and internal customer service. In 1987, ChemBank purchased a Rockwell Industries Galaxy automatic call distributor (ACD) and established ServiceLine in its operations facility in Jericho, Long Island. The ServiceLine system allows customer service representatives to give information to customers and bank employees through the ACD, which is a standalone switching system to automatically distribute incoming calls to ServiceLine customer service representatives without going through any operator. If all customer service representatives are busy, a call is held and music is played in the background until a customer service representative is available to handle the call. By constantly polling the available lines, the ACD knows exactly when a customer service representative completes a call. It immediately passes along any call it has been holding.

The ChemBank ACD receives approximately 15,000 calls per business day from trunk lines with approximately 130 circuits and from a microwave system at 55 Water Street in the Wall Street area. The ACD then distributes these calls to over 300 customer service representatives, who are the backbone of the ServiceLine system. Since

ServiceLine was created, calls going into the branches are automatically switched to ServiceLine. Many times customers think they are calling their local branches when in fact their calls are being answered at this centralized facility. Branch staff give customers only the four ServiceLine telephone numbers, and are no longer allowed to give out their direct numbers. Regardless of where the customer representative is located, the resident in the New York City area can call for the price of a local telephone call. Customers outside the tristate area can call the toll free number.

To its ServiceLine system, ChemBank added ServiceXtra based on an automated voice response unit that gives information to customers without human intervention. In August 1988, ChemBank rolled out ServiceXtra to its customers. This service uses a Rockwell Industries automatic voice response unit (VRU) to give customers account balance and rate information. The customers call a local number then enter their account numbers and social security numbers using their telephone keypads. Almost 24 hours per day, any information available on a statement can be supplied by ServiceXtra. ChemBank installed a tie line between the VRU and the ACD so that a customer who chooses to speak with a customer service representative after receiving information from the VRU will automatically be connected. Exhibit 11.1 diagrams the entire system. A current pilot program forces selected ServiceLine calls into ServiceXtra. Since many customer requests are for demand deposit account and transaction information, the goal is for ServiceXtra to filter these calls and reduce the volume into ServiceLine. By April 1989, approximately 40 percent of the callers into ServiceLine who were switched over to ServiceXtra ended up using ServiceXtra only. Of those who called ServiceXtra directly, practically all used ServiceXtra only. The call volume into the ServiceXtra system increased steadily after the program was rolled out. By the end of 1989, ChemBank was handling between 10,000 and 20,000 calls per month through this system.

The ChemBank customer service representatives were also given direct access to the corporate IBM mainframe in Manhattan and the account database through personal computer terminals. This enabled them to retrieve the information necessary to handle the wide variety of calls they received. To ensure security, customers were required to give

Exhibit 11.1

Chemical ServiceLine and ServiceXtra

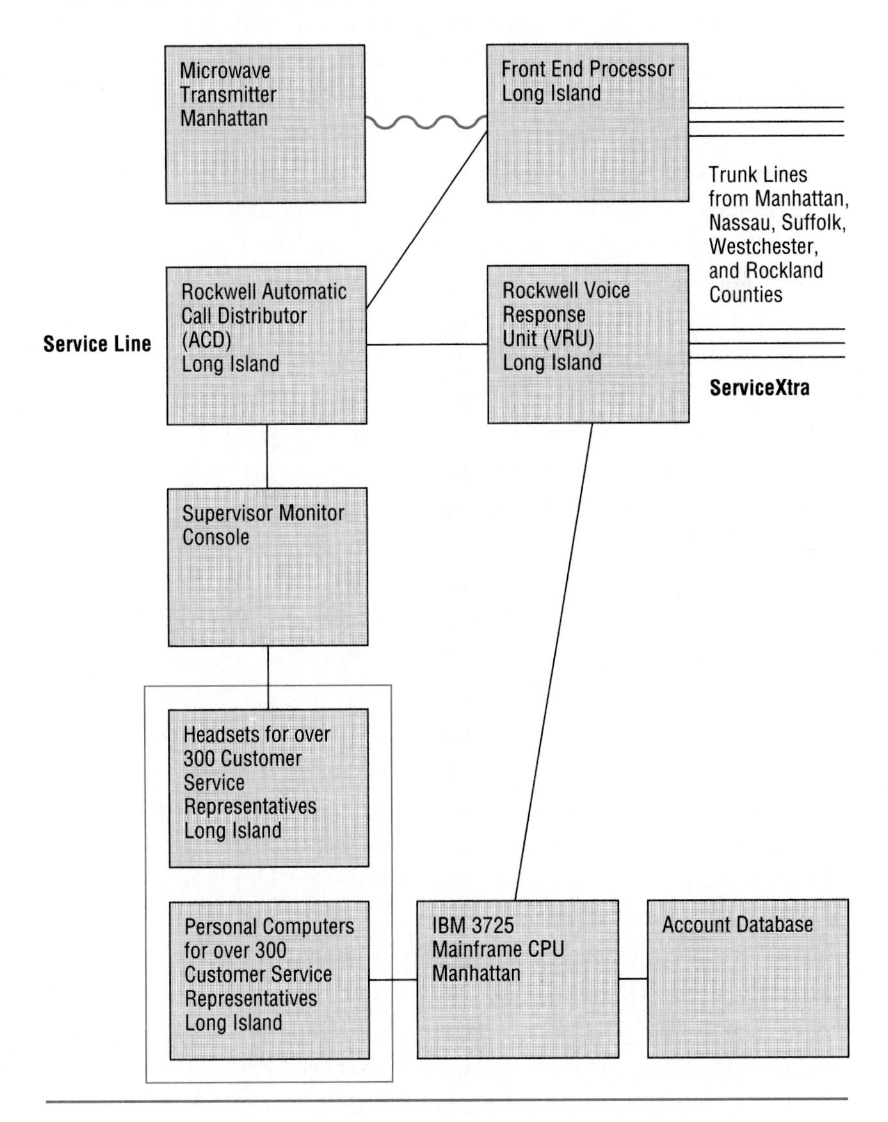

their names, addresses, and social security numbers as means of identification before information could be given out. Once customers had properly identified themselves, the customer service representative could carry out banking services that could previously only be performed in the branches. Services included account history reviews, address changes, interest/float calculation explanations, rate quotations, BankCard™ (credit card) information, check verifications, credit inquiries, check stop payments, debit memorandum information, deposit information, error corrections, product descriptions, photocopy requests, service fee information, etc.

Although customer service representatives could process most customer requests, they sometimes had to transfer customers' calls to other areas of ChemBank that could provide more detailed information or bilingual services. So that they did not have to memorize or look up numbers, their telephone sets were set up to transfer calls by speed dial to the branches (using the branch numbers) and product help desks.

Chemical Bank also set up a telemarketing area to handle sales of new services to potential customers. Potential customers who called the advertised toll free number or who were forwarded from the ServiceLine requesting sales information were handled by salespeople in the Telemarketing Department. Telemarketing sales representatives helped identify the best product for customers and then gave them the option of either opening accounts over the phone, having applications sent to their homes, or being assigned a contact person if they preferred to go to a branch. A basic part of the telemarketing sales representative's job was to arrange a follow-up call on an agreed-upon date with the customer. At that time, the sales representative would ask whether the customer had received the requested information and answer any question the customer may have had. Sales representatives also called customers to cross-sell additional products or services. Telemarketing maintained two Digital Equipment Corporation VAX 8550s separate from the main Chemical Bank database, because many of telemarketing's contacts were not yet actual customers of Chemical Bank.

To support the new telecommunications-based strategy, ChemBank initiated intensified training for the customer service representatives. Because the new ServiceLine system pooled all questions from customers, the customer service representative was required to handle a remarkable variety of subjects. The customer service representatives had to be trained not only to understand all of the products and services provided through the bank, but they also had to master their personal computer terminals and know how to access and interpret their numerous information screens. Each customer service representative received $3^1/_2$ weeks of classroom training and another 3 weeks of on-the-job training. Team leaders and supervisors were responsible for being readily available if a customer service representative needed assistance in answering a customer's question.

On a routine basis, operations personnel listened to a randomly selected, statistically significant number of calls to measure the customer service representatives' ability to handle calls effectively. Each call was rated on the employee's opening (i.e., proper greeting), closing, and overall manner of handling the call. Other factors, such as uncontrollable system downtime, were accounted for in a calculated final score. Weekly meetings reviewed the problems and incorrect procedures that were observed to teach all employees how the calls should be handled. This constant review of operations enabled the bank to keep a high score for customer service quality.

Chemical Bank launched an intensive testing campaign to compare its call handling service and speed against those of Citibank, and eventually found it was doing better, as measured by the amount of customer calls that received busy signals. In addition, by the fall of 1988, external surveys showed that virtually all of the survey respondents would recommend Chemical Bank based on their experience with ServiceLine. Furthermore, ServiceLine ratings exceeded those of competitors' similar services.

Yamaichi Securities' Sun Line Home Transaction System

Yamaichi Securities Company is one of the "Big Four" securities dealers in Japan and one of the largest in the world. Yamaichi, founded in 1897, is ranked as the world's tenth biggest securities company in terms of assets controlled. Through its 130 domestic Japanese branches and 26 overseas offices located in 21 countries, the company provides many kinds of investment and financial services.

Yamaichi faces considerable competitive pressure, and in order to maintain and increase its market share, the company must devise tactics to lure investors and hold them. Securities companies in Japan, particularly the Big Four, Nomura, Daiwa, Nikko, and Yamaichi, compete strongly with each other. The major income of securities companies comes from commission fees. In order to increase its commission income and maintain or raise its ranking, each company has developed many tactics to offer better services to its customers and try to receive more orders from them.

Because of the boom in the Tokyo stock market during the late 1980s, many new investors entered the market, especially housewives, who traditionally control the household finances in Japan. Since 1985, due to the appreciation of the Japanese yen and the stable economic growth in Japan, many foreign investors followed Japanese investors into the Tokyo stock market. Stock prices increased very rapidly and trading volume also greatly increased. Because of this boom in the Tokyo stock market, many occasional investors entered the market since the potential rate of return was much greater than that available through the traditional postal savings system.

The problem was that housewives, because of their other commitments in running the home, particularly raising children and taking care of family members, found it difficult to visit the securities houses regularly to conduct investment business. Although Japanese society has been westernized in recent years, it still maintains many conservative traditions. Women, particularly housewives, must give their priority to their family. Therefore, even though many were interested in the stock market, it was almost impossible for them to go to the brokerage house.

Unfortunately, in order to get real-time market information or to place orders, investors must go in person to the securities company. Although investors can receive real-time market price information over the radio, without going to the securities company they cannot get other market information, market analysis, check on their own portfolios' market values, or benefit from many other services associated with buying and selling securities. These inconveniences greatly obstruct housewives' entry into the market.

As the Tokyo stock market boom progressed, the number of customers increased and the waiting lines at branches of the securities houses grew. This increased the customers' frustration and impatience.

As a result, service quality dropped and the relationship between the customers and the companies became worse.

In order to handle the increasing number of customers, Yamaichi put more counter attendants in each branch. According to data from the Securities division of the Japanese Ministry of Finance, labor costs accounted for 30 percent of the working expenses in the Japanese securities industry. More counter attendants boosted labor costs further. Because of the capacity of current branches, it was almost impossible to increase the number of counter attendants rapidly because there was not enough physical space. To solve this problem, some branches moved to larger offices, but this did not reduce all of the inconveniences faced by the customer.

An additional barrier to hiring extra employees involved Japanese tradition. Once a company hires employees, it is very difficult to dismiss them, even if the company is facing problems in the market. Therefore, most securities companies were unwilling to greatly increase the number of their employees.

The competitive pressures increased, and companies were forced to continue to compete for customers. Companies that found it difficult to settle their problems might find their customers flowing to other securities companies with shorter waiting lines and better services. According to the Ministry of Finance, 221 securities companies maintain about 2,500 branch offices throughout Japan. Investors can easily choose the one that provides the best service.

In order to settle these problems, Yamaichi invented a convenient and low-priced system called "Sun Line" in 1988. Because of its convenience, low price, and ease of use, it became very popular, especially to novices and the occasional investor. The Sun Line system used the Nintendo home game system to extend the telecommunications linkages directly from Yamaichi to the home of the customer.

There are some comparisons between the Yamaichi Sun Line system and home-banking systems such as that of Chase Manhattan. The objective of these two systems was to increase the quality of the relationship between the customer and the company, reduce operational costs, and increase the company's profits. The Chase system required customers to use their personal computers to connect into the banking computer. Since a personal computer is very expensive and

Exhibit 11.2

Yamaichi Securities' Game Computer-Based System

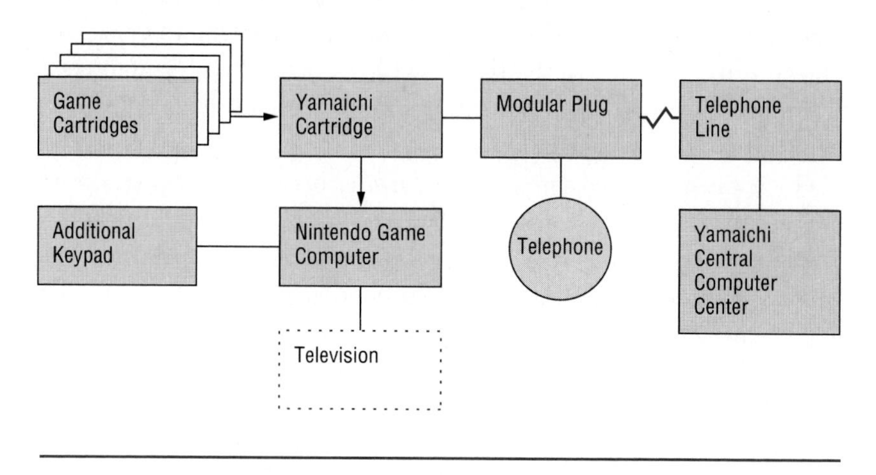

hard to use, and the number of people having personal computers is relatively small, the Chase experiment ended in failure.

Yamaichi's Sun Line system was designed to be used by both a personal computer, which some Japanese had, as well as by the family game computer produced by Nintendo. The Nintendo family game computer was already in almost 40 percent of all Japanese households, and the price of about $75 was very inexpensive compared to a personal computer. These facts helped the Sun Line system become very popular once it was introduced into the market.

As Exhibit 11.2 illustrates, the Sun Line system transferred data through the telephone line supplied by Nippon Telephone and Telegraph (the Japanese version of AT&T). The total fee for using the system was relatively low. The type of phone line used, called *DDX-TP,* was a digital line. Using this line kept the communications fee lower for long distance customers than ordinary long distance telephone calls. The Yamaichi computer was located in Chiba Prefecture, a city adjacent to Tokyo.

The cost to the consumer was set up to be relatively modest. The initial charge, a contract fee of 12,360 yen (approximately $88), was waived if the customer had a Yamaichi credit card, had more than 3 million yen in a Yamaichi savings account, or had traded at Yamaichi at least once every 6 months. In addition, there was an apparatus fee of 19,500 yen ($139) for the communications cartridge, the keyboard, and a special integrated circuit card inside the communication cartridge.

Once the customer was linked into the Yamaichi computer system through the Nintendo game unit, it was possible to display a great deal of information on the home television screen, including real-time stock prices and currency rates, market information, information on Yamaichi's new products, and many other services. Sample screens included prices for convertible bonds, individual stock information, market news, a list of the customer's portfolio, rankings of stocks, listings of Yamaichi-recommended stocks, and order information.

In addition to receiving information, customers could use the Sun Line system to place orders without going to the securities company in person. This "home dealing system" included a five-step ordering process:

- The customer first read the available information on the screen of the home television, then called up the order screen and placed the order electronically using the keypad attached to the Nintendo game unit.
- The information was transmitted at 1,200 bits per second to the Yamaichi computer center.
- The information was entered into the Yamaichi market division's terminal linked to the central computer of the Tokyo stock market.
- The Tokyo stock market computer received the order.
- The order was executed on the Tokyo stock market.

Since the customers were able to place orders for buying and selling stocks through the Sun Line system, Yamaichi felt reduced pressure to hire more counter attendants. In addition, the waiting lines for customers at the branches were reduced and the relationship between Yamaichi and its customers improved. Ultimately, the counter workers were in part shifted to other tasks, reducing labor costs for Yamaichi.

Because of the Sun Line system, customers became more willing to open accounts with Yamaichi and less willing to switch to other companies' services due to high switching costs. The low contract and application fees encouraged customers to use the Sun Line system. However, once they opened their accounts with Yamaichi, they became less willing to switch to a competitor and tended to become permanent customers of Yamaichi.

Since the orders were transacted by computer, Yamaichi could handle more orders than before. Consequently, the company's market share increased and profits rose. The effects of the introduction of the Sun Line system were felt throughout the company. Yamaichi's employees totalled about 12,000, including its computer center, research institutions, investment company, real estate investment company and many other subsidiaries. Salesmen and counter attendants, the major part of employees, included about 60 percent of the total employees, a total of approximately 7,200 persons.

Assuming that each of these employees covered 50 customers who opened accounts with Yamaichi, the total number of customer accounts would be approximately 360,000, not including inactive accounts. Nintendo's family game computer is in about 37 percent of all Japanese households, available to about 133,200 customers. As of 1989, about 52,000 customers used the Sun Line system. Therefore, the percentage of people who had Nintendo family game computers or personal computers that had contracts with Yamaichi to use the Sun Line system was approximately 52,000/133,200 or 39 percent. This number clearly shows the popularity of the Sun Line system; because of the popularity of the Nintendo family game computer, the percentage of customers using the Sun Line system is likely to increase in the near future. There is a good potential that each of these customers will become a permanent customer of Yamaichi, expanding its market share and increasing its profits. In the end, Yamaichi's market share increased approximately 8 percent as a result of the Sun Line system.

Ronzoni Foods Corporation

The General Foods Corporation is a conglomerate of food companies with sales of over $10 billion. (See Exhibit 11.3.) It is the largest U.S.-based company operating solely in the food business. Currently,

Exhibit 11.3

General Foods Corporate Organization

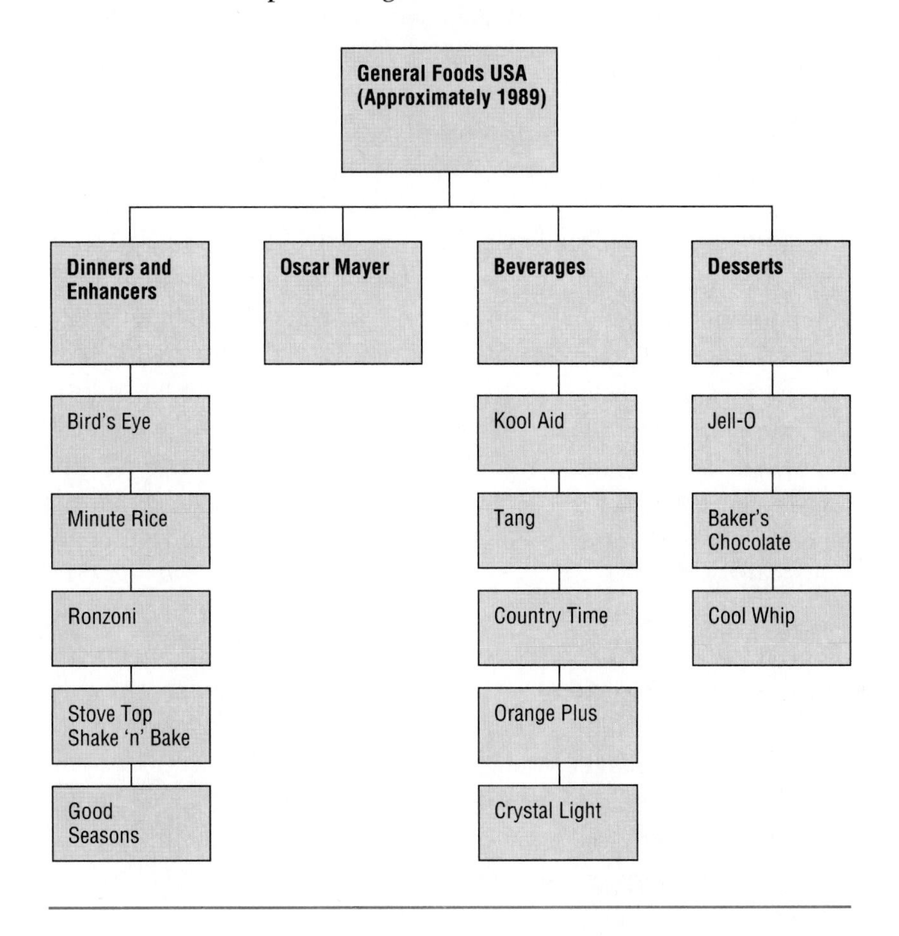

General Foods employs over 50,000 people worldwide, operates more than 100 manufacturing plants and distribution centers, and has operating subsidiaries and joint ventures with other firms in more than 20 countries outside the United States. General Foods exports to more than 100 nations worldwide. It has continually strived to be a market leader in the food and beverage industry. It has had a long-standing, high-quality reputation and has long been a leader in new product development.

Ronzoni Corporation has been a long-time market leader in the northeastern United States, with a 35-percent share of the Italian foods market in the metro New York area and an 8-percent national share. It was founded in 1915 by the Ronzoni family in Queens, New York and built its reputation by providing authentic Italian foods. Well-known for its high-quality products, Ronzoni generates over $90 million in annual sales revenues. It operates two manufacturing plants and employs nearly 400 people.

In 1984, General Foods Corporation acquired Ronzoni for approximately $70 million in order to capitalize on the high-quality reputation of the brand and diversify into the national Italian foods market. General Foods' primary objective was to generate profits in a business that was losing over $13 million annually. By slashing costs and improving sales volumes, GF expected to turn a profit in 5 years. In previous years, manufacturing costs had been reduced nearly 30 percent, but volumes had been dropping as well. New product introductions were expected to boost volumes 30 percent and generate significant profits.

General Foods and Ronzoni faced some formidable problems. By 1986, customer service levels (defined as the percentage of orders filled) were at 95.4 percent and falling due to frequent out-of-stock situations. These stemmed from a number of factors. Competition in the Italian foods market began to intensify and customers began to demand just-in-time shipments in smaller quantities. In addition, Ronzoni had to balance inventories on over 80 different products with a manual inventory control system. These two factors resulted in frequent changes in the production schedule (as often as once per day) and a lag of 2 to 4 hours before these changes were communicated to the manufacturing organization. Manufacturing often did not get up-to-date information about urgent needs as communication was done by word of mouth.

Difficulties in interdepartmental communication resulted in the shipping of 213 incomplete orders in 1986. Each department had several personal computers, but no formal method of relaying information to other departments. Therefore, because the Quality Control department did not have access to information regarding which products were needed for shipment, it customarily ran a day late in releasing finished product for shipping. Also the manual handling system in place was unable to handle rush requests from GF Distribution Centers and retail warehouses. Manufacturing received day-old production schedules as no one had access to timely information about finished goods needs.

Another problem was that finished goods inventory levels were at 1.1 million units, equivalent to 17 percent of annual sales, and increasing. Because of the manual inventory system, Logistics often received information on finished goods inventory with a 1- to 2-day lag. This resulted in frequent over-production to minimize the risk of finished goods outages. Eventually, three outside warehouses were leased to store the product. Inventory turned over only five times per year and total inventory carrying costs for 1986 were $589,000.

In addition, with the manual system, transaction processing and reporting was highly labor intensive, requiring the equivalent of eight full-time employees, over 14,000 labor-hours, at a cost of $215,000 per year. The Packaging department spent 600 labor-hours annually at a cost of approximately $6,000 to manually reconcile discrepancies between production counts and finished goods receipts. The Quality Control department logged an estimated 1,000 labor-hours per year at a cost of approximately $12,500 to maintain a manual filing system for all embargoed product and report to the organization on its status. Warehouse personnel conducted daily physical inventory counts, manually fed this information into a personal computer, and updated files daily. The error rate was estimated to be about 2.5 percent, or about 22,500 units of inventory despite 1,560 labor-hours (at a cost of $24,400 per year) spent to track inventory.

The Shipping department relied on manual ticketing and log books to track outgoing goods and log completed orders at a cost of $158,500 per year. All paper work was prepared manually, and information was fed into a personal computer at the end of the day to record shipments. A report was prepared and distributed 12 to 24 hours later. The error rate for manually tracking shipments was around 0.3 percent and was reconciled at significant cost to the firm. In addition, five people were needed to process shipping transactions, load data into personal computers, and prepare reports. Finally, the Accounting function handled all transfers, returns, adjustments, and

DC claims by manually feeding this information into the department personal computer and generating a specialized report at a cost of $13,800 per year.

Reports on finished goods shipments, inventories, and production counts were sent to corporate headquarters 1 to 2 days after actual production or shipment, and to higher management levels with a 1-week time lag. The corporate office began requesting this information to enable division managers to keep in touch with the plant and to make key business decisions involving Ronzoni's customers and suppliers. However, Accounting experienced internal delays, receiving final production counts, and shipment and embargoed inventory counts, a day late. Still, these reports were inaccurate. The final report was sent via messenger to the corporate office with a 48-hour lag, providing information of limited utility to upper management. This inability to communicate key information in a timely manner hampered top management's ability to control operations.

All inventory systems within Ronzoni ran on an IBM AS/400 minicomputer located in Avon, New York. It ran at 85 percent of CPU capacity and 75 percent of disk storage capacity. The system also linked six other General Foods plants, from Boston, Massachusetts to Dover, Delaware. It was experiencing significant response delays. Surveys indicated that the user response time had averaged about 10 seconds. The industry average is about 6 seconds. During peak working hours, response time often went as high as 40 seconds.

The solution was to implement a fully automated finished goods inventory system in February 1987 for a cost of $300,000 as illustrated in Exhibit 11.4. It was planned to be fully operational two years from installation. The system relied on a local area network and leased lines tying together the different Ronzoni sites. The system consisted of a bar code transaction processing system for finished goods to provide on-line, real-time information to users throughout the organization via the local area network, which linked all of the departments to an on-site minicomputer.

Initially, another IBM AS/400 minicomputer was installed at the site at a total cost of $180,000. The on-site AS/400 gave Ronzoni the ability to expand applications development internally, including the finished goods inventory system. In addition, the site gained the ability

Exhibit 11.4

General Foods Comprehensive Information System

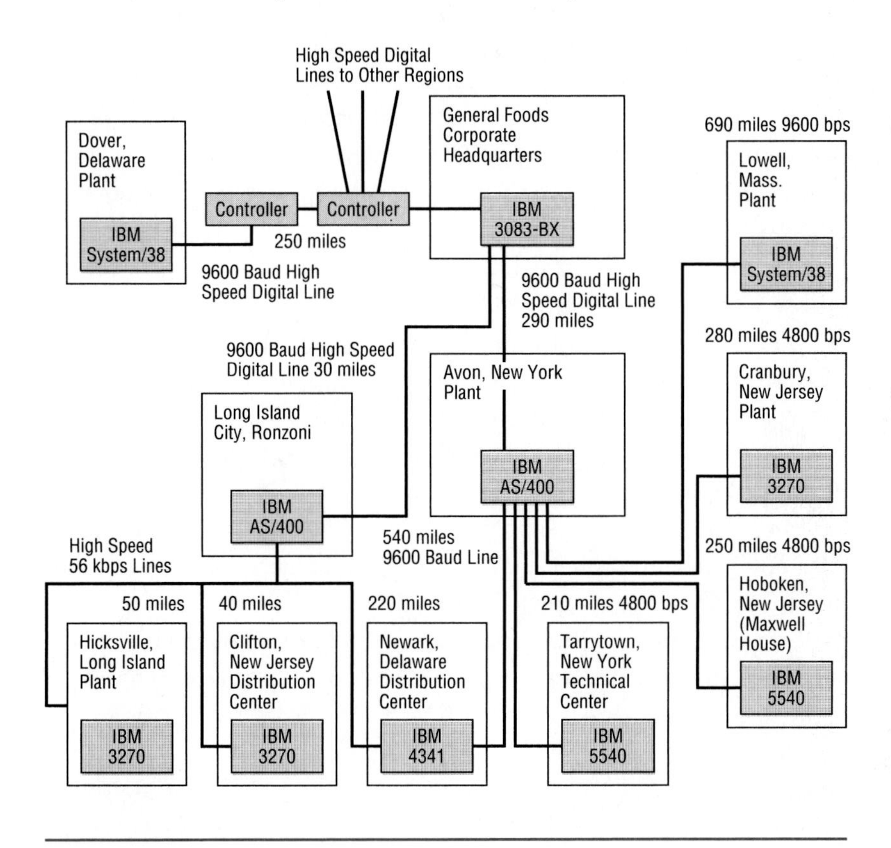

to network all personal computers to integrate all office and manufacturing applications. The system eliminated the plant's dependence on the over-utilized computer at Avon, thereby allowing real-time inventory information to become available.

Ronzoni leased a private 9,600 baud coaxial cable from NYNEX to link the on-site AS/400 to the corporate office's IBM 3083 mainframe in White Plains at a cost of $7,000 per year. The cost of leased lines was reduced due to shorter distances; Queens was 30 miles from White Plains, but 320 miles from Avon.

A second part of the telecommunications solution was to install a Xerox Ethernet local area network with a data transmission capacity of 10 mips to link the personal computers, terminals, and workstations in the various departments. (See Exhibit 11.5.) The high-speed network could handle on-line transaction processing and data acquisition and linked the personal computers, workstations, and printers of the Accounting, Manufacturing, Shipping-Warehousing, Logistics, and Maintenance functions. The local area network is flexible enough to add users and has the capacity to handle electronic mail and video transmissions. Finally, the local area network provides users with access to a shared laser printer.

The third part of the solution was to install a $77,000 bar code transaction processing system to handle production reporting and inventory control applications. (See Exhibit 11.6.) The bar code system operated a Crossbar network from within Intermec, Inc., which allowed the creation of interactive bar code reader applications. A series of bar code laser scanners collected finished goods transaction data and transmitted them to transaction managers, which processed and transmitted the completed transactions to the host AS/400. Thus, the host computer was not directly involved in data collection, only in the processing of data. Data were buffered in the transaction manager if the system went down.

The completed Finished Goods Inventory System streamlined the tracking of finished goods, reducing the time to position goods on pallets, wrap them, and inventory and store them by 35 percent. A fixed position bar code scanner read the bar codes on boxes as they moved along the conveyor. The data were transmitted through RS-232-C communications links to the AS/400 where they were used for production counts. Parameters, psychologies, and function modes could be defined and entered through the ASCII CRT terminal. The case of product was palletized automatically, and a bar code applied to the pallet was read with a laser scanner for input into the terminal. A keyboard available nearby enabled users to add supplemental data for special shipments, products on hold, and other exceptions. The terminal also provided information on usable storage space to minimize finished goods handling.

Exhibit 11.5

General Foods Local Area Network

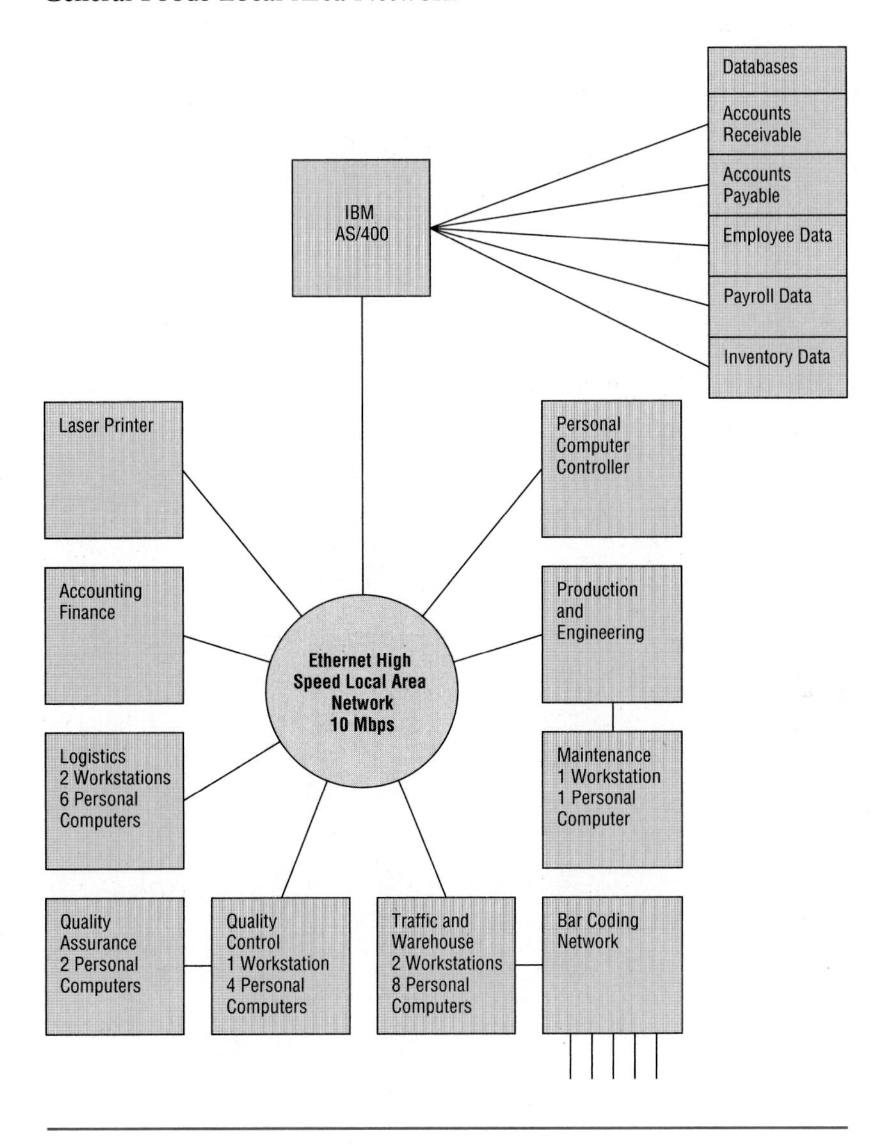

Exhibit 11.6

Ronzoni Bar Code Information Network

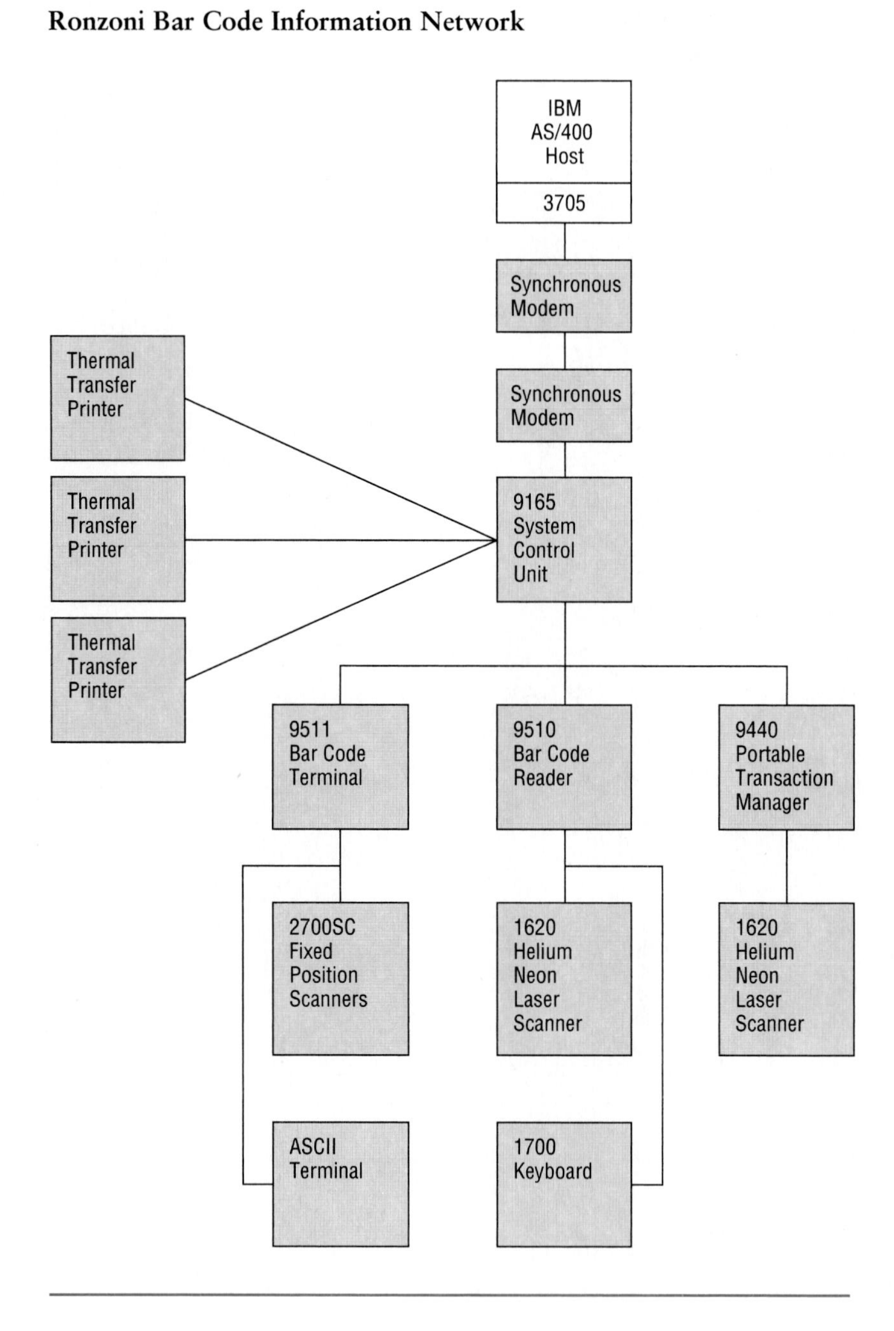

Inventory tracking and loading time was reduced by 20 percent. To fill an order, the user accessed the system and typed in the product code. The information was downloaded from the mainframe to the terminal, including age of stock, location, quantity at each location, quantity on hold, and total quantity available for use. The worker then pulled the product, scanned it before loading the truck, and completed the transaction with a few keystrokes. Again, information was available on-line regarding cases shipped, product codes, carrier, destination, etc., for use by Accounts Receivable personnel. In addition, bills of lading were generated from the information.

Two years after implementation of the Plant Finished Goods Inventory System, customer service levels improved significantly to 98.1, surpassing the industry average of 97.5 for the first time. Individual departments had on-line access to finished goods inventory data and were able to react more quickly to changing demands. Manufacturing received real-time information on low-inventory items and could modify the production schedule in time. Quality Control was able to schedule product inspections and releases more efficiently to ensure that product was available for shipping when needed. The result was a reduction in the number of out-of-stock occurrences from 213 in 1986 to 96 in 1988, and to 16 for the first quarter of 1989. Moreover, the ability to provide better customer service improved customer relations. It is estimated that discretionary sales to retail chains increased 60,000 units (approximately $680,000) due to improved reliability.

The completed system running off the Ethernet local area network and the AS/400 greatly improved efficiency of inventory tracking and reduced labor costs by $29,300. Instead of manually ticketing finished product for storage, the bar code scanner read the UPC symbol on each case of product and the program calculated total cases and total pallets, performed statistical analyses, allocated storage space, and stored data for report generation. Installation of the AS/400 reduced user response time by 60 percent, thereby increasing employee productivity. Average response times were slashed down to 2.5 seconds. This represented a reduction in waiting time of 370 labor-hours per year, or 7,000 percent. The system reduced the cost of reporting inventory information, tracking quality control information, and reconciling production discrepancies, saving an estimated $45,800 annually. Quality Control was able to disposition product electronically and generate historical and statistical reports without creating massive paperwork files. Finally, the warehouse conducted physical inventories on a monthly basis rather than daily to ensure accuracy.

Improved inventory control reduced inventory levels 29 percent to 776,000 units and reduced inventory carrying costs $162,200. The availability of real-time information enabled logistics to plan production more effectively and order materials more efficiently. Reliable information resulted in the ability to forecast sales and inventory positions and reduced uncertainty.

Finally, the system gave division managers at corporate headquarters real-time access to inventory, accounting, and production information. Although difficult to quantify, the benefits of this timely information were many. It improved the competitive advantage of the business unit and enabled tighter management control over the operation. The estimated payback of the system, including the 10-month implementation time, was just over 2 years.

Apple Computer's AppleLink™

Apple Computer is one of the most famous computer manufacturers in the world. Its line of Macintosh computers has swept through universities and businesses and gained market share in many important markets. Some of the most famous accounting, consulting, and investment banking firms use only Macintoshes. All of this was caused by the Macintosh operating system, which has an icon-driven user interface that makes many tasks, such as manipulating files, extremely easy and intuitive for the user. The icon environment allows the user to perform complex manipulations of data and information with no programming or computer experience. The Macintosh system has also benefitted substantially from the growing use of graphics and desktop publishing, which is accomplished in conjunction with a laser printer.

Apple Computer is rumored to have started in the garage of its founders, and was one of the fastest companies of its time to achieve Fortune 500 status. As this growth took place, the founders left the company and it has been taken over by more professional managers who have powered Apple into another wave of expansion. Apple appears to have successfully managed the transition from a startup company to a larger company.

At all times, Apple has faced severe competition from IBM and the makers of IBM-compatible equipment (called "clones"). It has been very difficult to remain price competitive against clones. Purchasers who have not experienced the simplicity of the icon driven environment may avoid Apple based on cost alone. Apple also faces many persons who are biased against the icon driven environment because they are accustomed to manipulating arcane commands which gives them a certain misplaced satisfaction. The result is that Apple faces a long upward battle to break into the business market, a battle it has yet to fully win.

The critical factor in the business market is service. Companies that have been highly successful, such as IBM, realized from their founding that service was critical. The cost of information technology is typically far less than the financial risk involved should a problem occur in the system. Apple, as a startup company, faced several major problems with its sales and service network:

- The number of general technicians who were able to work with the Apple operating system and equipment was small.
- Coordination of technical information was difficult in the absence of established procedures.
- Many of the problems being encountered were software related (because of the complex operating system) and this called for a particular type of expertise.
- There was a need to get reports on emerging problems from the field as quickly as possible so that manufacturing could make changes in product it shipped.

The solution adopted by Apple involved setting up an electronic mail system to link the various dealers and technicians around the United States into a system through which they could exchange information. The system works by linking Apple personal computers to modems which then dial up the network for connection with the host. The AppleLink system has gone through several major transformations, but one of the most significant has been the replacement of line driven command interfaces with the menu driven and icon driven methods so familiar to Macintosh users. This type of interface with the electronic mail system makes both sending and receiving messages extremely easy, and many bookkeeping and related tasks are automated to a great extent.

This automation extends to signing onto the system and collecting the newest messages, as well as sending a block of messages. Some of the features of AppleLink include:

- *Electronic Mail*
  Users can exchange telex-like messages with the engineers at Apple and field service representatives.
- *Bulletin Boards*
  Users have access to public bulletin boards in which people carry on a public dialogue (in the sense that everyone can read it) regarding various topics.
- *Chat Capability*
  Users can interactively type messages and have the information appear on the screen of the recipient, and receive replies.
- *Database*
  The system maintains large databases which can be downloaded (electronically transferred from the host to the user's microcomputer) when needed. Examples would include notes on repair procedures, updated manuals, technical information, etc.
- *Interest Groups*
  Users can have electronic discussions between the members of major interest groups, for example, the Eastern Marketing region, the Sales department, the Engineering department, etc.
- *Closed User Groups*
  Users can partition the E-mail system so that special interest groups organized around specific questions and problems can communicate without having to review information regarding other topics.
- *Scheduling Capabilities*
  Users can schedule meetings and conferences through an electronic system, often called a "calendaring capability."

As a result of creating the AppleLink system, Apple continued to build up the sophistication of its service and repair capabilities and was able to demonstrate the credibility it needed to go into the business market. It achieved better market penetration in specialized niches (i.e., desktop publishing, then called "Desktop Media"). Service levels improved. Also, information regarding problems being encountered in the field came back to Apple quickly and the engineers were able to make the necessary changes in product specifications in order to respond to chronic problems. This, it might be argued, helped to increase the turnaround time for Apple in making needed technical adjustments.

Apple has gradually improved the quality and comprehensive services in AppleLink. The capability of operation in an unattended mode with an icon-driven user interface was added. Also, technical diagrams and drawings, even color graphics, can be sent through the system without any special programming or operating requirements of the end user. All of the formatting commands and specifications of each file are also transmitted.

IBM Digital Packet Devices

IBM is the leading vendor of computers and related equipment in the world. By the late 1980s, however, its market share had been slipping as a result of the onslaught of personal computers, minicomputers, workstations, laptops, etc., into the market which it once dominated. IBM remained in a commanding position in the mainframe business, primarily because of its installed base in the commercial world.

From the very beginning of its existence, its founder Thomas Watson, Sr., placed a very high premium on service. Of the many famous stories of his emphasis on service, the most famous might be about a time when an early 360 machine broke down in New York on a Friday evening. He made sure that all of his service technicians and related executives showed up in their suits to work 24 hours per day until the service on the machine was reestablished. This set a style for IBM, and IBM has remained one of the best service organizations in the world.

The problem with information processing in general, and especially with a company as large as IBM, is that there is a great variety of equipment, and it is difficult to train technicians to handle all

Exhibit 11.7

Sequence of Computer Service Tasks

On-site service operations are the most critical step in achieving quick service turnaround times.

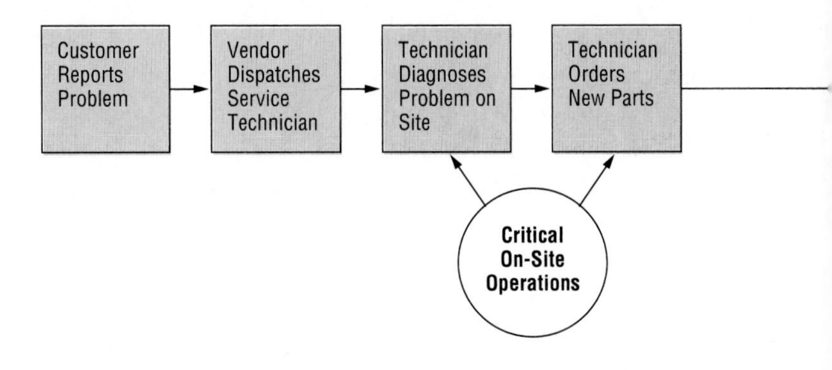

problems. Specialization is needed so that technicians can perform at optimum levels on their specific ranges of problems, but specialization has its disadvantages. The chief disadvantage is that specialization leads to difficulty in having correctly trained personnel available at every given moment of the day.

There are two ways out of this general problem for the vendor. First, it can hire more trained personnel so that the load is leveled, but over time this may raise costs in comparison with competitors who might have simpler product lines. A more simplified product line would mean less problem in training persons, and a lower cost per sales call. The second option would be to increase the skill and ability to handle complex sales calls of each of the service representatives.

An additional factor in computer servicing is time. Generally, repairs must be made as quickly as possible, with little, if any, exception. The repair process can have several components, as Exhibit 11.7 shows. The most critical step in the service process takes place when the service technician is operating on-site with the customer. At that time, the service technician is cut off from the expertise available in the office, and must survive alone to carry out the efficient repair

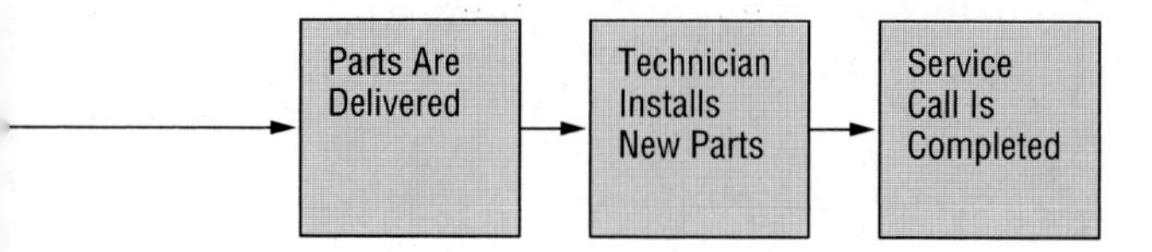

operation. This is when the training received by the technician either works or fails to work. In the past, the technician would have had to diagnose the problem, then use the customer's telephone and call the vendor office either to get advice on making the repair or to order the spare parts that were needed.

Many problems with this type of service process are obvious:

- If the service technician is unable to determine the problem, then other expertise is needed.
- If the service technician orders the wrong part, parts have to be shipped again.
- If the service technician discovers a problem for which he or she is not trained, another service technician must be dispatched.
- Double tripping may occur if the service technician has to get parts and return, or comes once with the wrong parts.

In order to solve part of this problem, IBM pioneered the use of digital hand-held terminals which are capable of keeping the service representative in touch with the office at all times. The digital terminal operates on the frequencies assigned to the cellular telephone network, yet it is optimized for digital transmission, i.e., for short digital bursts of information, rather than for voice.

This small unit enables the technician to accomplish several very important tasks:

- The unit can automatically transmit parts orders to the vendor, and confirm the delivery time, without having to make telephone calls.
- The unit can help diagnose the problem by accessing routines on the vendor's service computers and guiding the technician step-by-step through the diagnostic routines. This vastly expands the capabilities of the service technician.
- The unit acts as a communications device for alerting the technician to pending service calls, and their level of urgency, regardless of the location of the device.

These benefits of the digital cellular telecommunications device have included the ability to deliver continued high-quality field service, and to hold onto market position. The excellent service provided by IBM has continued to attract customers as they upgrade equipment, add new applications, and rearrange their information technology architectures. This type of service has reinforced the common saying among users: "No one ever got fired for buying IBM." IBM has set the standard to which other vendors must aspire in this service industry, and IBM has done it largely through the innovative utilization of advanced telecommunications technologies.

Honeywell HVAC

Honeywell is a large maker of industrial air-conditioning equipment for building complexes. The units are typically placed on top of large buildings, such as malls, and provide millions of BTUs of air-conditioning power. These air-conditioning units require regular maintenance, and for many service calls it is critical that the system be fixed as soon as possible: a retail complex could lose a great deal of money in sales as customers are driven away by heat and bad air; in a hospital, patients could become very uncomfortable, or even worse, due to heat. Furthermore, in many large office buildings, the air conditioning is tied in directly with the basic air circulation system, which is definitely a critical subsystem in most situations.

In order to deliver the type of service needed, Honeywell maintains a highly trained force of service technicians who are radio dispatched to service sites to effect repairs. Since the systems are often highly complex, Honeywell must invest much in training its field service technicians so that they can quickly get results when placed in the field. Honeywell has done this, and as a result, is able to deliver superior service in its many locations in the United States.

One of the chronic problems faced by Honeywell (and by many companies with similar service regimens) is that the service technicians often leave the company and form their own firms that compete with the service being provided by Honeywell. Since they have already received their training from Honeywell, they are able to accomplish the work. However, since their organizations are typically small they do not have the overhead of training employees. This makes their general price for maintenance lower than Honeywell's. Also, often they have built up close relationships with customers, establishing personal bonds between the customer and the service representative. This bond is used to "steal" the service business customer away from the mother company, Honeywell.

What could Honeywell do? It was essentially providing the training and marketing expertise for its competition. The answer for Honeywell was to develop artificial intelligence routines in its information system for diagnosis and repair of its equipment. The service representative would connect to the service computer through a telecommunications linkage. The system would work as follows:

- The service representative shows up on the site.
- The representative connects to the Honeywell computer through a telecommunications channel (typically a dial up line).
- The Honeywell computer takes the service person through a series of tests based on an artificial intelligence routine.
- The AI system diagnoses the problem and then the service person is able to effect the repair.

The advantages to this system are very great for Honeywell:

- Service personnel can be sent out into the field with a great deal less training than in the past because the knowledge for servicing is locked into the computer.
- Service personnel can work on a wider variety of equipment, since they just follow the instructions of the computer.
- Labor rates can be lower, because there is much less expertise needed on the part of the service personnel.

In addition to providing better service for the company as a whole, this system helped to curb the formation of rival service companies composed of ex-employees. How could the employees use knowledge that had never been given to them? Honeywell would not provide access to the artificial intelligence routines to any group outside its own organization, therefore the competitors would have no way to efficiently service the equipment. Also, training costs were greatly reduced because the entire process was essentially one of deskilling, purposefully downgrading the role and knowledge base of the service technician. Although there are ethical questions about deskilling, it has undeniably been a way to drive down the cost per service call and also keep down the cost in general as compared to competition.

American Airlines Sabre

American Airlines, Inc. was one of the largest United States airlines by the late 1980s. It served 149 airports in 42 states, the District of Columbia, Bermuda, Canada, the Caribbean, England, France, Switzerland, West Germany, Mexico, Puerto Rico, Japan, and Venezuela. Its headquarters was located in Dallas, Texas.

The company president, Robert Crandall, was a major figure in the airline industry's competitive wars. He set American Airlines firmly back on course after violent turbulence in the late 1970s. Faced with the deregulation of the airlines industry, price competition, and unprofitable operation of the airline, Crandall sharpened the computerized reservation system Sabre (Semi-automated Business Research Environment) into a powerful marketing tool.

Before Sabre, American Airlines was suffering from a terrible communication problem. Before the development of a computerized reservation system, travel agents had to call American Airlines for information concerning flight availability, route destinations, time schedules, and confirmation. In addition, they ran into the trouble of making phone calls to discover the fares for certain flights. All too often they would just receive busy signals. This was complicated by the agony of making phone calls for flight cancellations, changes, etc. Also, American Airlines had the difficulties of having to answer phone calls and keeping up records about commissions, etc.

Without a uniform, centralized database, immediate corporate decision making was almost impossible. It was difficult to get strategic information generated from the transactions at the lower levels of the corporation. There were also significant time delays in processing information on operations. In 1973 alone, American Airlines lost $50 million on revenue of $1.3 billion. In the early 1970s, American Airlines operated at a loss.

When then-president George Spater altered the flight plans in the early 1970s to concentrate on resort destinations and leisure travel, this eliminated the heavily traveled schedules and convenient connections used by business travelers. The fortunes of the airline began to sink. Operating at a loss, it found itself in great danger. It began to lose the patronage of its frequent customers, particularly business clients.

In addition, government deregulation of the airlines in 1978 left American Airlines facing low-fare price competition. Competitors offered ticket prices 10 percent or more below American. The deregulation of airlines changed the long-term structure of the airline economy and traditional methods of running the business. The action by the U.S. government in 1978 left the airlines in a state of confusion. American Airlines was undecided on how to operate in the new laissez-faire business. New airline companies were attempting to compete with major airlines such as TWA and American by introducing low ticket fares, which American could not match without losing profit. In addition, the competitors hired labor at much lower cost. Government deregulation had given the airlines freedom to manage and operate their businesses without government imposed fares, flight route schedules, and profit limitations. On the other hand, deregulation suddenly put pressure on management to rely on strategy in the newly competitive free market.

By the mid 1970s, American Airlines, along with a few other airlines, had developed computer reservation systems for internal uses. American Airlines began to market the system to travel agents under the name Semi-automated Business Research Environment or SABRE. Sabre uses five mainframe computers (three IBM 3080s and two IBM 3083s) and hundreds of IBM 3380 disk drives for storage of information concerning not only airlines, but also hotels and other services. Sabre operates as a hierarchical computerized reservation system.

Sabre linked in many travel agents around the country. The data communications from the travel agents to the hosts were based on asynchronous dial up through public packet switched networks. The Sabre system served as the principal distribution network for flight schedules. In the schedule displays, each flight listing is a type of electronic shelf. Sabre lists codes for requesting information about origination and destination, flight time, departure and arrival time, and so forth. Several other airlines displayed their flights through Sabre. However, in order to be listed on the system, other airlines had to pay a fee to American composed of a basic capital fee plus a bookkeeping fee. The travel agents see flight schedules in their computers six to eight lines at a time. To request flight schedules, they punch in the desired dates, times, etc.

In order to build the system, from 1976 to 1982, American Airlines had invested approximately $114 million in hardware plus another $46 million in expansion of central site facilities and equipment. The result was a dramatic transformation of airline ticket distribution methods. Between 1977 and 1982, American Airlines had installed more than 20,000 computer terminals for agents across the nation and some in the international markets. The installations of the reservation system were intended to boost service revenue through ticket sales. In the short run, American Airlines lost money. Between 1976 and 1982, the company experienced an estimated cumulative net loss of $123,888,000.

In the longer run, however, American Airlines was able to use Sabre to change the basis of competition. Some 500,000 regular business customers account for 40 percent of American's airline traffic and 41 percent of the market for automated reservation systems. The installation of the computer reservation system terminals linked through telecommunications networks to travel agents influenced the sales of the tickets. The agents were given easy access to the information they needed to process ticket sales, cancel reservations, or to book the previously reserved airline seats vacated by cancellations. The agents were able to provide much better service to their customers, since they could find out immediately about confirmations of reservations. Agents were also given all the schedules of airline flights without having to call the airlines to check on or verify flights. (This type of system is still used in countries such as the People's Republic of China.) Additionally, the air travelers themselves were given the option of seeing the various flight schedules.

Competitive Effects

By using an extended telecommunications system to give access to its mainframe computers, American Airlines used the Sabre system to lock in the customer relationship and raise the cost of entry against competitors. American Airlines developed an ability to market and distribute the terminals to travel agents. After this, the agents would be transformed into customers, because they would use Sabre to reserve tickets and flights, etc. Since American Airlines was the first to develop this product, Sabre provided an additional way to differentiate its product in the market. In addition, since it represented a considerable investment to build and operate the system, any competitor wishing to compete would have to make the same investment. This essentially raised the cost to enter into the market for the competitors. Also the development of a competitive system would take time, and during the intervening period, American Airlines would continue to have the advantage.

In addition, American Airlines developed a cohost system to get additional revenue and potentially run the telecommunications-based reservation system as a separate cost center. American provided the ability to make reservations for other airlines through the Sabre system. As a result, only about 10 percent of travel agent bookings made through Sabre are for American Airlines flight segments. Approximately 58 percent of all messages through the system do not involve bookings on the airline. The American Airlines cohost system collected monthly capital fees from other airlines for installation of the system to designated travel agency locations. Booking fees were collected for each reservation made on the cohost from other Sabre locations.

There were other great secondary benefits from the telecommunications strategy of American Airlines. The telecommunications linkages allowed American to sense the pulses in the market ahead of its competitors. Information on various levels of activities, such as sales volume, busy flight destinations, and the type of people who travel frequently—all of this information and more is collected. In addition to being a reservation system, Sabre can be thought of as a very large point of sale terminal for the airline.

Seeds of Destruction

Unfortunately, the success of the Sabre system attracted a spate of lawsuits from the other airlines operating in the cohost mode. They claimed that the system was biased against them and pointed as evidence to the way in which the information regarding their flights was presented on the Sabre screen. American Airlines' flights were listed first, they argued, with unreasonable messages introducing the flights of competitors. Because more than 75 percent of all flight bookings were made off the first batch of listings, the Sabre system was biased, they argued, since American Airlines flights were listed first. Although American Airlines argued that this type of activity, such as structuring and controlling the way in which information was distributed, was part of the competitive arena, and that Sabre was just another tool in the competitive battlefield, the legal challenge eventually forced American to modify the way in which it offered its information regarding the flights of its competitors.

Prior to 1984, computerized reservation systems were not regulated and airlines developed the systems to give themselves competitive advantages in the distribution process. This created political conflicts over bias and ultimately antitrust lawsuits against biased information display structuring. However, as a result of a Department of Transportation ruling adopted in 1984, all U.S. computer reservations systems are now totally free of bias favoring individual carriers. Flights are listed by time, not by carrier. These systems are funded by a combination of equipment rentals paid by travel agents and booking fees paid by the airlines and other transportation suppliers whose services are sold through the systems. Government regulation eliminated monopoly and removed a natural advantage of the owners over the cohosts.

At the same time, other airlines began marketing their own systems to travel agents. United Airlines started marketing Apollo and British Airways started marketing Travicom™. Once this element of competition entered the market, other airlines started offering better financial deals to travel agents who adopted their systems over Sabre.

From 1977 to 1987, since Sabre entered operation, American Airlines has increased its operating profit almost 400 percent. In 1977, it profited $94,767,000 and 10 years later, it earned a profit of $473,184,000, a difference of $378,417,000. It might be argued that Sabre pulled American Airlines out of the economic valley it was in. It helped attract more business and increase the total operating revenue dramatically.

The reader is warned, however, never to attribute such dramatic success to one factor. Economic factors, the larger picture of competition in the industry, and many other factors no doubt come into play in accounting for the profits made. It would, however, not be wise to ignore altogether this telecommunications strategy which seemed to work very well for a few years.

Saks Fifth Avenue Folio Fill System

Saks Fifth Avenue is one of the largest specialty retail companies in the United States and one of America's leaders in fashion with 58 stores of different sizes located throughout the United States. Saks has maintained a tradition of excellent service and high-quality merchandise.

The store in Manhattan was opened in 1928 and serves as the company's flagship store, particularly during the holiday season when its window displays attract tourists from all over the world. The Manhattan store is Saks' largest store, having ten floors. Nine floors are used for selling purposes while the tenth floor is used for office purposes. The store employs over 800 sales personnel (not including office workers, maintenance people, security personnel, cafeteria workers, and stock people). The Manhattan flagship store sells a very wide variety of merchandise.

Folio Fill is the name of the system used by Saks' Catalog Order Department. The Folio Fill system helps the store fill back orders that the Saks Fifth Avenue Catalog Service cannot fill. Under most circumstances, customers who receive Saks catalogs are able to order items they desire with no problem. Saks periodically mails out catalogs, which are called "Folios," to its customers. To order, the customer dials a toll free number and places the order using either a Saks Charge Card or any other major credit card. After the order is received, the Saks Fifth Avenue Distribution Center located in Yonkers, New York, sends the merchandise to the customer.

In the past, if the distribution center had the ordered merchandise in stock, there was no problem. If, however, the distribution center did not have the merchandise, it had to contact other Saks stores. The missing item could either be a back-ordered item or simply out of stock at the distribution center. The distribution center contacted the various stores in the area near the customer to locate the item. It then gave the customer identification information and the item was shipped from the remote store to the customer.

The problem was that the process of supplying the customers with the merchandise was all done by telephone. The distribution center would notify the remote store in the morning, usually around 9:30 a.m., reporting which items were needed. The items were listed by department on a printout. Each department manager would look for the merchandise or assign a salesperson or stockperson to look through a department for the merchandise. Once found, the merchandise was taken to the Private Line office where workers would call the distribution center to say that the merchandise was available and request that they be the store to fill the order. The more orders filled, the more business for the store.

The distribution center would then send the store a special form that contained the customers' information. Once the form was received by the store, the merchandise was rung up and sent out to the customer. This special form would be sent back to the center along with part of the sales check that was used to ring up the sale. This return of the form with the sales check was a way of confirming to the center that the process was carried out.

Since this order fulfillment operation was carried out over the telephone and by mail, several problems arose, including mistakes over the telephone, problems with the mail, wastage of time, and mistakes filling out the required forms. Many problems went along with having this operation done manually. Many mistakes were made by employees in the store talking with the distribution center, including mistakes about merchandise information. Saks workers in the Private Line office had to read off the vendor, style, size, color, and department information concerning the available merchandise to the workers back at the center. Any mistake in reading off the merchandise information would slow the whole process down, eventually forcing repetition, if the order was not lost altogether. Many times the store and the center were slow in mailing forms to one another. This would delay the process. There were many instances of forms with wrong customer information being sent to the store, or forms being sent to the wrong store.

This process took a considerable length of time to serve the customer. The customer often had to wait 3 to 4 days. This did not include the length of time the customer had to wait while the item was in the mail. If there was a delay in serving the customer, it was difficult to estimate the waiting time.

To solve this problem, the Electronic Data Processing Department at Saks used telecommunications to link the distribution center with each of the various stores throughout the country. The Folio Fill system was put into operation in June of 1987. The system used IBM terminals, and personal computers within the stores that emulated terminals, for telecommunication with the host. The Private Line staff and the departmental managers could use the system to send and receive information. A store's Private Line staff was made up of six people supported by the 30 managers who headed the store's departments. Trained staff and managers could work the terminals to request that the store be allowed to fill a customer's order.

The Folio Fill system replaced the old methods of operation. The new process had several major steps, as outlined in Exhibit 11.8. Under the new system, when a customer could not get the desired item through the distribution center, the center gave that customer a

Exhibit 11.8

Saks Fifth Avenue Folio Fill System

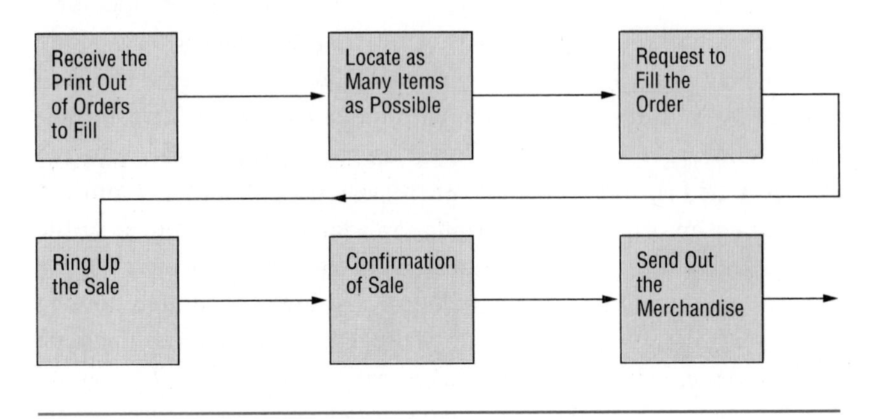

nine-digit ID number. The center then sent the stores in the customer's vicinity the merchandise information to see which store had the item. If two stores had the item, the store that notified the center first would be directed to fill the order. The New York store received a printout of the merchandise information.

Merchandise, if available, was found and brought to the Private Line office. At 10:30 every morning the center opened up the system, and all stores could try to request the chance to fill an order. Once the store requested all orders that it could fill, the Private Line people punched into the system to receive customer information in the form of a printout. Once customer information was given, the merchandise was rung up and sent out of the store to the customer. Before the merchandise was sent out, a Private Line worker punched confirmation information into the system to notify the center that the operation had been carried out.

As a result of this telecommunications-based solution, Saks Fifth Avenue has achieved many benefits that the old methods of filling orders could never produce. These benefits include less time for servicing, more money for the stores, and fewer opportunities for mistakes. The system shortened the service time from 3 to 4 days to 1 working day. Saks can receive an order at 10:30 in the morning and have the customer's items ready to be shipped out by 4:40 to 5:00 p.m. on the same the day.

The Folio Fill system appears to have helped cash flow, as well. In 1987, Folio brought in $1,599,591 in sales for the Manhattan store. In 1988, it brought in $2,067,550, an increase of about 30 percent. The figures for 1989 were even higher. Before the telecommunications linkages, this type of increase in volume could not have been handled.

In addition, the system reduced opportunities for mistakes and clerical errors. The entire process of sending forms between stores by mail was eliminated. All of the information flowed through the telecommunications systems to the terminals.

Nissan Motor Corporation VSAT Network

In introducing its new line of luxury cars to the U.S. market, the Nissan Motor Corporation built a VSAT network to enhance its automobile dealership network and provide superior customer service. The new Infiniti™ line of automobiles was supported by a telecommunications system that linked the firm's host IBM 3090 and 3084 processors running MVS/XA. The mainframes used the standard 3725 front-end processors that connected into the satellite network. Each Infiniti dealership has an IBM AS/400 minicomputer.

The new Infiniti telecommunications network handled the following applications:

- Interactive data processing for order entry, parts and vehicle location, and back-office billing
- Broadcast video
- Facsimile
- Audio transmission
- Teleconferencing

Nissan found that the use of the Scientific Atlanta VSAT system aided in the very rapid introduction of its dealership network. The days of spending dozens of years to set up an automobile dealership network are over. With the introduction of the Infiniti line of automobile, Nissan has shown that it is possible not only to set up dealerships quickly, but to have very advanced telecommunications-based applications to give superior customer service.

Summary

Providing good customer service is an important way to get competitive advantage through telecommunications. Chemical Bank demonstrated that advanced voice processing technologies can be used to provide a very wide range of informational services to customers. Ronzoni Corporation demonstrated how telecommunications can be used to prevent out of stock situations for product deliveries, thus helping to maintain market share. Apple showed how an electronic mail network could be used to help companies that are struggling to maintain good customer contact and service. The speed of the network helped feed information back to the manufacturing process, as well. IBM and Honeywell demonstrated how telecommunications can be used to deliver advanced artificial intelligence and mainframe computing to even the most remote field site, thus speeding up customer service. The American Airlines example is similar in that the mainframe-based reservation systems were being extended out to the remote locations (travel agent offices). The Saks Folio Fill system demonstrated how telecommunications can be used to fill orders under limited supply conditions. The Nissan VSAT example shows how telecommunications can be used to create a very sophisticated service and accounting network.

Questions

11.1 a. Discuss why telephone answering service for a company is important.
b. What are some ways to improve it?

11.2 a. What is the role of telecommunications in just-in-time delivery systems for inventory?
b. How can telecommunications be used to aid effective inventory management?

11.3 a. How can telecommunications be used to give remote sites access to sophisticated services and capabilities of mainframes?

b. What can this mean for customer service?

11.4 Discuss how satellites can be used to set up a customer service network quickly and tell why that might be important.

Chapter 12

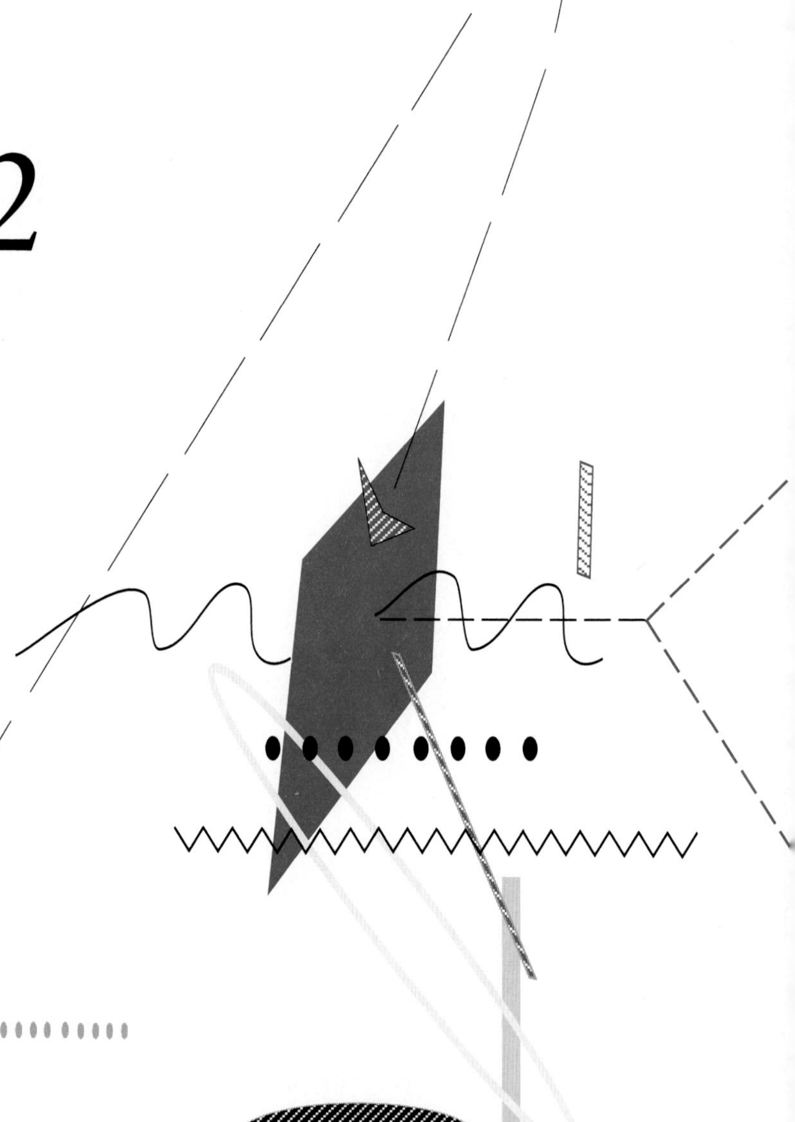

Operate a Distributed Business

After the rise of the multinational and the national corporation, it is no longer possible to think of a business as existing in only one location. Businesses must be operated in a distributed manner, and this poses several problems for information resources management.

What is the role of telecommunications in operating a distributed business? We will see in this chapter that telecommunications is critical. Without it, businesses would not be able to operate at all. Telecommunications becomes the nervous system through which all of the critical information regarding the firm flows.

The 24-hour Market at Morgan Guaranty

As the end of the 1980s approached, Morgan Guaranty's foreign exchange department found itself in a leading position. It held a commanding presence in the foreign exchange markets through its careful management and quality information and telecommunications technologies linking together its foreign exchange traders. The newer Morgan traders were encouraged to take short-term trade and positions in foreign currencies, leaving longer-term positions for the more experienced senior traders.

The foreign exchange trading room employed a series of trading stations linking together several critical telecommunications functions, illustrated in Exhibit 12.1:

- Reuters and Telerate information services were accessible from each trading station.
- A bank of telephone circuits giving access to customers, other brokers, and others was accessible from each desk.
- A terminal was also linked to the Internal Rate System, which kept track of rates, particularly cross rates.

Morgan realized that access to information through telecommunications plays a vital role in the foreign exchange market. The job of a foreign exchange trader is to determine the direction of movements in the American dollar against other currencies, buying currencies at one rate and selling them at a higher one. Since many world events instantaneously influence, and in turn are influenced by, the dollar's rise and fall, the trader needs to have up-to-the-second information relayed through telecommunications linkages by news services and word of mouth. Any delay in getting the critical information can result in either lost opportunities for making deals or in outright losses.

Therefore, the trader's success is a function of individual ability and the quality of information received. If sources of information are suddenly cut off, the trader is vulnerable to changes in rates, which can swing dramatically in the space of seconds. This can cost millions of dollars and a person's job if a trader is caught with a wrong position. Improper telecommunications linkages deprive the dealer of important news sources and price information.

Exhibit 12.1

Morgan Guaranty Foreign Exchange Information System

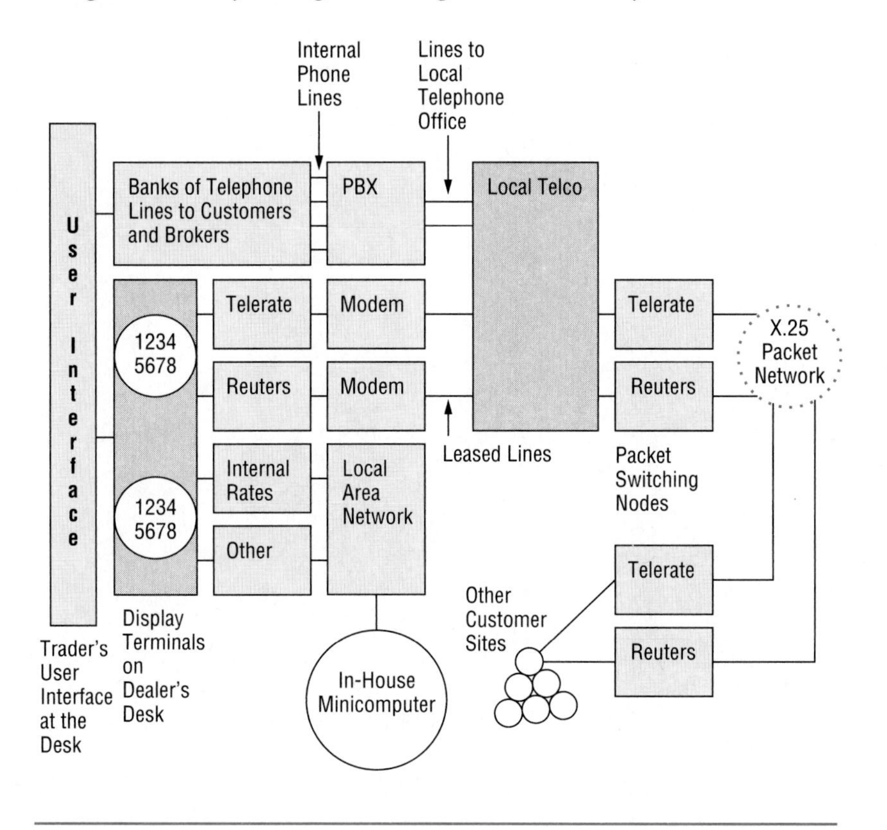

Since a trader's price is only good for five to ten seconds because the market moves so fast, the telecommunications linkages must be fast enough to allow the supporting information technology to update information in only a few seconds. Morgan knew that, although its almost decade-old system was good, it deteriorated in efficiency after trading volume reached about 600 deal tickets per day. (See Exhibit 12.2.)

Morgan was playing an essentially unregulated market. Since it was unregulated, new players in the market could and did enter every day. Each new player was typically loaded with the newest information technology equipment and the fastest telecommunications

Exhibit 12.2

Trading Volume and Response Time

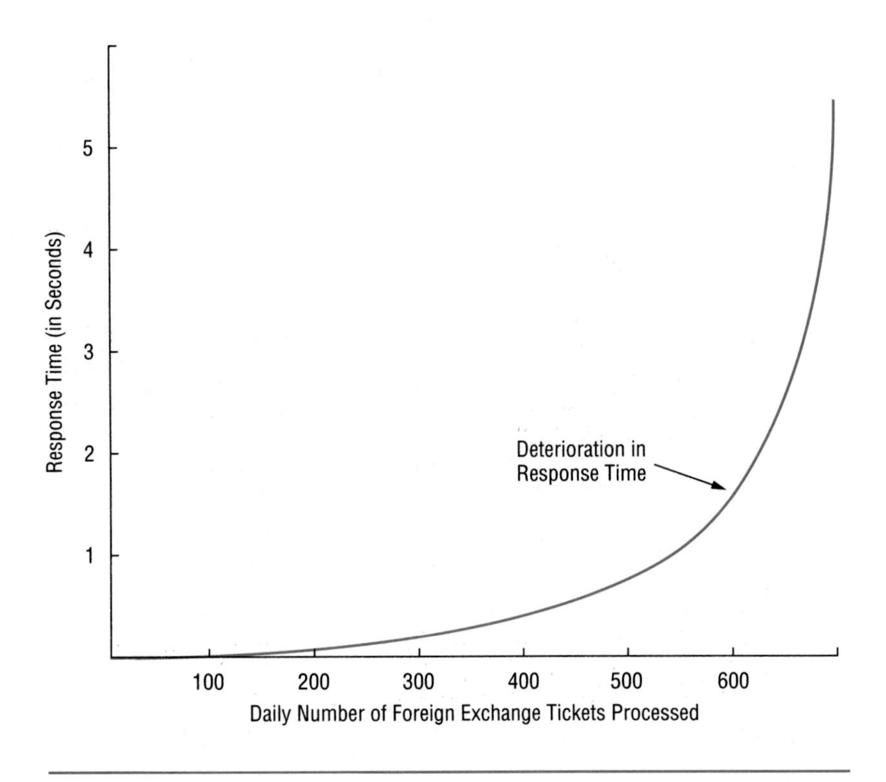

linkages to shave seconds off the time for each foreign exchange transaction. This response time advantage, although sometimes not apparently significant, would translate into competitive advantage.

Morgan therefore faced the need to completely redesign its foreign exchange system in response to competition. It had to consider the various sources of response time delay for the foreign exchange trader, and what type of technological options would aid in giving a competitive edge to the foreign exchange (FX) desk. Unlike most environments, Morgan was not limited in its expenditure of money. In relation to the amounts of cash involved in foreign exchange transactions, the amount spent on the computers and telecommunications links serving the FX traders was minimal. But having essentially unlimited money to

Exhibit 12.3

Information Network Options

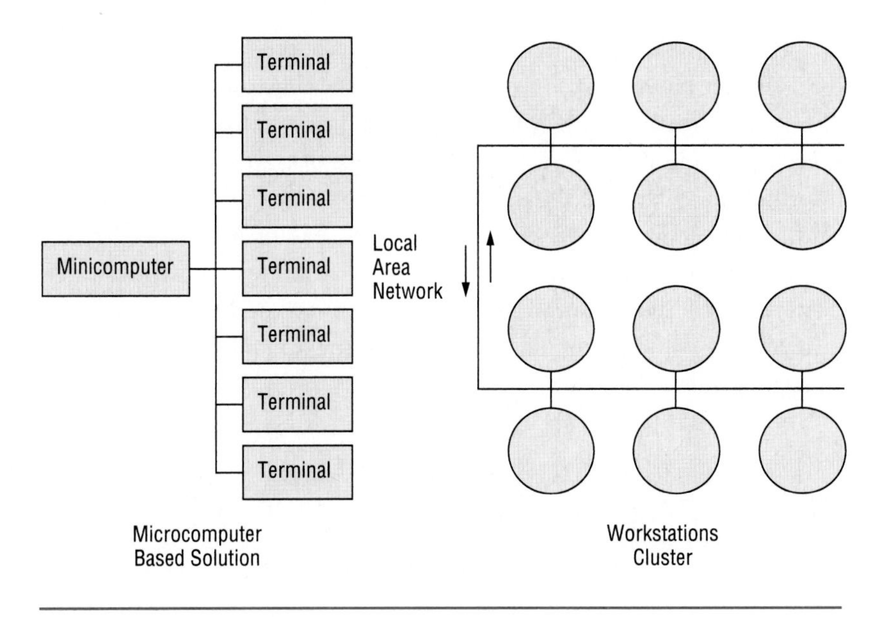

buy information and telecommunications technologies did not solve the critical design problem.

Morgan identified one of the most critical points to rectify in its system. It observed that the internal rate system was having difficulty keeping the price updates available with the needed response time. Also, it constituted a single point of failure for the entire system.

The design it considered consisted of a movement away from a terminal/minicomputer-based solution to a networked workstation approach utilizing a high-speed local area network. (See Exhibit 12.3.) This solution made possible several key advantages:

- The single point of failure at the minicomputer was eliminated. If one of the workstations broke down, it would not harm the entire system of workstations.
- In addition to getting basic information from the news and quotation services, the trader would be able to perform various analyses that in the past had been done manually using a hand calculator.

- Faster microprocessors available on workstations could be used instead of the older generation of processor available in minicomputers.

This possible movement from a minicomputer-based solution to a workstation-based solution was typical at the time of a larger trend in the same direction in the industry. The faster telecommunications speeds possible through the local area network allowed much more information to be made available to the workstations simultaneously. The relatively more advanced microprocessors on the workstations allowed multitasking. Multitasking means that several windows of display on the workstation screen could be updated simultaneously, information could be copied from one window to another, and sophisticated graphics could be utilized.

Distribute Critical Activities Where They Are Most Needed

Telecommunications can be used to distribute critical business activities to where they are most needed, where they are performed, or wherever they do the most good for the competitive success of the business. Telecommunications allows linkages with remote locations, and also between critical decision making centers of a firm. By providing instant communications between remote locations of the firm, telecommunications shrinks the logical distance between locations and provides a seamless web of interconnection. An occurrence thousands of miles away can appear as though it is no farther away than the next room.

Manufacturing can be coordinated in many different locations simultaneously, documents can be printed in several locations simultaneously, and hardware and software systems which in the past have been closely linked together can be separated and still work connected together only by the thin nerve fibers of the telecommunications network.

Wan Hai Steamship Company

Wan Hai is the second largest nonstate-owned shipping company in Taiwan. It is a full-container shipping company with more than 20 years of history. It rose up during the years after World War II when

East Asia began to obtain startling productivity gains through international trading and commerce. Wan Hai maintains a fleet of 12 full-container ships and will receive 4 new vessels by the end of 1990. Wan Hai moved itself into the full-container shipping business approximately 10 years earlier. It is the market leader on the routes among Japan, Taiwan, and Hong Kong—routes that carry the bulk of the East Asian shipping traffic. By the late 1980s, Wan Hai had started service to destinations in Korea, Singapore, and Malaysia. Exhibit 12.4 diagrams the route network.

Use of information and telecommunications technologies began at Wan Hai in 1982. The main hardware used included the Hewlett Packard HP-3000 minicomputer series and several personal computers. Wan Hai developed its own software through its information system department. Although large information and telecommunications activities take place in departments such as shipping affairs, administrative, agent, auditing, and operations, the bulk of computerization took place in the sales and finance/accounting departments.

By their very nature, businesses such as shipping are highly distributed. Although shipping firms typically have central headquarters offices, their operations are distributed among distant ports. It would be fair to say, historically, that the way in which shipping documentation is handled—the frames of reference, the forms of papers, and the procedures—are all based on assumptions about technology and telecommunications that no longer hold true. As shipping companies such as Wan Hai build sophisticated information and telecommunications systems, they must take the frames of reference for documentation and procedure developed in the past, and then transform them into modern automated systems. There is absolutely no doubt that if shipping systems, and many other systems, as well, today were completely redesigned with references to telecommunications and computer technologies, the world of business would be radically different.

Before introducing the telecommunications system, Wan Hai's method of operation in sales differed for export and import transactions. Manual processing of export shipping documents involved approximately 13 steps (see Exhibit 12.5):

1. The client (the shipper or customs broker) came to the sales department's documents processing section (DPS) to get the shipping order (S/O) issued.

Exhibit 12.4

Wan Hai Steamship Company Route Structure

Wan Hai is the market leader in East Asia.

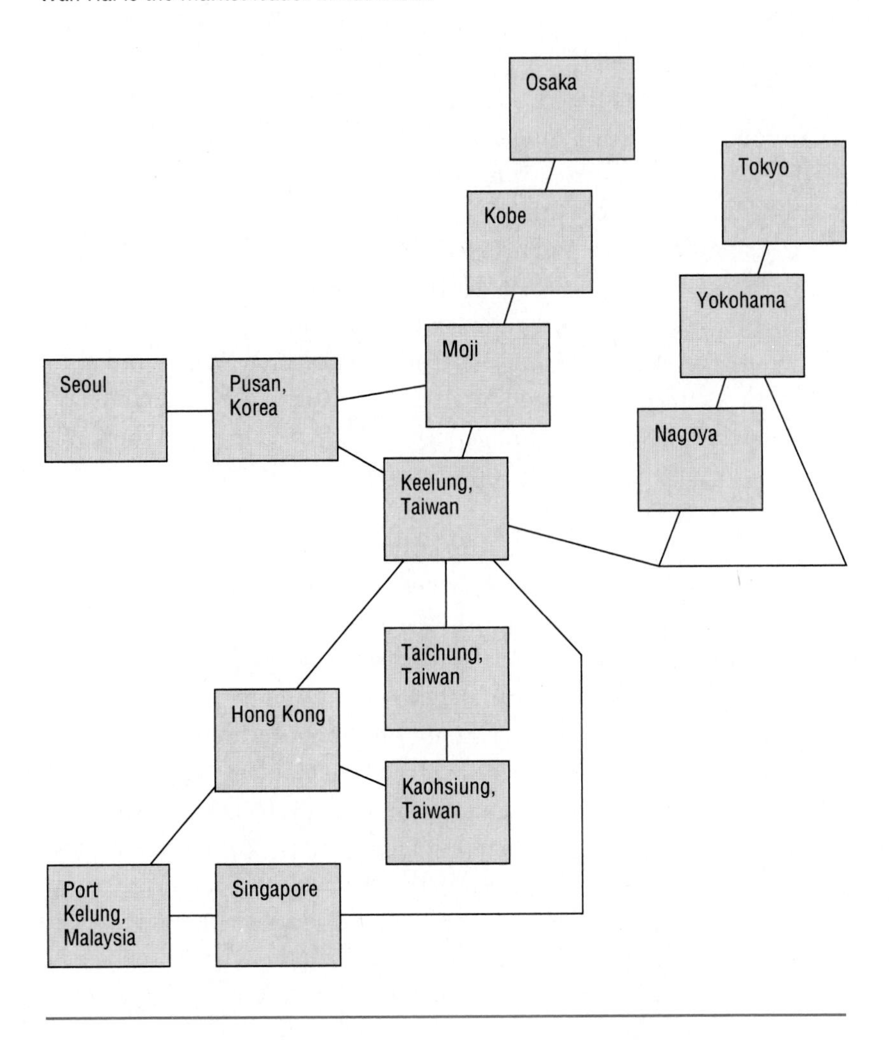

2. One copy of the S/O went to the sales area in charge of the account. The salesperson agreed on the freight rates with the client following either a predetermined schedule or a different fare structure. At the same time, another copy of the S/O went to the shipping agent at the harbor.

Exhibit 12.5

Wan Hai's Cumbersome Manual System

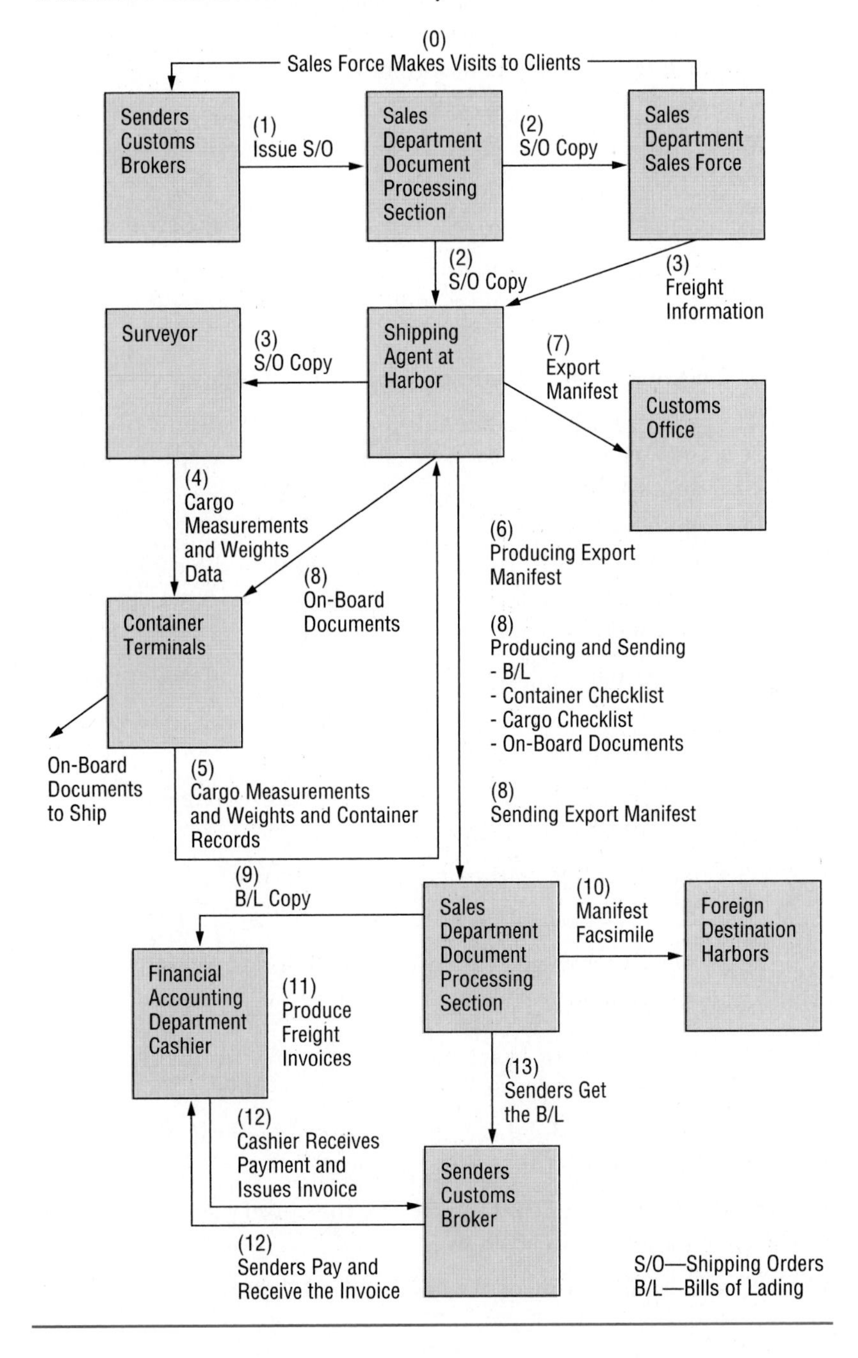

3. The shipping agent filed the application for export with the customs office at the harbor. The agent also sent the S/O to the surveyor for the cargo measurements and weights survey and recording. About the same time, the sales area determined the freight fare and sent this information to the agent.
4. The cargo measurements and weights data were sent to one of Wan Hai's container terminals at or near a harbor by the surveyor.
5. The same data were sent back to the agent by the terminal.
6. The agent produced manifests according to the previous data from the container terminal and the freight information from the sales department.
7. The agent sent the manifest to the customs office.
8. The agent produced other shipping documents (bills of lading, container check lists, cargo check lists, on-board documents, etc.) and sent them back to the DPS in the sales department at headquarters in Taipei.
9. A copy of the bill of lading (B/L) was sent to the cashier in the finances/accounting department (F/A). An invoice was produced by the cashier.
10. The export manifest was faxed to the offices at the destination harbors.
11. The cashier issued the freight invoices.
12. The sender made the payment and received invoices at the cashier's office.
13. The sender received the original copy of the B/L from the sales department.

The manual information processing procedures for import operations had a similar complexity (see Exhibit 12.6):

1. An overseas agent or branch sent the manifest to the DPS of the sales department at the headquarters by facsimile.
2. The overseas agent or branch also sent the arrival notice to the DPS at the headquarters by express mail.
3. The on-board documents carried by the vessel were given to the shipping agent at the harbor.
4. The agent sent a copy of the B/L to the DPS at the headquarters.

Exhibit 12.6

Wan Hai Manual Information Network before Telecommunications

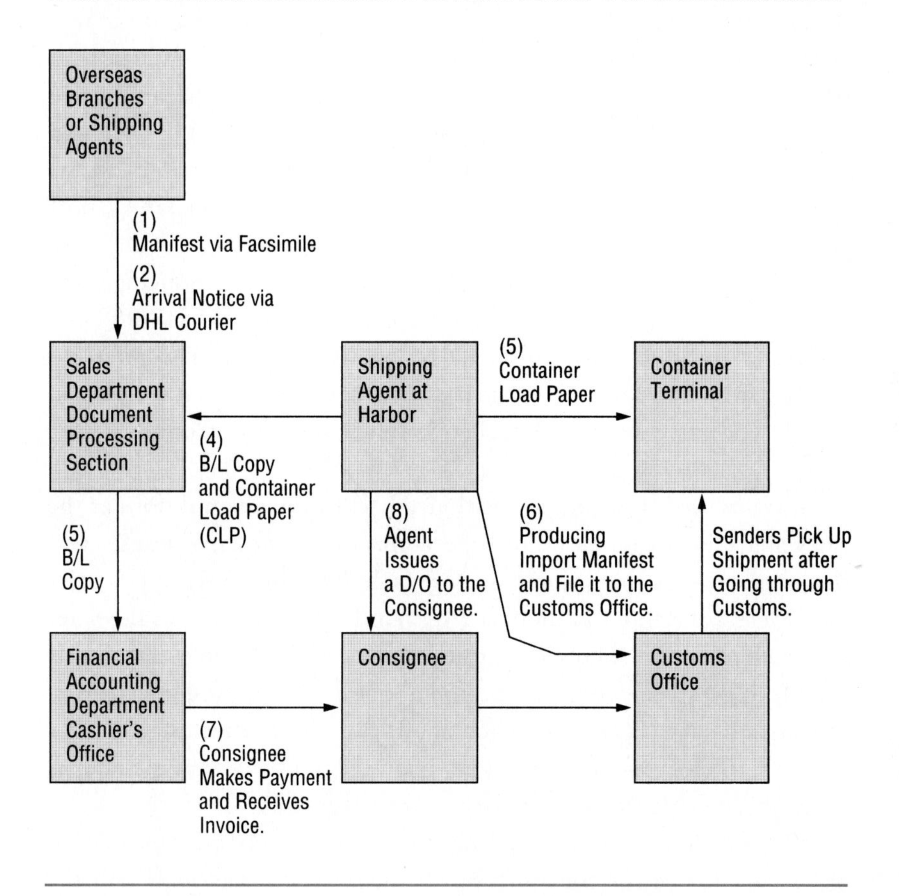

5. The agent sent the container loading paper (CLP) to the container terminal. At the same time, DPS sent a copy of the B/L to the cashier's office at the F/A department.
6. The agent produced the import manifest and filed it with the customs office at the harbor.
7. The consignee made the payment (if the freight was not paid by the senders) and received the invoice.
8. The agent issued the delivery orders (D/O) to the consignee so that the consignee could pick up the shipment.

The problems experienced by Wan Hai with these procedures were very significant. The shipping documents were usually late, and the error rate of the documents was 1.6 percent. Correcting errors was very expensive. The clients complained about the long document processing time. The delay of the shipping documents became more serious during peak seasons. The error rates were high, 1.5 percent and 1.7 percent, respectively, in 1980 and 1981, according to a 1981 study. The mistakes were found mostly in the items such as vessel names, port container numbers, item numbers, package units, freight calculations, exchange rates, B/L and S/O numbers, etc. The cost of producing these documents was high. The shipping agents charged NT$13,000 per voyage to produce the shipping documents in 1982. The price increased from 1982 to 1989. The cost of giving the job to the agents in 1989 could easily have approached NT$1.5 million per month, as the number of the voyages exceeded 75 per month.

An additional problem was that the sales force could not get the sales analysis information until the 10th of the *following* month, and this resulted in a sluggishness in responding to the market. The sales data were collected and analyzed by hand by the people in the sales department in the following month. As a result, the sales force was unable to get the newest information about the clients and the trend of the business at the end of each month. This resulted in a slower reaction to changes in the market. The variety of the analysis and its accessibility were not satisfactory to the sales staff.

Within the finance/accounting department, there were problems, as well. The cost of transmission of information was high and the accuracy of the information was not satisfactory. The transmission of the financial data among branch offices, shipping agents, and the headquarters relied on express delivery services and facsimile. Express delivery was not fast enough. The cost of express delivery and facsimile were both high. Additionally, the accounting and financial reports had to be done by hand. This was not quick or accurate enough. The data were sometimes found to lack the required uniformity. The people in the F/A department had to spend a lot of time correcting and reconfirming the information they were working with in producing their reports.

Overall, the lack of easy accessibility of financial data complicated financial planning and effective management of the company. The tardiness of information had prevented easy accessibility of financial data in the beginning of each month. This caused inconvenience in financial controlling and planning, particularly in such areas as budget control, working capital management, etc.

In order to solve these problems, Wan Hai began, in phases, to build a telecommunications and information technology system to link together the many facets of its operations. Wan Hai started the computerization of its information and telecommunication system in 1982. The information system department was formed in January 1982 with only four persons. These included a department head, two programmers, and one data controller. The number of personnel in the information systems department as of June 1989 was 18 persons, including the manager of the department. Wan Hai started with the first CPU in June 1982. It was an HP-3000/40 which was subsequently upgraded to an HP-3000/950 (see Exhibit 12.7).

The important first step in the process was setting up the Transaction Processing System (TPS). Most of the daily transactions and paperwork in these two departments were taken over by the computers, and the telecommunications infrastructure was essential in supporting the movement of documentation throughout the system. The documents that were computerized included the following:

- *Sales department*
  Shipping documents (B/L, S/O, D/N, etc); sales performance statistics and analysis; freight control; cargo control; client data; space booking system.
- *Finance/accounting department*
  The entire bookkeeping system; cash control system; invoice system; accounts receivable collection system; commissions processing system; freight correction system; budget control system; assets control system; branch offices; and shipping agents network.

The telecommunications supporting the Wan Hai system is a combination of local area networks, dial up personal computers operating as terminals, digital data networks, and satellite networks

Exhibit 12.7

Phases of Computerization at Wan Hai Steamship

HP-3000/40	Head Office, Taipei	June 1982
HP-3000/42 (upgrade)	Head Office, Taipei	May 1984
HP-3000/48 (upgrade)	Head office, Taipei	August 1985
Local PC System set up	Kaohsiung Branch Office	December 1985
HP-3000/42	Dawin Inc., Hong Kong Agent	March 1986
HP-3000/37	Kaohsiung Branch Office	September 1986
HP-3000/42	Osaka Branch Office	January 1987
HP-3000/70 (upgrade)	Head Office, Taipei	June 1987
Local PC System set up	Swire, Agent in Hong Kong	September 1987
Local PC System set up	Taichiung Branch and Agent	November 1987
HP-3000/950 (upgrade)	Head Office, Taipei	February 1989

all supporting the linking together of the different processors at each of the major Wan Hai locations. In Japan, the Nippon Telephone and Telegraph company provides the domestic linkages, including dial up and dedicated leased circuits. Kokusai Denshin Denwa, the international carrier of Japan, transports the information to Taiwan using the Venus-P packet switched network. At Taiwan, the information switches to the Pacnet packet switched service network. At the headquarters, a high-speed local area network serves both the various departments and the communications controllers linked to the overseas locations. Exhibit 12.8 (on pages 310-311) presents a detailed diagram of the system.

The benefits to Wan Hai from adopting this telecommunications and information technology strategy have been great. The transaction processing time has been reduced for handling the large amount of paperwork. The computer processing of daily transactions reduced the operational time of most of the documents in the two departments. Within the sales department, for example, shipping documents can be finished in 3 to 4 days less than was required when the process was done manually. At peak performance, it takes only 1 or 2 days to finish the documents, with a substantial number being processed the same day. Also, the sales performance analysis can be generated 1 to 5 days sooner than by the manual operations of the past. In the finance/

accounting department, all booking work and invoice processing can be done 2 to 5 days earlier than by manual methods. Financial statements can be finished 5 to 8 days more quickly than by manual operations.

The sales force has benefited substantially by the shorter processing times. It is now able to get the first-hand information it needs at the earliest moment. The sales staff now can review the sales performance and adjust their strategies and tactics as soon as possible. The ability to do this, it is argued, gives Wan Hai a competitive edge in the market.

In addition, the error ratio for documentation was trimmed as a result of the telecommunications and information technology system. Within the sales department, shipping documents errors were reduced 25 to 45 percent. In the finance/accounting department, errors in invoice processing were reduced 30 to 35 percent, errors in taxation documents were reduced 10 to 15 percent, and errors made on financial statements were reduced 5 to 10 percent. The result of this was the reduction of the overall document error ratio to less than 1 percent throughout the company! This lower ratio of error certainly is another important factor improving the attitudes of clients. The salespersons can spend less time in dealing with mistakes, and this extra time can be spent within the company, with their clients, or in getting more business and better sales performance. As a result, the morale of the sales force has been enhanced.

The lower error ratio has meant a lower overall cost in processing documents. The savings have been substantial. The system has reduced the cost within the finance/accounting and sales departments alone by an average of NT$23.6 million per year. Over a 7-year period, the cost of the system was approximately NT$77,268,000.

Net cost savings of NT$165,192,000 was realized in the sales and the finance/accounting departments over 7 years: the total savings of NT$242,460,000 minus the cost of the computer system of NT$77,268,000 (see Exhibit 12.9). This NT$165.2 million savings amounts to NT$23.6 million per year. Furthermore, this is the benefit only in the sales and the finance/accounting departments (including the same units at all the branch offices).

Specifically, Wan Hai has processed all of the shipping documents by its own computer systems. This saved approximately NT$1,500,000

Exhibit 12.8

Wan Hai Steamship Company Telecommunications System

The Wan Hai telecommunications network has won several national awards for its sophistication and efficiency.

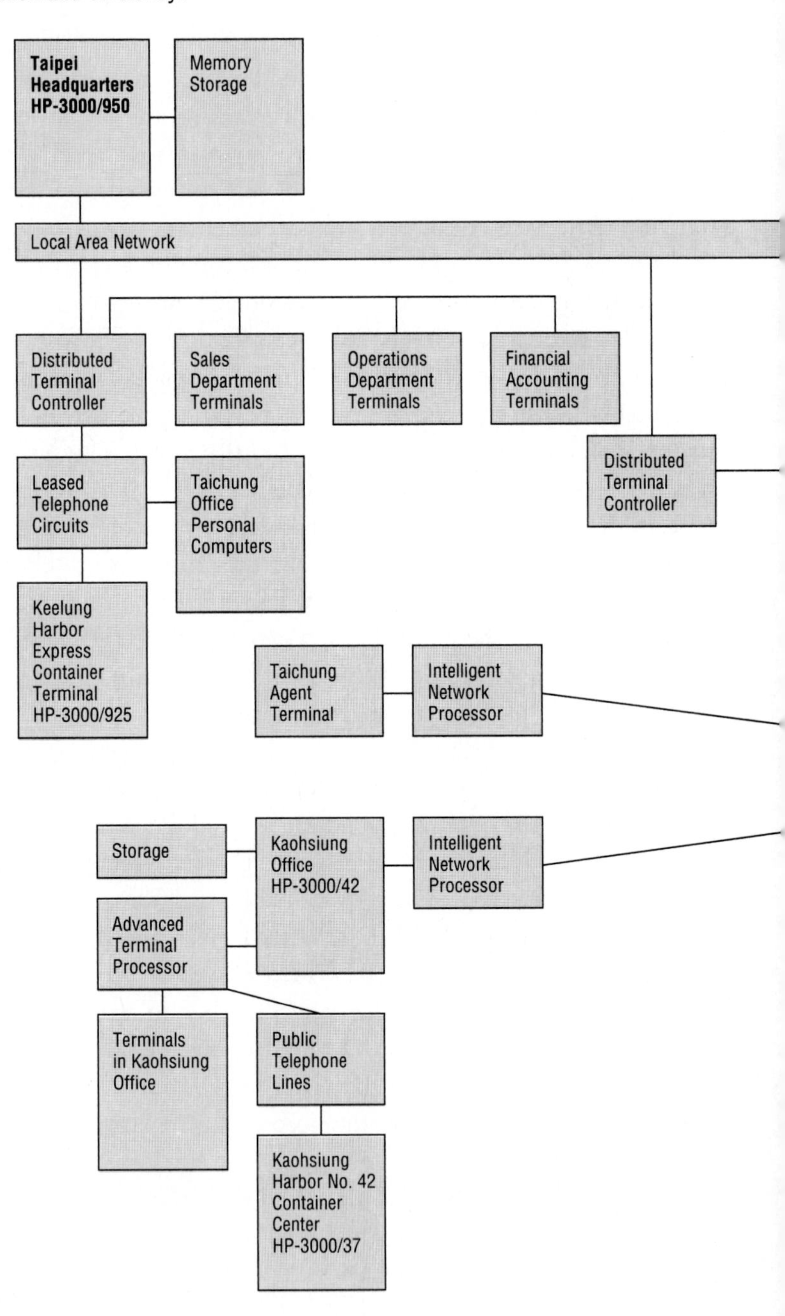

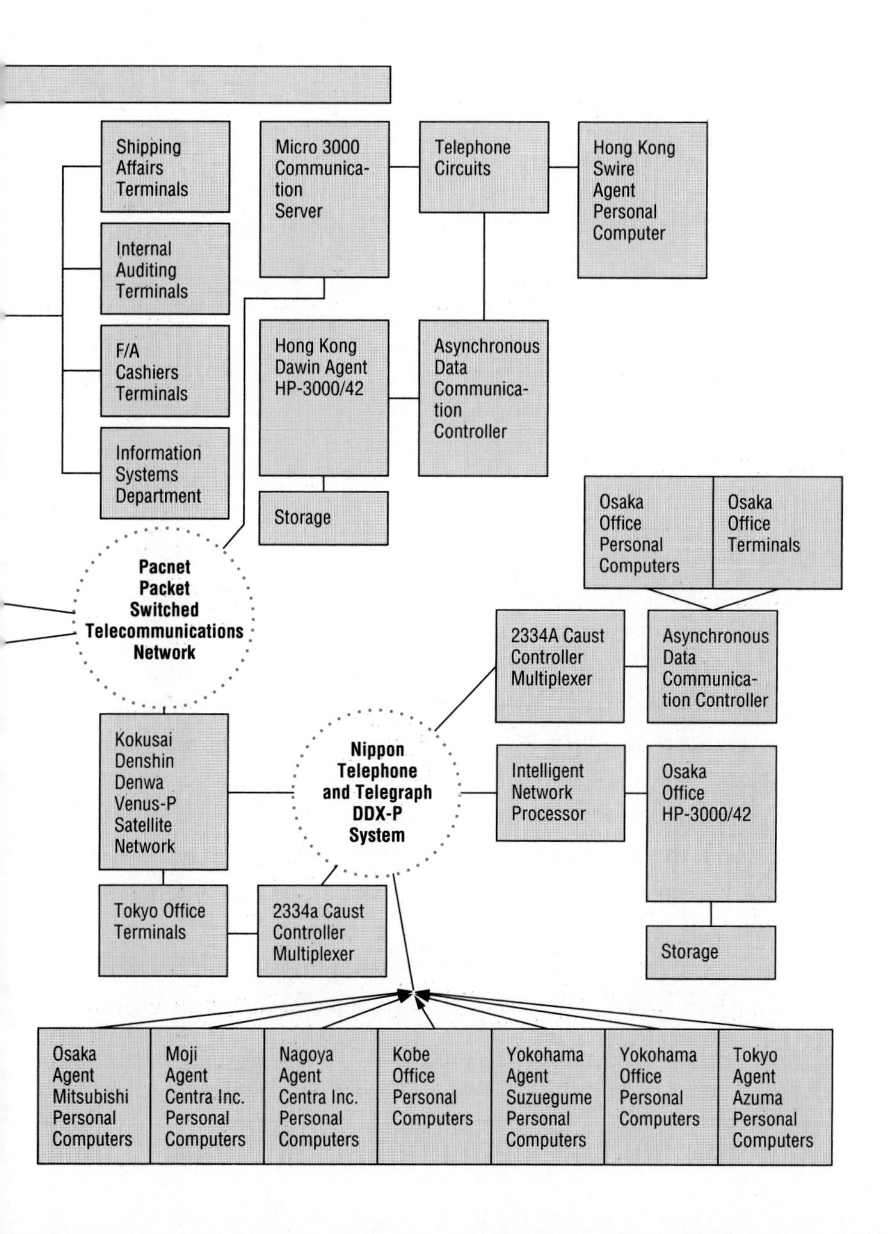
Shipping Affairs Terminals
Internal Auditing Terminals
F/A Cashiers Terminals
Information Systems Department
Micro 3000 Communication Server
Telephone Circuits
Hong Kong Swire Agent Personal Computer
Hong Kong Dawin Agent HP-3000/42
Asynchronous Data Communication Controller
Storage
Osaka Office Personal Computers
Osaka Office Terminals
Pacnet Packet Switched Telecommunications Network
2334A Caust Controller Multiplexer
Asynchronous Data Communication Controller
Kokusai Denshin Denwa Venus-P Satellite Network
Nippon Telephone and Telegraph DDX-P System
Intelligent Network Processor
Osaka Office HP-3000/42
Tokyo Office Terminals
2334a Caust Controller Multiplexer
Storage
Osaka Agent Mitsubishi Personal Computers
Moji Agent Centra Inc. Personal Computers
Nagoya Agent Centra Inc. Personal Computers
Kobe Office Personal Computers
Yokohama Agent Suzuegume Personal Computers
Yokohama Office Personal Computers
Tokyo Agent Azuma Personal Computers

Exhibit 12.9

Cost of the Wan Hai System (in thousands of NT dollars)

	Average Cost/Year	Number of Years	Total Cost
Hardware		7	NT$40,000
Personnel	NT$4,224	7	29,568
Training	500	7	3,500
Maintenance	600	7	4,200
Total cost			NT$77,268

Savings from the Wan Hai System (in thousands of NT dollars)

	Average Savings/Month	Number of Years	Total Savings
Document processing	NT$1,100	6	NT$ 79,200
Document transmission	45	5	3,240
Salary savings	1,890	7	158,760
Error correction savings	15	7	1,260
Total savings			NT$242,460

per month in 1989. Document transmission costs of NT$455,000 per month were incurred with the manual system. Without the information system, it would cost NT$520,000 per month in 1989. This is a 12-percent or NT$65,000 saving per month. The salary savings are estimated based on the average number of employees saved at the sales, finance/accounting, and counterpart units at the branch offices over 7 years. The sales and workload increased 600 percent during those 7 years. If the telecommunications and information system had not been used, the old system would have required 210 extra employees. Additional cost benefits were realized in all other departments, especially the ship affairs and operation departments.

Wan Hai clients are much happier than before. Complaints from clients are significantly reduced. There are other benefits as well:

- A better image is presented to the clients. As the company provides more efficient service to clients, it enhances its image. The wholly computer-processed documents and invoices help with making a good impression. The firm has experienced a much lower rate of complaints by clients.

- Upper-level managers can now utilize more and quicker information to help them make decisions.
- The data on clients enables sales to better prepare before they visit their clients.
- The operation data such as the cargo lifting statistics, shipping operations, and so on are very useful for the sales staff in making decisions on sales tactics.
- A faster operation has greatly reduced the cost of interest for Wan Hai's clients, as well as its own cost.
- All other cost savings and benefits at other departments also add up to a very significant amount.

Publishing: Xyvision's Distributed Printing Architecture Meets Financial Deadlines

Xyvision, a medium-sized company on the outskirts of Boston, produces publishing systems. Its systems and telecommunications linkages are a good example of advanced workstations being linked together through telecommunications systems. The Xyvision systems are aimed at the professional publishing operation. They perform complex layout, formatting, and composition functions, and are designed to send information directly into typesetting equipment. Xyvision faced a great deal of competition in the field of producing publishing systems, including old-style composition systems, desktop publishing with small laser printers, and well-established direct competitors such as Compu-Graphic and Linotype that had been in the business for a very long time and held very strong positions with the customer base. Xyvision was a startup, and even within the field of startups, Xyvision faced competition from companies such as Interleaf, which had built its strategy upon interfacing with IBM mainframes.

Xyvision developed an architecture for its equipment, shown in Exhibit 12.10, that would allow a completely transparent interface between the composition terminal and the storage device where the document information was held. If the file was located on the storage device for the local workstation, then the composition terminal had

Exhibit 12.10

Xyvision Workstation-Based Networks

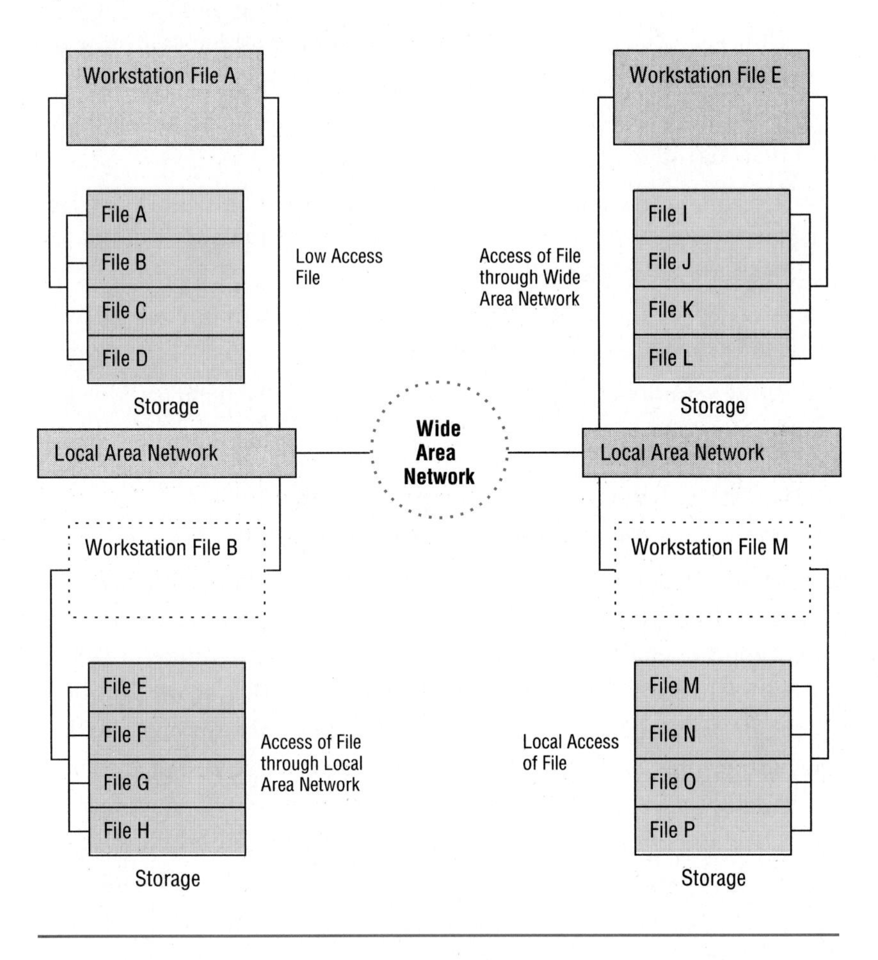

direct access. However, the file might be located on a nearby machine connected through a local area network. Under those circumstances, the workstation would transparently access the file through the local area network. To the user, this transparent access would appear as though the file were located locally. There would be no practical difference in the performance of the machine or in what the user had to do to get access to the file. If the editing terminal was in a different location from the storage device, then a telecommunications linkage could be used to access the remote file.

What distinguished the Xyvision approach was the ability to connect workstations and storage devices through both local area and wide area networks. This implied that a user located in one city could immediately access a file located on a storage device in another city. The telecommunications link was transparent.

Xyvision successfully introduced this type of system to financial printers who work under very tight deadlines, particularly when a new issue is involved. Using the Xyvision technology, the financial printers are able to edit documents and print them in several locations simultaneously. They can make changes to a file remotely, for example, if teams working on an issue are located in two different cities. This ability to produce and edit documents in a distributed fashion has served Xyvision well. It has been, however, under other financial and commercial pressures from its competitors for a variety of reasons. There is a good chance that the particular nature of the Xyvision system will eventually be absorbed by stronger players in the market. This is often the fate of innovators.

Telecommunications at Nippon Life Insurance Company

In Japan in the mid 1980s, the life insurance sector began work on an extensive nationwide telecommunications network to support information processing. The overall trend of deregulation and internationalization in the financial arena in Japan brought dramatic changes to the financial sector in the early 1980s. The life insurance industry, as an important part of the financial sector, began to compete with domestic banks, securities companies, and property insurance companies, in addition to heated competition against foreign corporations. As a response to this changing environment, the Japanese life insurance industry started to build what it called "The Third Generation On-Line System," a general, nationwide telecommunications network aimed at developing productivity savings, improving customer services, and getting more sales performance in a fiercely competitive market.

One of the largest players in the Japanese life insurance industry is Nippon Life. Its response to the new competition was to build the Nippon Life System 100 at a cost of 70 billion yen ($483 million), from 1985 to 1988.

Nippon Life Insurance is the largest life insurance company in the world. Its assets amount to 18 trillion yen, equivalent to $120 billion at an exchange rate of 150 yen per dollar. It had issued total contracts (policies) worth 4.47 trillion yen ($29.8 billion). It has 1,900 satellite offices and 121 branch offices under 2 headquarters, employing more than 92,000 people.

The problem faced by Nippon Life involved the need to cut the cost of routine clerical work involved with processing new contracts and handling the regular payments of premiums. This system of bureaucracy involved all three levels of Nippon Life's organization. The organization of most life insurance companies is divided into three levels: the headquarters, which has the major information processing computer center, the branch offices, and the satellite office. This forms a pyramid, with many more satellite offices than either branch offices or headquarters offices. Routine clerical work originating from ongoing contracts was related primarily to either issuance of new policies or regular payment of premiums.

New contracts at Nippon Life were initially input on computers at the branch offices, while regular payments were input at the headquarters. Exhibit 12.11 diagrams the process. These actions took place almost exclusively on a monthly basis. However, in terms of customer service, it took Nippon Life approximately 1 month to complete the process of registering a new contract, and approximately three weeks to calculate premium payments each month. The clerks at branch offices were spending most of their time inputting data into the computer, with little time available for other activities, such as customer service. The salespersons spent time selling life insurance, but without the advantage of clerical workers' support for selling activities. Nippon Life was clogged up with too much clerical work.

In the Japanese life insurance industry, a firm's greatest advantage comes from having a large sales force. There are over 400,000 persons involved in life insurance sales. As the ratio of salespersons among total employees increases, so does sales performance, and statistical

Exhibit 12.11

Nippon Life Information Network before System 100

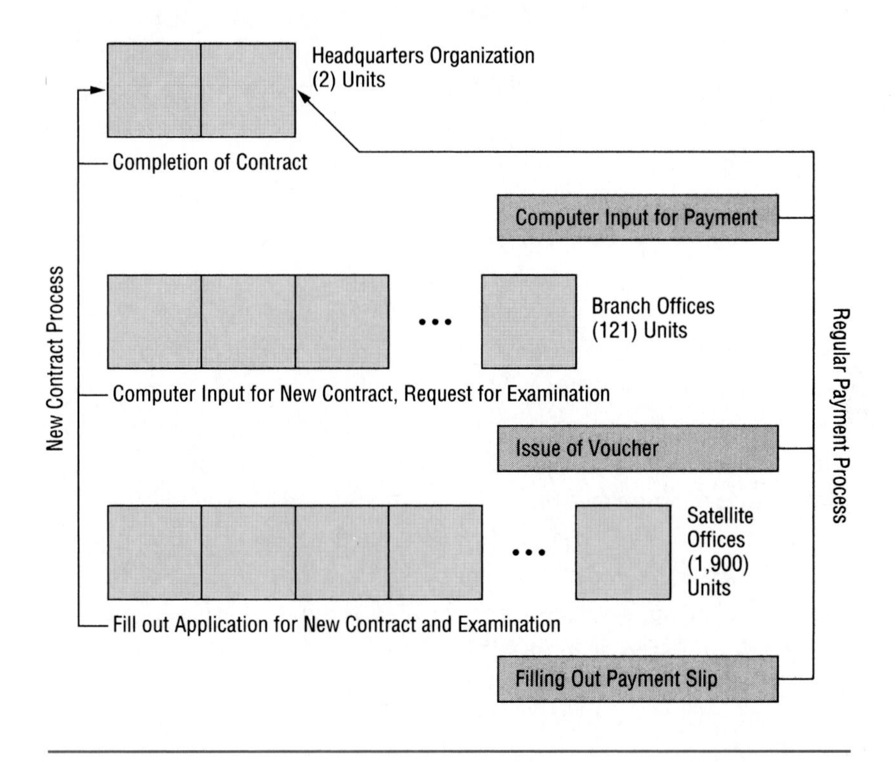

analysis has shown a direct correlation between the numbers of new contracts issued and the number of salespersons placed in the field.

Each organization aimed at taking advantage of this correlation, trying to decrease the amount of clerical work and strengthen the marketing and sales functions. Nippon Life Insurance Company estimated that it should be able to process the same work with 4,000 fewer clerical workers and save 6 billion yen ($41 million) per year in the process.

Another problem was that customers were seeking detailed information regarding financial investments, but typical life insurance salespersons were unable to offer the information. The Japanese are great savers, and in the mid 1980s disposable income was increasing. Since November of 1986, the bank rate for savings accounts stood at only 3.5 percent. The average Japanese therefore lost interest in typical savings accounts being offered by commercial banks. This shifted demand toward higher-risk investments, such as equities. In this type of environment, offering clear financial information on investment strategies became crucial to salespersons for the life insurance industry.

Unfortunately, this required a shakeup in the way life insurance was sold in Japan. Life insurance was typically sold by young, inexperienced women who depended on maintaining close contact with the customers. The traditional salesperson was not accustomed to effectively showing the advantages of each financial product for the consumer in comparison with the financial products being offered by banks, securities firms, and other financial institutions. It was realized that even if the information became available, the salespeople were not trained to use it effectively. At the same time, in order to get competitive advantage, it became urgent to give better information to each salesperson to support more convincing explanations of why customers should invest in Nippon Life contracts.

In order to solve these problems, Nippon Life introduced System 100, which was designed to expand its telecommunications network on a very large scale. For System 100, Nippon Life spent 70 billion yen ($484 million) including the cost of hardware, software development (including 8.2 million programming steps), and training.

Nippon Life Insurance Company has two computer centers, located in Tokyo and Osaka, with 121 branch offices and 1,900 satellite offices located throughout Japan. The headquarters computer center houses three IBM 3090 Model 400s and one 3084 mainframe. Each of the branch offices, with approximately 10,000 employees, has an IBM System 36GP and 716 megabytes of disk storage. At satellite offices, which together employ some 80,000 salespersons, Nippon Life has 1,700 System 36SXs with 120 megabytes of storage each. Approximately 5,000 IBM 5540 16-bit personal computers (with displays in the Japanese language) were added to the existing base of 2,000 personal computers, and 1,700 optical character readers controlled by Hitachi 2020 workstations were located at both branch and satellite offices.

The telecommunications infrastructure supporting this is equally large. The offices in seven major cities—Tokyo, Osaka, Sapporo, Sendai, Nagoya, Okayama, and Fukuoka—are connected by high-speed digital lines; branch offices are linked by 9,600 baud leased lines, while satellite offices use the NTT DDX-P packet switching services. Exhibit 12.12 illustrates the system.

The Nippon Life Insurance system links distributed processing and optical character recognition equipment through the telecommunications system so as to reduce the amount of clerical work needed to conduct business. The use of telecommunications for distributed processing allows processing of information at remote sites without losing contact with the advantages of centralization. Whereas in the past the vast majority of clerical work was done by clerks at the regional sites, who supported the large number of salespeople, after the system was introduced, telecommunications allowed the salespersons to input information regarding new contracts, or premium or coupon payment information. The IBM 36SXs and the distributed database allowed each satellite office to share, or take over completely, the clerical work. The new contracts and the records of premium payments are now input by salespersons and sent directly to headquarters through NTT's DDX-P packet switching service.

The optical character recognition equipment is used to read the handwriting and convert it into direct data entry for the information system. The Hitachi HT4171 interprets the handwriting of salespersons and automatically inputs the data into the IBM System 36. This process is completed without a single keystroke. The effect of this is that even without any specific computer training, every salesperson can input data from the very first day of starting the job. This reduced need for training is very important because the turnover among the salespersons is very high. Every year, life insurance companies hire 15,000 salespersons in total, while 12,000 to 13,000 resign.

Once the telecommunications system was in place, it allowed delivery of up-to-date information through user-friendly software to customers who were making financial decisions. In addition, since the central databases were now available to the satellite offices, the customer information system was able to deliver overall information on each individual and any previous contracts. This allowed the salesperson to approach a customer with a whole history, instead of

Exhibit 12.12

Nippon Life System 100 Telecommunications Linkages

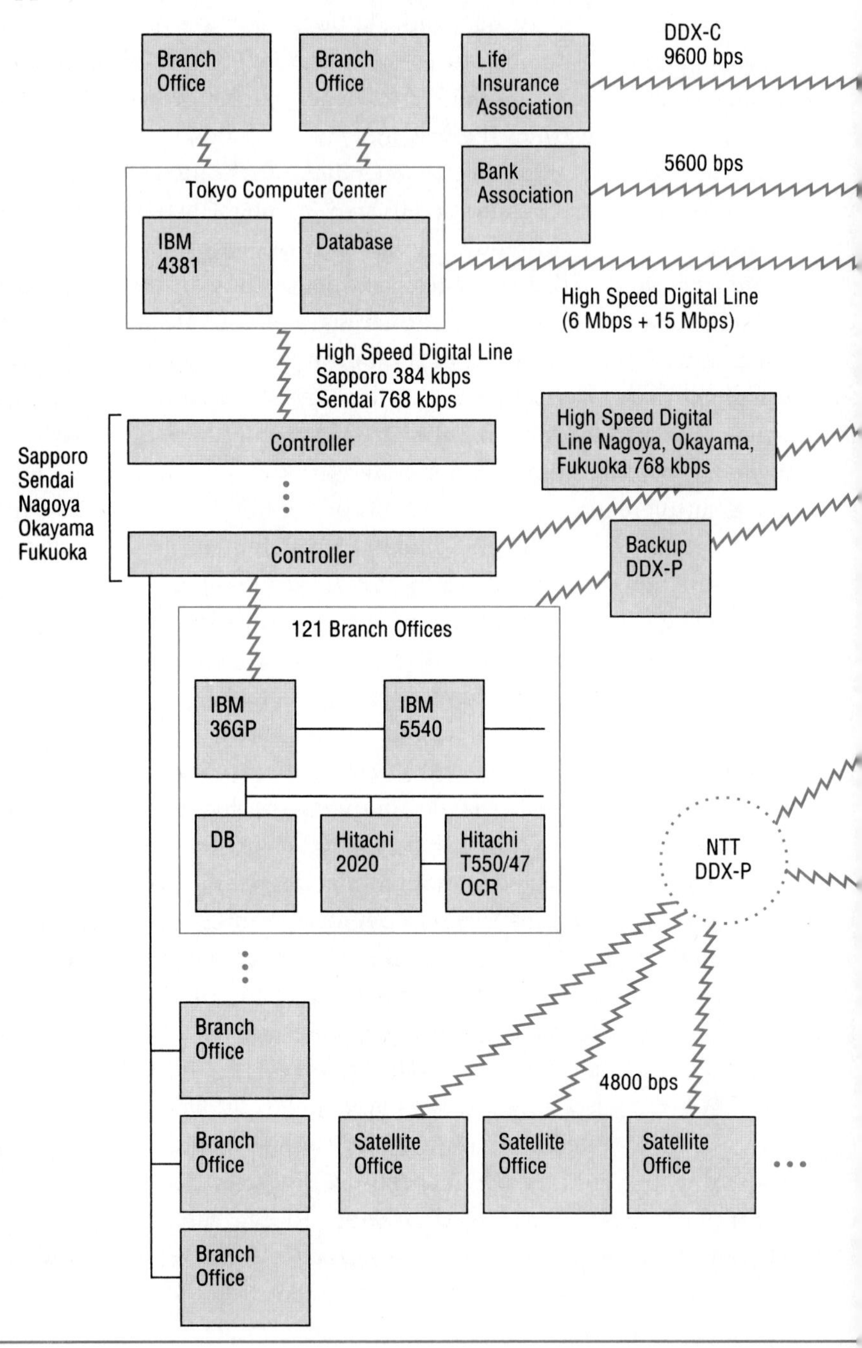

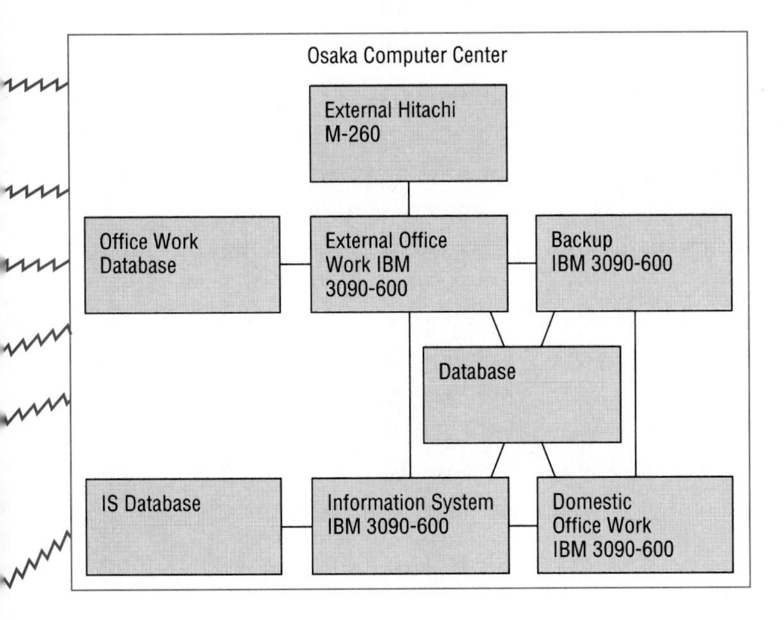
Osaka Computer Center
External Hitachi
M-260
Office Work
Database
External Office
Work IBM
3090-600
Backup
IBM 3090-600
Database
IS Database
Information System
IBM 3090-600
Domestic
Office Work
IBM 3090-600

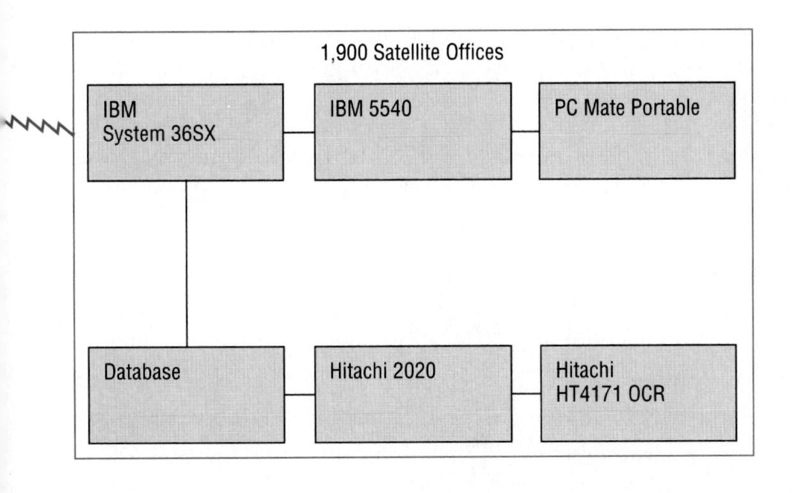
1,900 Satellite Offices
IBM
System 36SX
IBM 5540
PC Mate Portable
Database
Hitachi 2020
Hitachi
HT4171 OCR

operating on a contract-by-contract basis. For example, since life insurance policies generally cover long-term time frames, some customers must think of purchasing other types of insurance to replace old policies. In the process, the customer might wish to compare a new policy and an old policy. In that situation, the salesperson can tell how much the customer can get as a return for cashing in old contracts. This allows a comparison with the total amount of money the new contract would provide. As a result, customers can easily get information regarding their overall coverage. In addition, this allows the salespersons to approach the customers systematically.

This type of customer information system is also used by major U.S. insurance companies, as seen with John Hancock Mutual Life Insurance Company's "Customer Oriented System" and CIGNA's "Marketing Office Support System."

In addition to general information, the system aids the salespeople by providing on-line calculations for the following:

- The amount of money needed for each year, calculated on the basis of the life cycle of each member of the family
- The inheritance tax (This is extremely important in Japan, where inheritance tax that many times takes 50 to 70 percent can cause many families to lose their property.)
- The allocation of all funds and expenditures in the household budget
- Other miscellaneous sales gimmicks, including horoscopes, analysis of compatibility of spouses, or readings of implications from blood types.

The general trend of increased emphasis on telecommunications and information systems has become evident in Japan. For the past 100 years, the Japanese life insurance industry was protected by the Japanese government from all foreign competition. Also, it was forbidden to differentiate various products among insurance companies. The result of this system was the development of a direct correlation between the number of salespersons and the market share of each insurance company. With deregulation, firms have realized that better information infrastructures would free up more clerical people so as to give better sales performance. In addition, new, more differentiated products could be created. Besides Nippon Life, other insurance companies have followed the same trend. For example, Daiichi Life Insurance upgraded its three IBM 3090s to Model 400s from Model 200s. Also, Meiji Insurance replaced 41 communication controllers with more powerful models and continued to add more so as to double the capacity of all of its telecommunications links.

The benefits for Nippon Life Insurance Company from its telecommunications-based strategy were clear. Clerical work, especially at the level of the branch offices, was decreased. This allowed emphasis of tasks of the employees to change from clerical information processing to strategic marketing. By late 1988, Nippon Life Insurance Company had already succeeded in decreasing the number of clerks by 2,400. The firm projected savings from decreasing clerks by 4,000 in total of 6 billion yen ($41 million) per year. Since the relevant data are input by the salesperson at customer visits, there is no time lag between everyday sales activity and the central database.

The real estate market, for example, creates difficulties because the price of land is fluctuating and increasing very rapidly in the Tokyo area. By accessing the database of satellite offices, branch offices, and headquarters, it is possible to greatly improve the marketing strategy. Japanese need to buy more expensive life insurance policies in order to cope with losses due to the inheritance tax. Since the salespersons are inputting data on a day-to-day basis, it is possible to track which geographical areas need extended coverage to correspond with rapidly escalating land prices. This makes it much easier for companies such as Nippon Life to adjust to dynamic financial changes.

Another benefit centers around providing qualified information to the customer for investment decision making. The Nippon Life salespersons were given Fujitsu FC Mate portable computers to take to the customers' houses during sales calls. This enabled the salespeople on the spot to get exact information regarding insurance, including premiums, coupon rate, annuity, interest rates, etc. This also enabled them to deliver other types of value-added information to the customer directly, such as household budget and horoscope information.

By the end of the decade of the 1980s, Nippon Life was attempting to move beyond the technique of personal calls by salespeople to customers' houses. They were studying the use of Nintendo game

Exhibit 12.13

Key Developments in the Japanese Life Insurance Industry in 1985 and 1986

January 10, 1985

Life Insurance Association officially announced a Life Insurance Industry Value Added Network with installation to begin in April of the same year.

January 11, 1985

Meiji Life Insurance Company completed an on-line system for overall accounting functions. It was the first use of this type of system in a life insurance company in Japan.

February 15, 1985

Daiichi Life Insurance Company announced that it would complete its "firmwide banking system," which would connect Daiichi to its commercial banks (city banks and mutual banks) as well as to its remote offices.

March 12, 1985

Nippon Life Insurance started work on Nippon Life System 100.

April 19, 1985

Meiji Life Insurance and Mitsubishi Bank agreed to issue cash cards to allow customers to deposit money.

May 8, 1985

Sumitomo Life Insurance started inquiry services by using the CAPTAIN videotext system available to consumers through a telephone line link to their television screens.

May 24, 1985

Daiichi Life Insurance completed a nationwide facsimile system, the very first among life insurance companies in Japan. The facsimile system allows very rapid flow of information throughout the company.

July 5, 1985

Mitsui Life Insurance expanded its network to include 64 local banks and all mutual banks. (Mitsui Bank was already connected to Mitsui Life.)

July 24, 1985

Sumitomo Life Insurance announced a plan to introduce videotext services at its satellite offices to be used by customers in making inquiries and getting information.

---

computers to give customers the ability to purchase stocks and get access to databases. Nippon Life Insurance and Nomura Securities were two of the several companies participating in the Study Group on Family Computer Networks. Exhibit 12.13 gives a chronology of significant developments in the life insurance industry's use of telecommunications.

August 12, 1985

Nippon Life Insurance Company started locating insurance centers at major cities in order to give insurance information, as well as financial information, to customers located throughout the country.

October 3, 1985

Nippon Life Insurance became affiliated with the JCB Card Company and the firms announced plans to establish a company together. The JCB Card was one of the most popular charge and credit cards in Japan.

February 18, 1986

Yasuda Life Insurance Company started to process application forms for new policies using optical character recognition technology, which links institutional customers and headquarters through telecommunications linkages.

April 8, 1986

Meiji Life Insurance Company completed a nationwide on-line system, designed by Toshiba. This new system used an IBM CPU and Fujitsu equipment in conjunction with distributed processing computers from Toshiba.

May 26, 1986

Meiji Life Insurance Company issued cards by affiliating with Life Inc., a card company, thus giving its customers access to Life's automated teller machines.

August 1, 1986

Nippon Life Insurance company started using an artificial intelligence system to process health examinations for new policies.

August 10, 1986

Variable life insurance became available in Japan.

Network Bridges for the Apple Laser Writer

When laser printers for the Apple Macintosh computer first came onto the market, they were relatively expensive and slow. Although the Macintosh was sold with the capability of working in a small local area network called AppleTalk™, the network had space and number

Exhibit 12.14

Hayes Bridges for AppleTalk™ Networks

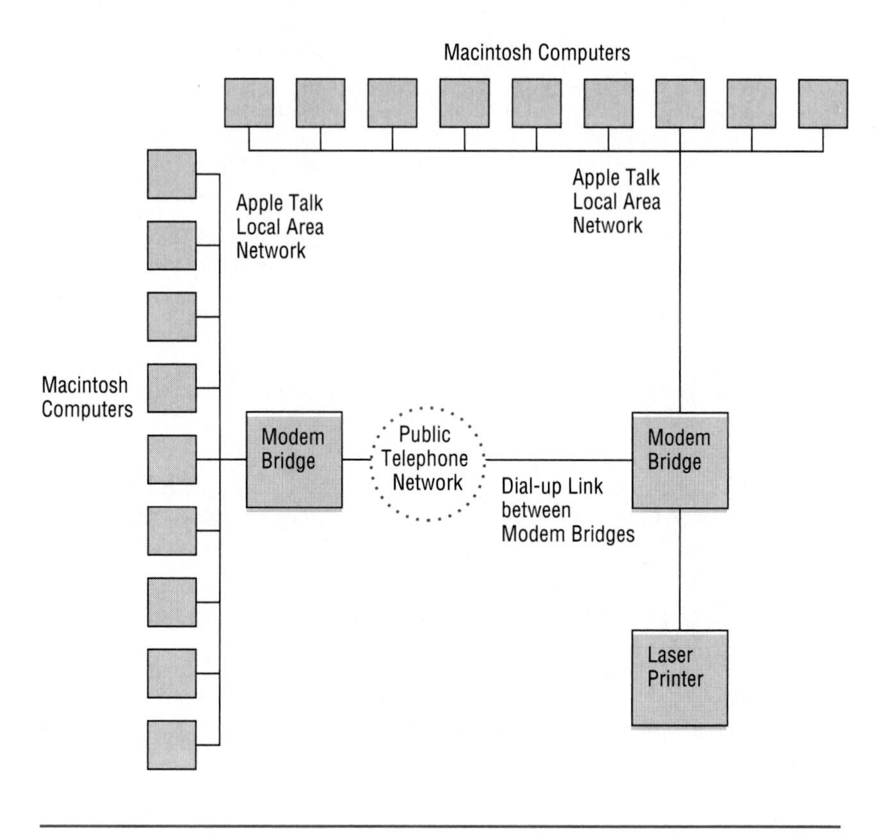

limitations. Only about 32 users could simultaneously work on the network without significant deterioration. In addition, the network could only be extended practically within a single office environment. At the same time, with the high cost of laser printers, an opportunity opened up for a company to create an extension of the AppleTalk network.

Hayes was one of the companies that pursued this opportunity. Using the Hayes Bridge system, illustrated in Exhibit 12.14, users were able to extend the reach of their AppleTalk networks to get access to Apple laser printers. This technology filled a temporary gap in the market for remote access to laser printers from other local area networks. However, things change. As the price of laser printers decreased by approximately 90 percent over a 5-year period, it gradually became practically as cheap to buy another laser printer, or an alternative technology such as an ink jet printer, as to install the telecommunications bridge. As the economics changed, the value of the telecommunications bridge faded, and it became a less and less viable option.

The history of telecommunications applications is full of such stories of technological fixes coming and going with the shifting tides of technological and economic change.

Summary

An examination of the 24-hour market in Morgan Guaranty's foreign exchange business demonstrates how price changes in one part of the world can have a great effect on other parts of the world. Telecommunications provides the essential connection that allows subsecond response times for responses to changing market trends, which is critical for competitive advantage. Wan Hai Shipping used telecommunications to build an information system for complex management of shipping. Xyvision showed how complex intellectual tasks can be accomplished by teams working in separate locations. Nippon Life showed that even in very highly complex matters such as life insurance, telecommunications plays an essential infrastructural role in operating the business.

Questions

12.1 How does telecommunications allow a business to respond better to changing conditions in the economy?

12.2 How can telecommunications help an organization provide complex services across international boundaries?

12.3 What is the role of telecommunications in global manufacturing operations?

12.4 What is the role of telecommunications in providing individual services to a mass market?

Chapter 13

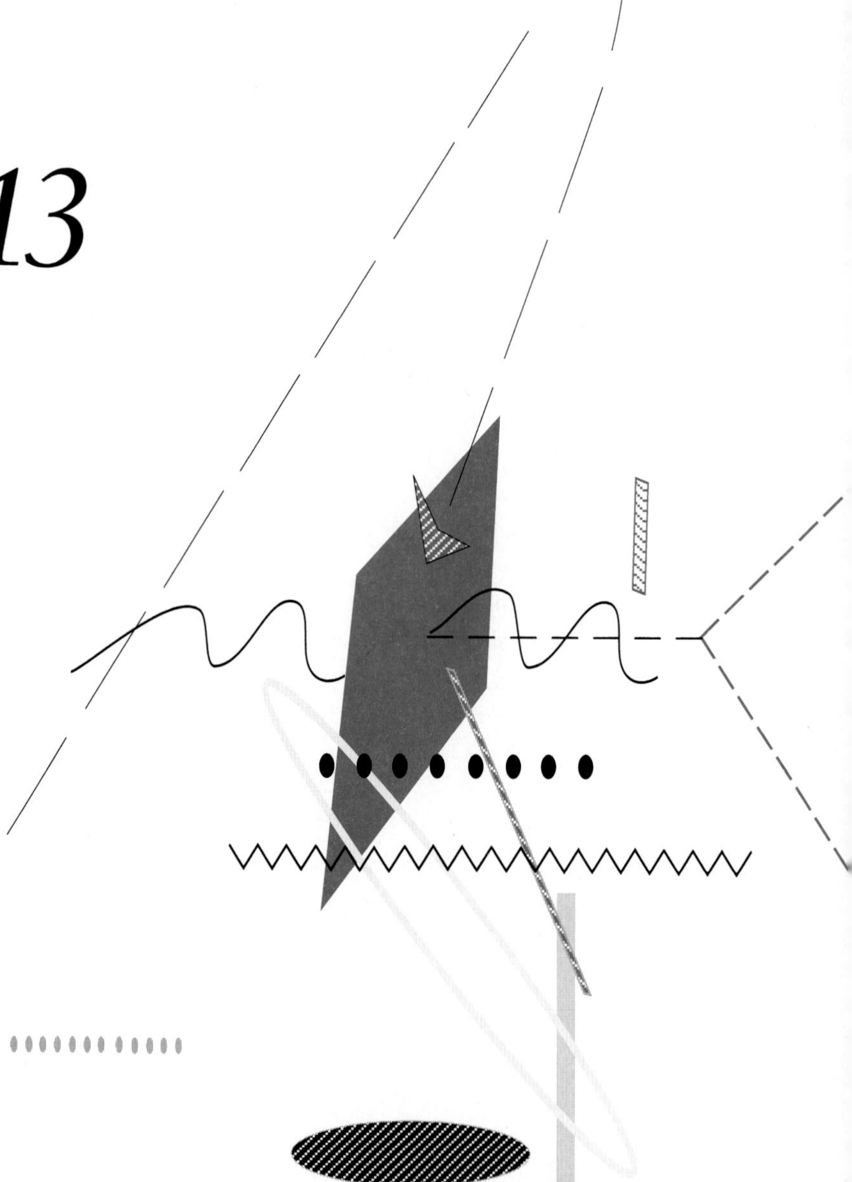

Slash Inventory Costs

Telecommunications can do much to facilitate the control and movement of inventory. Using telecommunications, firms can learn their exact inventory exposures much faster than in the past. When the cash register becomes a data terminal, the moment an item is sold, it is recorded in the store's minicomputer, then the information is quickly transmitted to headquarters over a telecommunications line. The firm can quickly identify items that are selling rapidly or slowly. This important information helps greatly in planning future purchases. In addition, it is possible to measure the sales response to a promotion or advertising campaign accurately soon after it is launched. In some cases, this information is sold to others at a profit.

This chapter goes into more detail regarding how telecommunications is used to control inventory and increase the profits of the firm. Large, integrated computer networks with centralized inventory functions and satellite networks that report the status of sales seven or eight times a day are all based on advanced telecommunications. These systems guarantee adequate safety stock levels and faster inventory turnover, and thus more profits and better shareholder value.

Sears SNA Retail Network

Although Sears is a conglomerate involving financial services, insurance, real estate, and retailing, the large retail department store operations have been its mainstay throughout its history. It has branches in almost every metropolitan area in the United States and Canada and also runs international operations in various countries. Sears works in many countries of the world to purchase the items it needs to fill its stores. In the late 1980s, however, Sears was doing the bulk of its shopping in East Asia, bringing into the United States competitively priced imported goods that it could then label with its own brand name and mark up handsomely.

Being a very large department store operation, Sears has always faced a horrendous problem of inventory control. For a store chain as large as Sears, small movements in inventory, or worse, lack of such movement, can rapidly add up to millions of dollars of unusable inventory. In addition, its wide product line includes everything from garden hoses to panty hose, so Sears had to maintain an inventory system capable of handling very fast-moving and seasonal items as well as slow-moving items. In order to do this, Sears needed to know quickly and efficiently how well different items were moving at its many different locations.

Its geographic range complicated matters, as well. While it was still selling snow plows in Wisconsin, it might be selling sun tan oil and beach chairs in southern Texas. It faced several problems:

- How to keep track of rapidly changing conditions in its inventory
- How to sense trends in sales, and to identify items that were not moving
- How to estimate sales and make forecasts based on accurate sales data
- How to keep track of inventory in warehouses and distribution centers

The solution Sears adopted turned out to be one of the largest Systems Network Architecture (SNA) networks in the world. The network and information control system envisioned by Sears would basically link up all sales terminals and inventory control points throughout all of its locations. The very rapid reporting system made the information on sales of each item from each store location available through the network every 24 hours. This enabled Sears to know much more accurately where it stood in terms of its inventory of goods.

As Sears began to implement this system, each of its cash registers was converted to a point-of-sale inventory terminal. They recorded information regarding the sales price of the items and the total cash volume sales, along with other information, daily. In the old system, the store had to take inventory to gather this information; the implementation of point-of-sale inventory control terminals automatically provided it at the end of each business day. This information was available at the individual store level. However, it was equally, if not more, important that the information was transmitted to headquarters on a daily basis. Just as the individual store was able to learn its position every day, the headquarters as a whole also was able to know its position every day. This aided greatly in planning and strategic decision making for the store.

Caterpillar Logistics Systems

Caterpillar is one of the largest and most famous companies in the world. It is headquartered in Illinois and is a major producer of very large diesel engines, earth moving and agricultural equipment, and other heavy equipment. Its bright yellow tractors and heavy earth moving equipment are recognized throughout the world at major construction sites. One of the oldest and most noble of American corporations, it had its start during some of the early agricultural developments in California, where very large harvesting machines were developed with special tracks that would not bog down in the soft soil. Later on, as the world was entering the First World War, Caterpillar invented the type of track-wheel that is commonly used on military tanks. It produced a great deal of equipment for use on the battlefield.

Some might have the impression that since Caterpillar makes large but essentially simple equipment, the type of operation it runs might be unsophisticated. Nothing could be further from the truth. The machines Caterpillar makes are large and bulky, but they typically represent a very large investment on the part of the customer. Some front-loading machines cost $750,000 to $1 million, but this is typically not a large cost in comparison to the work they do. In a coal mining operation, for example, a large Cat may be responsible for moving a third or more of the entire output of the mine every day. Therefore, should a problem take the machine out of service temporarily, the output of the mining operation can be seriously affected.

Under those circumstances, the amount of money required to get a spare part is relatively inconsequential; the critical factor is how long it takes to get the Cat up and running again. In other operations where Caterpillar equipment is found—oil drilling operations, diesel power generators for small towns or hospital complexes, construction projects at remote locations in the developing world—speedy repair of a broken machine is usually critical.

Caterpillar has recognized this problem over the years and has developed a unique system of spare parts logistics to cope with it. Caterpillar promises to deliver a part anywhere in the world within 48 hours or the part is free. That guarantee depends on keeping spare parts available at short notice from a local or regional Caterpillar parts depot.

In order to accomplish this type of performance, Caterpillar uses a sophisticated telecommunications system that supports a completely centralized operation in Morton, Illinois. The telecommunications system links the regional depots into a seamless web of logistics control. Dealers are also linked into the Morton central computer complex by telecommunications lines, although many use dial up linkages.

It is only with the use of a large telecommunications system that Caterpillar can operate with its stunning efficiency. The linkages with its logistics depots enable Caterpillar to know the location of every spare part in its inventory within seconds. Should parts not be available in one depot, they can be quickly rerouted from other locations to where they are needed. As parts are sold from each of the remote locations, including from the various authorized dealerships, reports are made back to the centralized operation at Morton immediately. This fast reporting system based on secure telecommunications linkages throughout the system allows the headquarters to calculate safety stock levels accurately and reliably. The parts inventory of the company is known immediately, not with a month or so delay common in many industries.

Caterpillar is able to gain competitive advantage through the use of this telecommunications system in several ways:

- It can offer a guarantee for time-specific delivery of parts that cannot be matched by its competitors.
- Its feedback system allows it to calculate safety stock levels quickly for each individual location throughout the world.
- It can hold a smaller gross amount of inventory than its competitors.
- Its dealers are linked in electronically to the system for electronic ordering and billing.
- It can find the location of an individual part anywhere in the world within seconds.
- It can very accurately manage spare parts inventory and quickly notice any digressions of true stock levels from what is reported on the system.
- Its costs are lower per dollar of spare parts delivered.

Although its business is tough and very cyclical, Caterpillar has managed to hang onto its very strong position over the years. It has been challenged on many fronts, even by makers of cloned spare parts. It has, however, continued to constitute the supplier of last resort. It is a well-tested truism in the industry that if you can't get the part you need somewhere else, you can always buy it from the Caterpillar dealer.

Ozeki Sake's Trial Facsimile Ordering System

In 1985, the Nippon Telegraph and Telephone Corporation (NTT), the largest telecommunication supplier in Japan, tried to apply a value added network facsimile system to the sake industry in the Kansai district of western Japan. Kansai is an area of Japan where many sake companies are concentrated. At that time, the industry had been

Exhibit 13.1

Sake Distribution in Western Japan

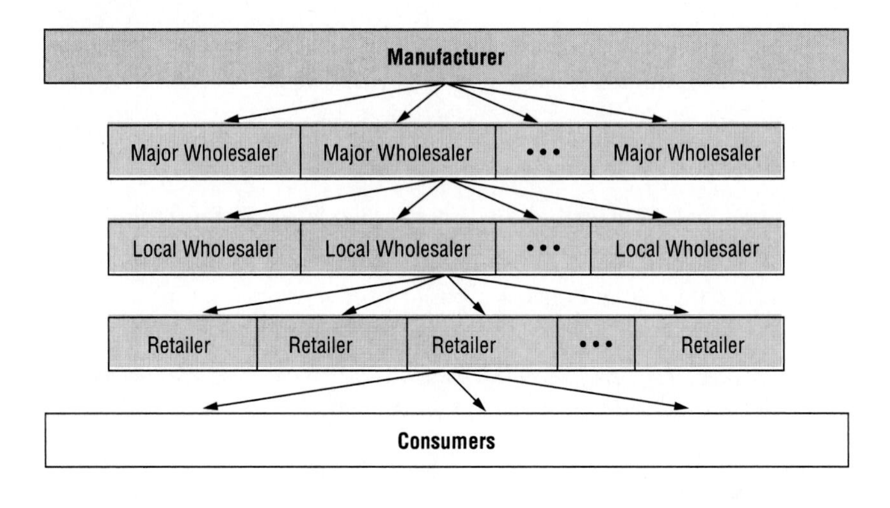

suffering from a decrease in sales. Primarily as a result of increasing consumption of whiskey, sake consumption had gone down by 17 percent in the previous 10 years. In addition, the cost of the rice used in production of sake had risen (sake is made from fermented rice) as had labor costs. As a result, the overall profitability of the industry had deteriorated.

In order to improve efficiency, the industry was attempting to reduce the cost of mass production and introduce more efficient inventory control. Another factor in the revamped strategy was the need to more quickly ascertain the changes in consumer tastes for sake in order to introduce more responsive marketing strategies. The strategy of NTT was to help the sake industry adopt a VAN system to improve its overall profitability.

The structure of Japanese sake distribution is composed of five levels: the manufacturer, the major wholesaler, the local wholesaler, the retailer, and the consumer. (See Exhibit 13.1.) The ordering and delivery of sake through this system were arranged primarily by telephone calls, which had replaced the use of walking messengers at the turn of the century. There is little doubt that, in a close society such as Japan, the people involved in ordering and delivering sake knew each other very well from their many telephone conversations over the years. For a large order from the manufacturer, the wholesaler called the manufacturer once or twice a month and delivery was made within a week. For smaller orders between the wholesaler and the retailer, calls were placed once or twice a week and delivery was accomplished within a few days.

Given this situation, Nippon Telephone and Telegraph attempted to become an intermediary between wholesalers, manufacturers, and retailers. For large orders between manufacturers and wholesalers, NTT proposed that data be automatically distributed through NTT's Dendenkosha Real-time Sales Management System, known as DRESS. For connection between the retailer and the wholesaler, NTT proposed to automatically distribute ordering data through a Facsimile Data Conversion and Interface Control (FDIC). The FDIC is a concentrator and conversion device that attaches to the network to convert handwritten information on a facsimile to digital information, which can then be transmitted as an order to a computer system. Exhibit 13.2 diagrams this system.

Using the FDIC, the retailer was able to fill out an order form by hand. Afterward, the form was faxed to the wholesaler. However, prior to delivering the facsimile message to the wholesaler, the FDIC unit would use optical character recognition (OCR) equipment to convert the handwritten information to digital data. The FDIC would then emulate a data entry terminal and pass on the order, now in the form of data, to the mainframe. To the mainframe, it would appear as though a person at the other end of the telecommunications linkage was entering the order data.

Under this arrangement, the wholesaler could save time and labor in taking the telephone orders. The orders for sake from the retailers would pass through the telecommunications system directly to the mainframe computer without human intervention.

NTT expected to be able to demonstrate several positive effects from utilization of its system. The automation of the ordering process would save time and ensure data accuracy. Orders entered from a facsimile machine or a retailer's host computer would pass through the

Exhibit 13.2

Facsimile-Based FDIC System

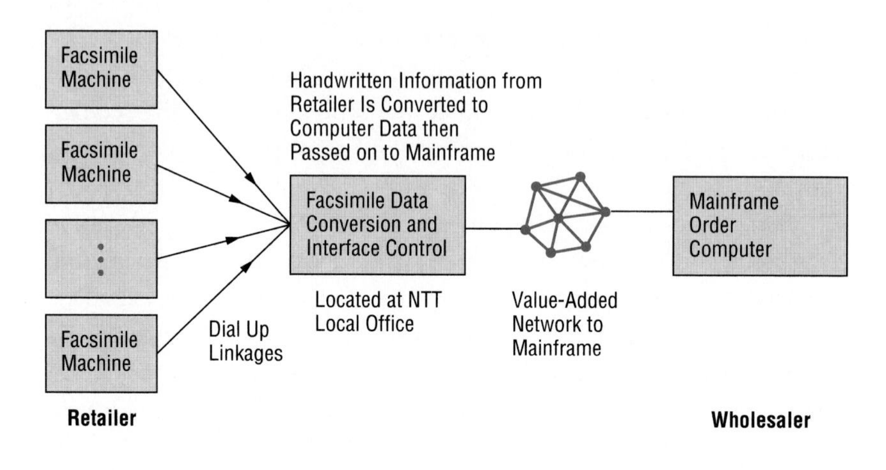

data network to be distributed to the related wholesalers automatically through the DRESS network.

Also, inventory cost would be reduced as the retailers could expect delivery within 24 hours of placing an order with the wholesalers. Furthermore, the order frequency from the wholesalers to the manufacturers could be increased, further reducing the amount of inventory kept on hand.

Finally, critical management reports would automatically be generated to help manufacturers plan production operations and back up the retailers. Information regarding product and also regarding sales to retailers would be vitally useful for manufacturers. (With the telephone-based system of the past, it was not easy to aggregate information regarding sales and product quality.)

In order to pilot this system, NTT reached agreement with five major manufacturers of sake in Japan: Ozeki, Konishi, Hakuturu, Kizakura, and Sawanoturu. In addition, agreement was reached with three wholesalers: Kokubu (one of the most famous), Nishuhan, and Koami. Twenty-five retailers agreed to take part in the system. (See Exhibit 13.3.)

Exhibit 13.3

Value Added Network for Sake Distribution

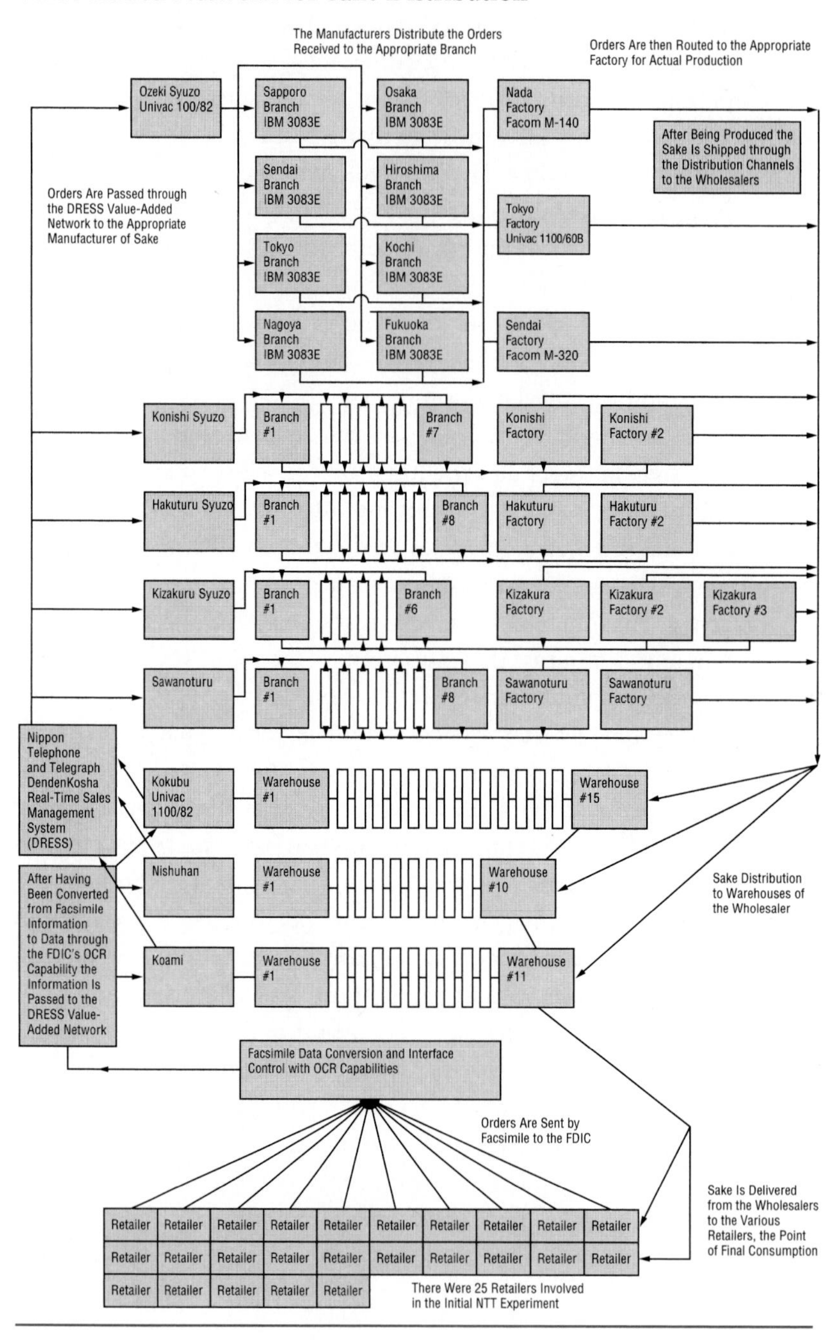

Unfortunately, the introduction of the system was not successful. It had many defects that were quickly realized by the retailers and wholesalers. For one, the optical character recognition (OCR) equipment needed much more time and caused more mistakes than ordering by telephone. Facsimile machines were used as data entry points for retailers who did not have host-computers and were not willing to purchase microcomputers. The facsimile machine required the use of many codes for the OCR equipment and it took the retailer a longer time to fill out the new-format data entry form. In order for the OCR to work properly on such a massive scale, standard forms were used, with requirements for neat handwriting in very specific areas of the form. The retailers felt the use of the form was a troublesome hindrance. In addition, the OCR equipment could not recognize incomplete data that might be supplied by the retailer: uneven marks or sloppy handwriting were a particular problem. The end result was that the OCR method, despite transmitting information faster and more efficiently, caused more mistakes and, compared with the former telephone ordering system, took more time.

When a consumer survey was made by NTT of the retailers approximately 2 months after initiation of the pilot, the results were disappointing. It found that 88 percent considered the facsimile-based data entry system less convenient than the former telephone-based system.

Another problem with the system was the willful suppression of data on the part of the retailers and wholesalers, making the creation of efficient management reports on the market situation impossible to obtain. The original intent of the system was to make it possible for participants to access information regarding other retailers', wholesalers', and manufacturers' sales reports and stock levels at any time. However, since the participants were afraid of competitors knowing their real sales and stock levels, they did not report the information accurately. There was a strong tendency to hold back information. (One wonders why this type of reaction was not anticipated by the initial planners.) The end result was that the management reports that were supposed to give participants precious information about such subjects as consumers' new taste preferences did not reflect the real

situation in the marketplace. Therefore, the information could not be used for the manufacturers' sales promotions.

For another problem, the system tended to increase the operational costs for the retailers. Of course, this was highly unpopular. In the old telephone-based system, the retailer had to pay approximately 10 cents per 3 minutes, and the typical length of time required to place an order was less than 3 minutes. However, using the new system, the retailer was responsible for paying both one-time and recurring fees:

- One-time fees included the system engineering fee, the software fee, the fee for the installation of the telephone circuit, and the NTT transformation control fee.
- Recurring fees included the information processing fee, the telephone circuit fee, the modem fee, the data file fee, and an NTT terminal fee.

For example, an order that had approximately 20 items might cost \$1.20 to \$1.50 to place instead of the regular 10 cents. When this was combined with the intended plan to increase the frequency of ordering, the ordering charge became substantial.

Ultimately, approximately 90 percent of the retailers who joined the project rejected the system.

K mart VSAT Network

In the early 1980s, K mart found itself facing a cutthroat battle with other discount merchandisers over market share and profits. K mart had developed a successful strategy based on selling discount items to middle class consumers located in smaller towns throughout the United States. Unfortunately, there was no end to competitive pressure. K mart was forced into using all possible avenues to cut costs and improve inventory control.

As part of this effort, K mart chose to install a private satellite telecommunications network based on Very Small Aperture Terminals (VSATs). The VSAT system uses satellite dishes a meter or so in diameter that can both broadcast and receive. The system K mart used is able to transmit both voice and data over the same channels.

Completely bypassing the local telephone companies, as well as long distance carriers, the K mart network is able to telecommunicate from every store in the system every day complete information about sales and inventory. It becomes possible for headquarters to know its exact position every day. This knowledge is critical for calculation of safety stock levels for inventory, and for monitoring cash flow, on a store-by-store basis.

The economics of VSAT technology generally do not compare favorably with terrestrial networks, particularly packet networks. However, in the case of K mart, the peculiar locational economics played a great part in changing the cost/benefit equation in favor of VSAT technology. Since most K marts are located in fairly remote or marginal areas, the cost of local telephone service tended to be fairly high. In addition, since the amount of transmission from each of the stores to headquarters in Troy, Michigan was clustered into only a few hours per day, the acquisition of a large leased line service was not economical. Leased lines are more appropriate when the amount of data telecommunications is fairly heavy over a good part of the day.

K mart used VSAT-based telecommunications to obtain competitive advantage in several ways:

- The immediate reporting of sales gave K mart the ability to spot trends or fads in sales of specific items quickly.
- Conversely, immediate sales reporting also allowed K mart to quickly spot items that were not moving quickly enough.
- K mart saved the cost of leased lines, which would be necessary for the same type of batch upload technology without satellites.
- K mart saved on its local telephone bills.
- Internal costs of distributing information were much lower, as a single announcement or broadcast could be sent to all store locations at once.

The K mart example is one of many involving businesses that have found it profitable to set up totally centralized information functions. They are able to get the advantages of extremely decentralized operations and market coverage, and yet administrative control over vital functions can remain in the center, thus increasing the power and effectiveness of executive decision making. The low cost of satellite

telecommunications enables K mart to eliminate many levels of bureaucracy that might otherwise characterize a national, regional, and local type of organization.

Johnson Controls Satellite-Based Inventory Control System

Johnson Controls, Inc., based in Milwaukee, Wisconsin, provides facilities management controls, systems, and services. It also manufactures car batteries, automatic seats, and plastic bottles. It is one of the old manufacturing firms that were the powerhouse of American production before the United States was eclipsed by other nations. Johnson maintains $^{1}/_{10}$ of its sites in Canada, and the rest are in the United States. Johnson's corporate information system was powered at the headquarters by an IBM 3090 mainframe. At its remote branches, it was using an Ethernet local area network with IBM AS/400 minicomputers. In order to connect all of the equipment into a comprehensive information system, Johnson considered using multidrop terrestrial lines. This strategy was complicated by the need to connect the Ethernet local area networks, which necessitated packet switching network technology based on the X.25 standard.

Johnson instead developed a very small aperture terminal satellite network with equipment supplied by Scientific Atlanta. In addition to providing for its data communications needs, the VSAT network supported the following applications:

- IBM AS/400 minicomputers at each remote branch processed customer service and inventory control data.
- Ethernet local area networks at each branch handled movement of the data locally, before the data were transmitted over the VSAT network back to the corporate mainframe.
- Broadcast video would also travel over the same network from a central studio located in Milwaukee, to transmit important video announcements for the Control Systems and Services Division.

The VSAT network provided a solid X.25 backbone through a shared hub site in Chicago. The network automatically shifted between interactive and batch traffic to respond to shifting applications loads as needed.

Summary

Telecommunications can be used in a variety of ways to improve management of inventory. It can aid rapid delivery of needed components, as illustrated by the Caterpillar example, a company with perhaps the world's best service in this respect. Sears typifies the difficulties of managing incredibly complex inventory on a very large scale, and how telecommunications can keep top management abreast of key developments. Sake distribution in Japan gives an example of how many different organizations can band together to create a highly complex, yet inefficient system. K mart and Johnson Controls used satellite technology to allow quick updating of inventory information and other critical data on a daily, or even hourly, basis.

Questions

13.1 What is the business effect of:
  a. Better inventory control?
  b. Faster inventory turnover?
  c. Better management of safety stock levels?

13.2 How can telecommunications help solve the inventory control problems of several different organizations handling the same product?

13.3 What is the incentive for a company to know its inventory on an hourly basis? Why can't it wait until the end of the month or the end of the year?

13.4 Explain the difference between satellite and terrestrial networks for inventory control.

Chapter 14

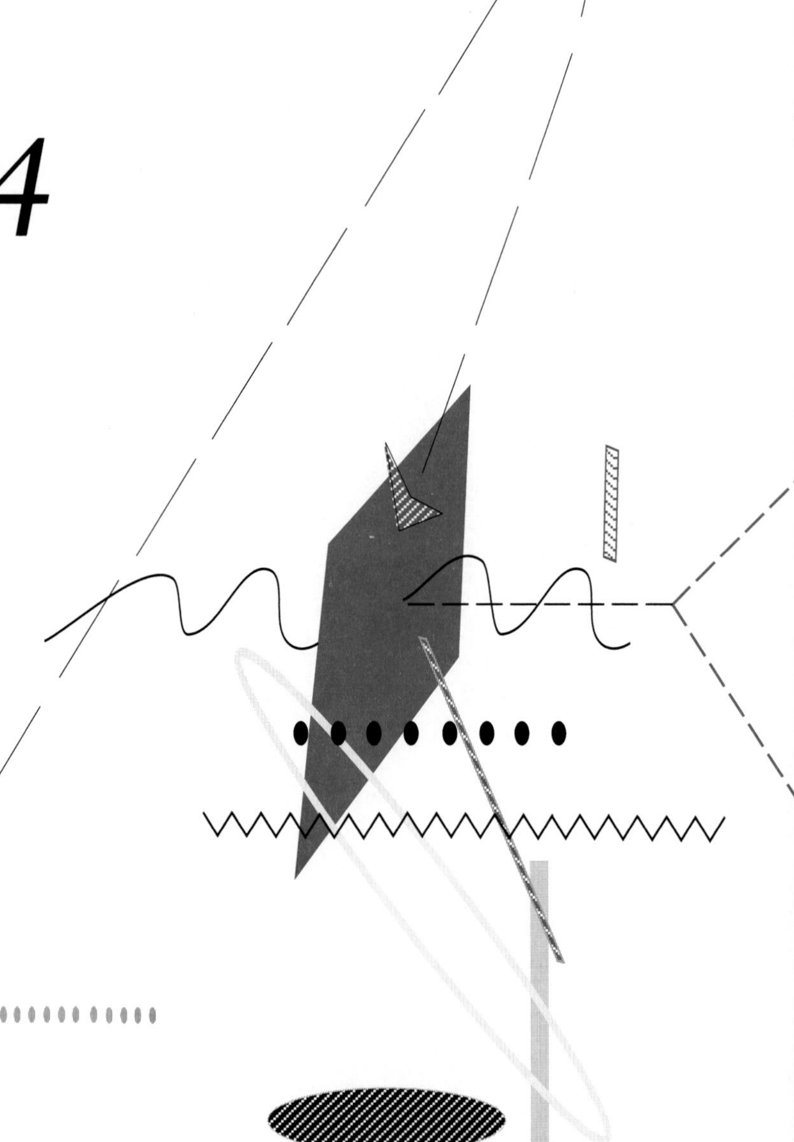

Slash Operating Costs

The use of telecommunications for cost reduction is another proven way to gain competitive advantage. Although some strategic moves using telecommunications, such as linking in suppliers, may give a great temporary competitive advantage, that advantage may erode quickly. Competitors may quickly copy your moves, and this means your advantage deteriorates rapidly.

On the other hand, the use of telecommunications and information technology to substantially reduce cost of production is more difficult for competitors to follow. Squeezing costs out of a process or business operation may be more difficult than adding linkages outside the firm.

This chapter focuses on various ways in which firms have used telecommunications to radically reduce their costs. As you will see, some very innovative technologies have been used to accomplish this purpose. The most fundamental aspect of this process is either finding substitutes for labor or wholesale export of labor-intensive processes to a location where labor is cheaper for the same quality.

Scan and Burn Technologies

One important way telecommunications can give competitive advantage lies in the adoption of scan and burn technologies. (See Exhibit 14.1.) The fundamental elements of scanning consist of optical character recognition (OCR) equipment and image capture. An OCR machine recognizes alphanumeric information that is either handwritten or machine written. OCR is a technology with a long history. The first attempts to produce such a system started in the early 1950s. Gradually, it has developed to the point where sophisticated OCR machines can read and understand even handwriting. The ideal application for OCR appears to be in high-volume situations involving highly standardized information formats. Credit card chit processing is one example. Sorting of mail, which uses sophisticated OCR technology, is another important and practical example.

Image transfer, another important component of this technology, is the ability to capture a bit-mapped image of a document and convert it to machine-storable form. Once the image has been converted to electromagnetic storage, it can be transmitted to another location, then decoded there and the image can be reconstructed. Facsimile is one such technology.

If the resolution is high enough, the scanned image of the document, when reconstructed at the other location after having been sent through the telecommunications channels, is impossible to tell from the original. In transmitting the image, the telecommunications system treats the information exactly as it would treat any other information. It is irrelevant, therefore, whether a sound, an image, or a piece of data is being transmitted.

American Express

American Express is one of the oldest and best-known credit card companies in the world. It has offices in virtually every country of the world, and issues cards on all six continents. By the late 1970s, American Express was facing competitive pressures from other companies. Visa and Mastercard were expanding rapidly, and were issued by banks, such as Citibank, which had extremely efficient credit card processing operations.

Exhibit 14.1

Scan and Burn System

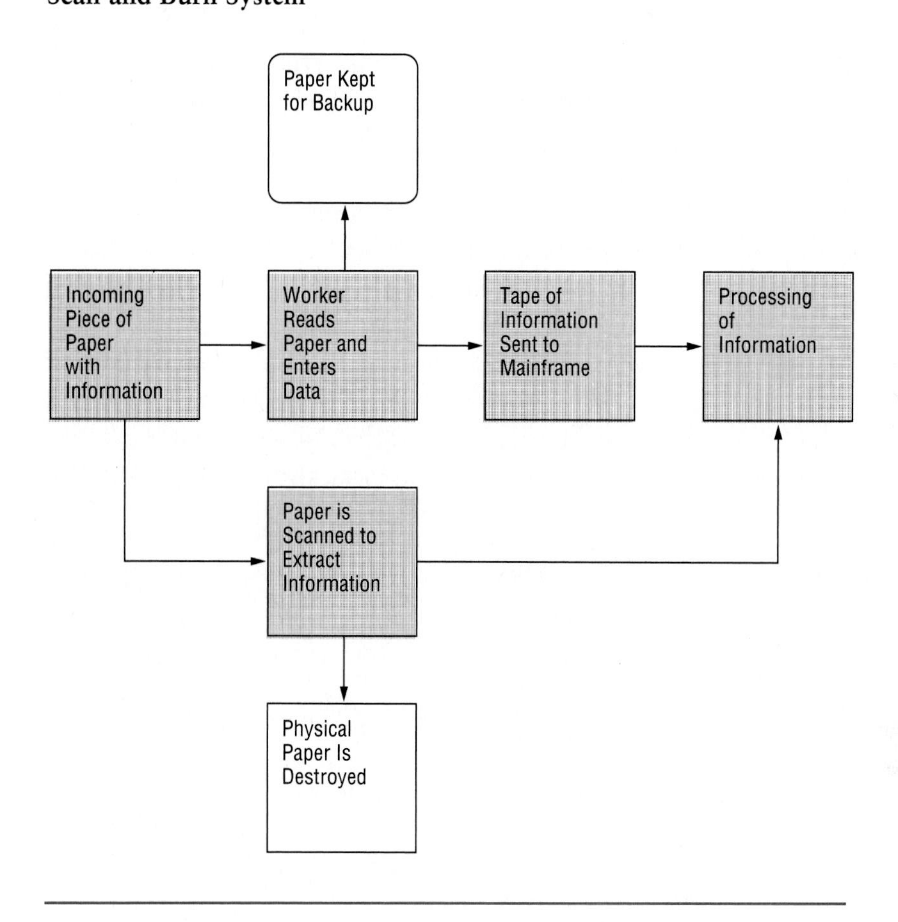

One of the unique benefits of the American Express card was that with each monthly bill, the customer received carbon copies of charge chits that had originally been signed at the point of purchase. Using this information, the customer was able to verify the signature, amount, and other details by comparison with the original copies carried away from the point of purchase. The problem for American Express was how to get the correct charge chits into the correct customer bills at the end of each billing cycle.

As the credit card business expanded rapidly, with increasing numbers of customers and transactions, the burden of sorting the chits and getting them back into the customers' monthly bills became correspondingly greater. It was as though American Express was running one of the largest private postal and sorting services in the world.

American Express adopted a telecommunications-based solution to this problem that has given it a tremendous cost savings. (See Exhibit 14.2.) Working with Integrated Automation, American Express developed an electronic scanner that could both read the account information from a chit and capture a complete facsimile image of the chit.

American Express then would destroy the paper chit and transmit its scanned image to the customer's billing center. The customer's bill included a laser-printed reproduction of the chit based on the scanned image, rather than the original chit. American Express saved in several ways:

- Much of the labor cost associated with data entry of the chit information into the billing system was eliminated by the optical character recognition technology.
- The cost of transporting and sorting the physical chits was eliminated, so American Express ceased running a large internal postal system.
- Cash flow and the cost of holding debits decreased because delays in the manual sorting process for American Express billing were eliminated by the much speedier electronic sorting and transmission system.

This strategic use of telecommunications technology has helped American Express to retain its position as a major player in the credit card business.

Time Inc.

Time Inc. is part of Time-Life, one of the largest publishing conglomerates in the world. As part of its on-going operations, it must process large numbers of orders and renewal notices for its books, videotapes, records, and journal subscriptions. One disturbing characteristic of the business is the highly variable format of the renewal notices. For

Exhibit 14.2

American Express OCR System

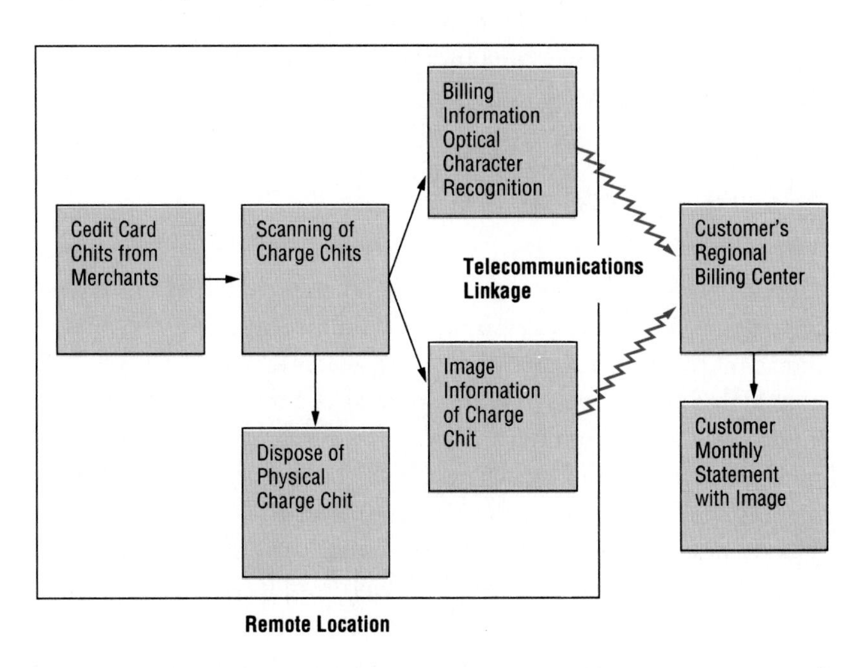

example, different magazines or mail order promotions may require different designs for the form on which the consumer either requests a new product or renews a subscription. This is true even for a single line of products, such as *Time* magazine. When this problem is increased to a scope of operations involving many different lines of subscriptions, the complexity becomes great.

This complexity, combined with the volume of incoming cards, which can reach more than 35,000 per day, makes this part of the process a very important target for automation. Just processing the cards requires Time to maintain a sizeable warehouse facility filled with people sorting, collating, reading, and doing data entry on all of the incoming cards. Ideally, Time would be able to have the scanner take in the card, lift off the required information, keep a record of the transaction on an optical disk for later recall if required, then send the information directly to its subscription mainframes. Perhaps one day Time will be able to solve this problem and introduce a great deal of automation linked to its mainframes through telecommunications systems, but Time is still looking for the ideal technology.

Remote Processing Services and Labor Cost Differentials

Telecommunications can be used to change the payoff of labor versus capital. Companies with information-intensive business operations have found it profitable, even very profitable, to export their labor-intensive activities to countries where labor is cheap. Under these conditions, the information-intensive work is done in a location where labor costs are substantially lower. The results of the information processing are telecommunicated back to the central information processing complex, typically in an expensive location.

We discuss here several examples of this trend. Banks have found that some of their most labor-intensive processing, such as credit card processing and customer service, can be more profitably done in locations where labor is cheaper, and perhaps more educated. Some insurance companies have found that the very labor-intensive processing of claims can be moved offshore to countries with cheaper labor rates and the results can be telecommunicated back to the headquarters mainframe, as easily as if the processing had been done down the hallway. Finally, some companies specializing in programming and software development for customized applications are finding they can achieve a competitive advantage by locating in low labor cost countries, but maintaining links to central locations through telecommunications.

The typical pattern is for businesses that face very high labor costs at their locations to export jobs to cheaper locations. This typically involves moving processing operations to developing countries, where labor costs are lower. Instead of having to move labor from one location to another, it is easier to simply move the work with telecommunications.

This exportation of information-intensive processing works by linking remote batch entry equipment or terminals to a central location via telecommunications channels. Most typically, this has been accomplished by using high-capacity satellite channels, but any type of telecommunications channel would work.

Some cases of this type of operation exist within countries, where an information-intensive process in a major metropolitan area is moved via telecommunications to a less developed, typically nonunionized, part of the country. However, the more publicized cases involve transfer of information-intensive work from a developed to a developing country. (See Exhibit 14.3.)

Economically, the cost savings from a lower labor rate pays for the investment in information technology and telecommunications infrastructure needed to set up the exportation of labor-intensive office work.

Printing *The Economic and Social Origins of Mau Mau*

David W. Throup's book on the Mau Mau is one of the better histories of this important African movement toward liberation from British imperial rule. The book is full of many interesting references to colonial documents and interviews with fighters in search of *Uhuru*. However, a look at the printing history of the book shows the incredible international division of labor which is now possible through the global telecommunications system.

According to the inside cover of the book, it was written in Charlottesville, Virginia, managed by Ohio University Press in Athens, Ohio, typeset by Colset Private Limited in Singapore, and printed and bound in Great Britain.

Only with a truly global telecommunications system can this type of distributed printing activity take place.

Citibank Visa

Citibank is the largest consumer bank in the United States, and among the 15 or 20 largest banks in the world. (The world's largest banks are Japanese.) It has been in the forefront of consumer banking, developing products and services for sale as retail products to the general consumer, small business, and also large business. Citibank pioneered the

Exhibit 14.3

Distribution of Jobs via Telecommunications

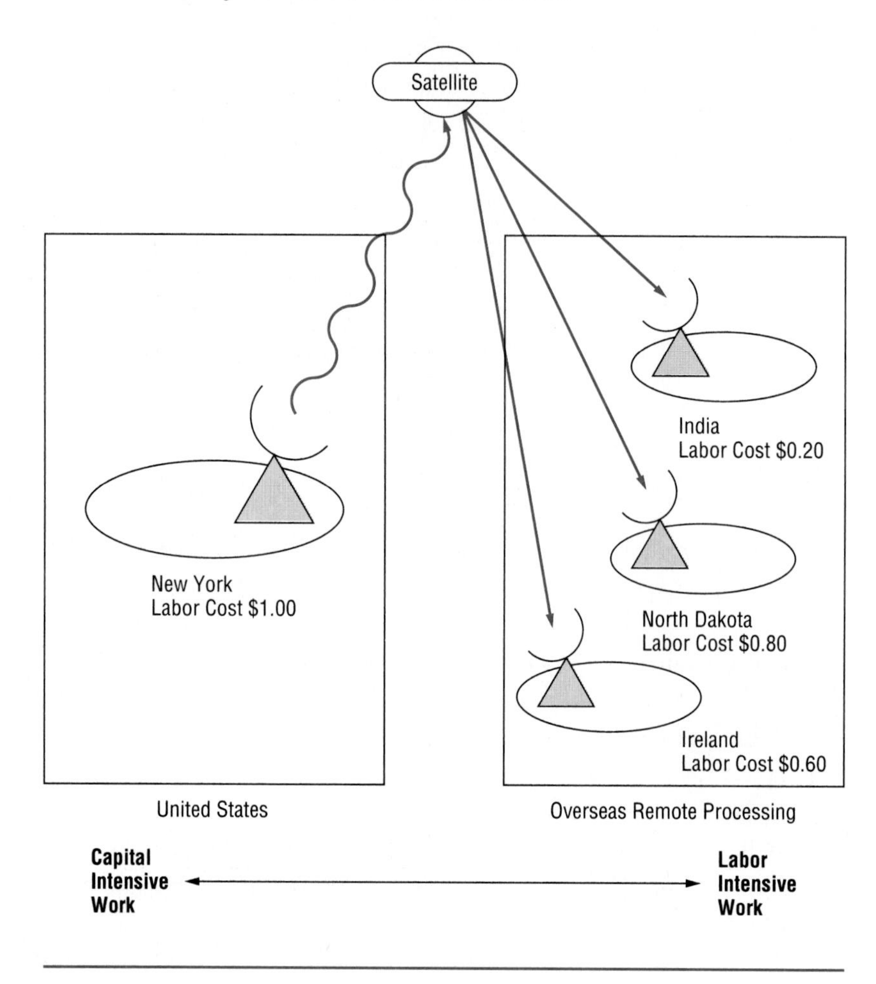

use of automatic teller machines in the early 1970s, and has introduced second and third generations of consumer banking technology.

In the credit card business, Citibank faced the problems typical of any institution. It is difficult to find responsible consumers, and the general rate of nonpayment is high. This makes relatively high rates of interest necessary. In the business of consumer credit card operations, several factors drive the cost per card or per transaction, which in turn determines how competitive a business is in the market. In order to attract more customers to reach the scale of operations necessary for a low cost per transaction, Citibank had to carefully screen its new customers and reduce its costs as much as possible.

Citibank faced the same labor and cost problems in New York City that have plagued many corporations and forced them to move to other areas of the country. In the 1970s and 1980s, New York faced several budget crises and a declining literacy rate among its workers, and most corporations were saddled with relatively expensive labor forces, in comparison with other parts of the country. Citibank in particular needed to lower its labor costs to achieve a low cost per transaction to meet the competition.

In addition, Citibank was facing a regulatory threat from the State of New York. The state was forcing it to cap the interest rates it charged for credit cards. Citibank concluded that if it were forced to place a cap on its interest rates at the same time it was forced to carry the inefficiencies of its workforce, it would eventually be driven out of the important credit card market.

To solve this problem, Citibank adopted a strategy of exporting the labor-intensive portion of its credit card operations to South Dakota. In fact, the credit card operation was virtually completely transferred to South Dakota. The transfer of operations was done by linking the New York headquarters with the operational center in South Dakota, relying heavily on telecommunications.

Citibank was able to use leased satellite channels between New York and South Dakota to connect to its remote office. This telecommunications bridge between the center and the periphery of the Citibank information processing system provided the capability for transmission of data, voice, facsimile, and other forms of information, including teleconferencing.

This location for credit card operations had several key advantages for Citibank:

- South Dakota's labor costs were generally lower than those in New York.
- South Dakota did not have and was not contemplating any cap on the interest rates charged for credit cards.

The results were generally successful for Citibank. Although it had to survive some typical derision from consumer groups who decried the use of technology to avoid the movement to put caps on interest rates, it accomplished that goal. Additionally, it did find the work force in South Dakota to be more capable and better prepared for efficient credit card operations. With a better educated work force, lower crime rate, and fewer economic opportunities for the average worker, South Dakota proved to be a better recruiting ground for quality labor per dollar.

The technical success of the operation led Citibank to move more of its credit card operations to South Dakota. Citibank purchased Diners Club, and moved those processing operations to South Dakota, as well. This further enabled Citibank to consolidate its operations and improve their efficiency.

Citibank's position now is that of a low-cost provider that at the same time is able to offer superior customer service. When calling for information regarding an account, a customer is usually satisfied by the quick response to inquiries from the Citibank employees located in South Dakota. The culture and psychological environment in South Dakota also paid off in the general attitude of the workers, which becomes apparent when the consumer calls for information. The South Dakotans provided warm and efficient customer relations. This additional perception has continued to help Citibank in its quest to gain market share across the broad spectrum of middle class America.

Citibank's position as a low-cost producer is caused by savings in several areas:

- Cheaper labor costs for information technology workers and customer service representatives
- Cheaper operating costs for the computer complex
- Cheaper rents and leases on property and office space, including suppliers, power, etc.
- Cheaper regulatory compliance and government red tape costs in South Dakota as compared to New York State
- Sustained high volume of credit card customers yields economies of scale in processing transactions

- Less constrictive regulations about denying credit card services and interest rates (Removal of the need for regulatory appeals has provided Citibank with pricing flexibility.)

The implications of this type of telecommunications solution are vast for the financial services industry. It means that work can be moved around to different locations in the country in search of the best labor at the cheapest prices. It also implies that other financial institutions that do not meet the challenge from Citibank will be forced into higher cost positions, and will ultimately not prosper in the credit card business. On the regulatory side, it means that telecommunications can be used to manage state-imposed regulations that interfere with business operations. Processing and services can simply migrate to areas where regulatory pressure is not as strenuous.

American Airlines

American Airlines is one of the premier airlines in the United States. It also operates in Latin America, Western Europe, and to some extent the Asian Pacific region. Like all airlines, it faced problems in processing the paperwork associated with billings for tickets purchased via credit cards. The chits, signed by the customers, had to be sorted and prepared for dispatch to the various credit card agencies and travel bureaus for billing purposes. The sorting of the chits delayed American Airlines' collection of payments for the services it had delivered.

American faced problems of cash flow and needed to collect its payments as soon as possible. In addition to facing the lost opportunity and interest cost of holding the chits for the uncollected cash and revenue receipt, American Airlines faced the relative inefficiency and high labor costs associated with manual processing of the chits. This back office operation had grown with the airline, but had never undergone the type of rigid, systematic operational study aimed at increasing efficiency that had been continually done in other parts of the business, particularly the basic airline operations.

American Airlines adopted a solution that used a telecommunications linkage to export the information-intensive part of its operations to a Caribbean island. Every day the chits from around the country would be transported to the Caribbean where they would be sorted and processed by local workers. Of course, the cost of labor in the Caribbean is much lower than the cost in the United States. In addition, American Airlines benefitted from centralizing what had been a scattered and ill-managed bureaucratic process for chit processing.

At the new processing center, when the chits arrive the workers begin the sorting process. After sorting, the workers begin to enter the required data into the computer. The computer is linked to the Caribbean operation via a telecommunications linkage. Logically, from the point of view of the system, the terminal locations seem as if they were next door. Once the information has been entered into the computer, the substantive part of the processing has been accomplished. The terminals are linked to the corporate mainframe via an international telecommunications satellite system. This provides another example of the exportation of an information-intensive process.

The effect was automation of the chit processing function, or at least creation of a much more efficient data entry and sorting operation. The turnaround time for chit processing was improved, and the cost per transaction was driven down, thus leaving more resources available for critical operations.

The long-term picture, however, is not entirely favorable for the type of solution developed by American Airlines. As with the American Express scan and burn example, eventually the information workers responsible for the sorting and manual data entry will be replaced by scanning equipment with optical character recognition capabilities. However, due to the high variability of chits, and to the differing quality of the images on them, it will be quite a while before application of artificial intelligence and other routines to scanning technologies will make it possible to completely eliminate the information worker. It appears that American Airlines realized this when it set up its facility.

CIGNA Insurance

CIGNA Insurance faced the same problem other insurance companies faced. It was very expensive to process claims. The job required a highly literate worker, who was able to work through a great many variations in the details of the coverage, applicability, and price payment schedules. Yet at the same time, the claims adjustment and processing functions were critical to the business. Claims processing is particularly important to the customer who is dependent, in sometimes unfavorable circumstances, on a speedy resolution. In addition, the determination of the exact benefits the company is required to pay is another critical function. Paying too much is waste, yet paying too little leaves the company open to lawsuits, which are very costly.

The problem was complicated by CIGNA's expanding case load and its difficulty attracting labor at a price that enabled it to operate responsibly without too much expense. Better-quality workers would have to be paid more and these higher costs would have to be passed to the consumer to avoid crossing the barrier of unprofitability. CIGNA also faced the problem of getting educated and literate information workers.

As a response, CIGNA moved its claims processing operations to Ireland. The CIGNA headquarters' information processing center is linked to its facility in Ireland using an international satellite telecommunications system. When claims arrive at CIGNA, they are sent by air to Ireland where adjustors and analysts do the processing. As the processing takes place, the appropriate information is entered on terminals linked to the headquarters mainframe computer through the telecommunications linkages.

The results and advantages of CIGNA's new telecommunications-based system are many:

- Labor rates for quality workers are lower in Ireland than in the United States.
- The general educational level and skill of the workers in Ireland is better than that available in many metropolitan areas in the United States.
- There is no appreciable decline in response time; in fact, the specialization helps to improve the response time from the customer's point of view.
- The workers in Ireland are generally more complacent and less captious than labor in the United States.

The effect of CIGNA's strategy has been to both increase the quality and lower the cost of its claims processing operation. Able to get good quality workers, CIGNA improved its turnaround time and customer service. The addition of toll-free telephone service from the United States to Ireland further enhanced the package. This type of transparent telecommunications system makes it impossible for customers to know the location of the customer service representative with whom they are speaking. Besides possibly fooling the customer, the tied-in international telephone service fools the host computer, as well. It does not know that its terminals are located in a distant country.

India's Programmers

Over the years, the cost of software maintenance for running information technology installations has greatly increased. In the early 1980s, software costs began to exceed hardware costs for the first time, and the trend was continuing. By the mid 1980s, salaries for analysts, programmers, or project managers with applications development experience were increasing rapidly due to the explosion in utilization of information technology and the shortage of trained personnel.

The cost of developing applications programs has become a major expense, and this helps explain the dramatic rise in the use of outside project management, consulting, and contract programming organizations. Firms have simply been unable to maintain the required skill base in house, and have had to resort to outside "console jockeys" to do their programming.

This change in the economics of development of applications programs has opened up the opportunity for the development of international services based on telecommunications linkages. Programmers and applications development specialists are able to live and work in India, yet telecommunicate their work to the developed countries where it is sold and marketed.

Again, this development is a type of exportation of information-intensive work to a geographic area where labor costs are relatively low. In India, the relative costs of labor are approximately 10 percent of those in the United States, and probably vastly cheaper than in the richer countries such as Japan. The result of this labor cost differential is the ability to produce advanced applications software at a fraction of the U.S. cost.

Digital Equipment Corporation Programming in India • In southern India, an affiliate of the Digital Equipment Corporation is linked directly through satellite channels to the Maynard, Massachusetts headquarters of DEC. Transmissions of details of assignments, source code, and messages flow immediately back and forth as needed. The only difference is that the Indian group is able to operate at a much lower cost than the U.S. group. In addition, India offers a highly educated workforce, schooled in English according to traditions that have their origins in the United Kingdom. The workers are able to communicate effectively, and the projects are managed by persons who have received advanced research and doctoral degrees at centers of learning such as the Massachusetts Institute of Technology and the California Institute of Technology.

Although this type of operation brings certain efficiencies, it is still a small drop in the bucket in meeting overall demand for programmers and analysts. However, the success of the DEC venture points the way to a great many other possibilities for using telecommunications to export information-intensive work in applications development and programming to areas with low labor costs.

Swiss Banks Processing in India • One of the highest cost labor markets in the world is Switzerland. In addition to its high cost, it has had to import much of the talent to operate its large data processing industry, a key component of its banking and financial services system. Many banks in Switzerland have also found that they can export their information-intensive jobs to India via telecommunications channels. By doing this, the Swiss banks have been able to develop critical banking applications more cheaply than they could in Switzerland.

Computerworld: Modems and Journalism

Computerworld is one of the largest weekly newspapers devoted completely to the computer and telecommunications industries. Published in Waltham, Massachusetts (as well as in other locations) *Computerworld* carries many product and service advertisements, covers most new product and service announcements, and runs regular

"In-Depth" specials on topics of enduring interest to the information technology community.

The problem with covering this topic is the high mobility required of correspondents. On the one hand, *Computerworld* is forced to follow developments in Japan, Silicon Valley, and the eastern seaboard; on the other hand, it is not a large-circulation paper, so managing costs is a critical function.

Computerworld's problem of having to cover a very wide territory was complicated by the need to have the most up-to-the-minute product announcements available. Getting scoops on product announcements was critical to *Computerworld.* Important announcements from major companies such as IBM and DEC could not be delayed one week because *Computerworld* was unable to get the information to print in time.

Altogether, challenges faced by *Computerworld* were great:

- Getting the most current information to the printing press up to the last minute for fast breaking news
- Delivering information to headquarters as quickly as possible
- Reducing the amount of labor required to take an article from first draft to finished, typeset copy
- Coordinating stringers and correspondents around the country (and the world) to ensure coverage of all major stories

Computerworld's solution was to equip many of its reporters with portable personal computers which could be used to write and compose stories on the spot then transfer them directly to the story preparation computers in Waltham. Correspondents were able to use modems and dial up the remote computer, sign on, then telecommunicate their stories. Once the stories were received in electronic form, they could be electronically edited by the staff editors at the headquarters, and sent to the typesetter without having to be entered more than once.

In addition, the stringers and correspondents were able to link into an electronic mail system through which to exchange information both with the headquarters staff and with each other. This electronic mail system was critical in helping management deploy its reporters around the country to the many different locations where the critical product announcements and other newsbriefings took place and coordinate their movements.

The benefits were numerous for *Computerworld.* It reduced both its cost and the interval for getting a story to press, which has helped it maintain market share and continue to provide some of the best coverage of events in the information technology world. Its correspondents have been able to deliver their stories at the very last minute, thus making it easier for the newspaper to cover fast-breaking events. All of these improvements have taken place through telecommunications, which enabled *Computerworld* to replace its old system of phoning in and double entry of stories with a highly automated system.

McDonald's Use of Telecommunications for Clocking In

McDonald's, the people who brought you the Big Mac, set up a system in the United Kingdom to process employee time clock records automatically. Each day, the employee punches a time card into the time clock where it is stamped when beginning and quitting work. At payroll time, the card for the employee is taken to accounting, the hours are calculated, and payment is made. McDonald's has changed this process by assigning its U.K. employees magnetic identification cards. The employees pass the card through the reader both before and after work. In the evening a minicomputer at the central payroll location in the United Kingdom systematically calls each of the remote, on-site clocks and polls for information. The remote clock unit transmits the information to the minicomputer. Thereafter, the information is transferred to the IBM mainframe responsible for calculating the payroll and cutting checks. This telecommunications-based solution completely automates the payroll calculation process, and betters the turnaround time for cutting checks. Now, instead of manually collating the time cards, then sending them to headquarters for processing, the process is completely automated.

Teleconferencing at Revlon

Revlon is a large cosmetics company catering to the vast middle class, primarily in the United States. Its cosmetics are widely advertised and sold through many different distribution channels throughout the United States. By its very nature, selling cosmetics relies heavily on salespeople, particularly when new products are introduced, and a promotion campaign is put into action. Depending on the product being introduced, a significant amount of training may be required to brief salespeople from around the country on the nature and uses of the product.

In the past, companies such as Revlon accomplished this briefing task by assembling the salespeople from around the country into a single location, typically a resort complex or hotel, and then conducting seminars and briefings before releasing the salespeople back into the field. Unfortunately, the rate of new product introduction of cosmetics has increased, and in many ways the natures of the products have become more complex as chemistry and research have pushed back the frontiers of cosmetology. More product introductions meant more meetings at more expense, ultimately reducing return on investment.

Revlon has, to a great extent, solved this problem by using teleconferencing. Teleconferencing allows products to be introduced quickly and in full color with sound in a remote, satellite-transmitted meeting, without either the time or resource utilization of the old travel-based system. Revlon uses portable teleconferencing units that can be rolled from room to room. The units use a 56-kbps line on a dial up basis, allowing Revlon to incur telecommunications costs only when it is using the teleconferencing system. The use of the 56-kbps lines just appears as an extra charge on the telephone bill.

The physical operation of the system is simple. A small modular plug fits into a socket on the wall. Revlon has saved much time and money by its use of teleconferencing in new product introduction campaigns.

Summary

Strategic use of telecommunications can greatly reduce the cost of some operations. Instead of moving physical goods, such as paper, telecommunications can move only the information, thus speeding up processes and cutting costs. In addition, telecommunications can be used to reach out to find cheaper labor to carry on the same work. These savings can be quite substantial. As the nontariff barriers to advanced information services are improved, this type of solution will become more prevalent. Telecommunications can also help in reducing travel costs, particularly through such technologies as teleconferencing.

Questions

14.1 a. How does scan and burn technology save costs?
b. What role does telecommunications play in this process?

14.2 How can telecommunications be used to change the fixed labor costs in information-intensive organizations?

14.3 a. Discuss how teleconferencing can help reduce travel costs.
b. Do you think teleconferencing can really work as well as face-to-face meetings?

Chapter 15

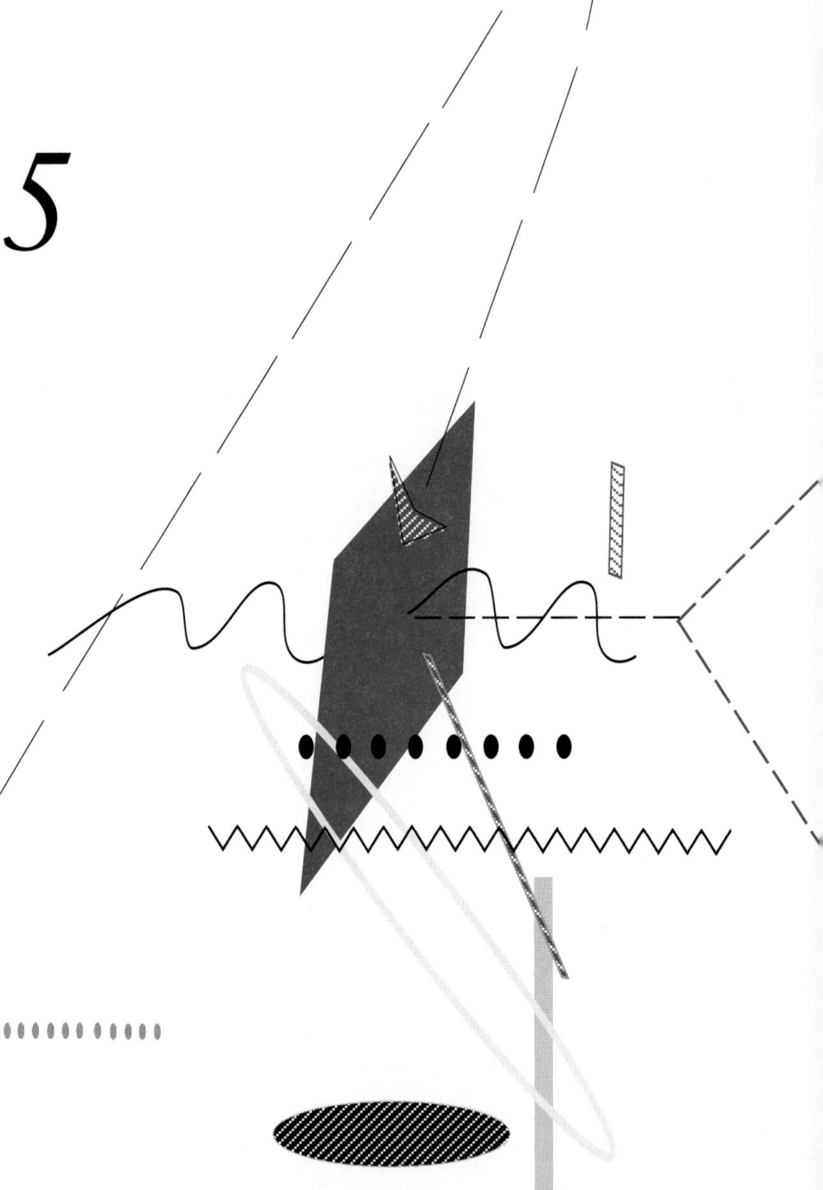

Deliver Innovative Services

Telecommunications can help an organization deliver a highly innovative service. This principle is particularly prominent in the financial services sector. In finance, such moves as home banking, automated teller machines, and newer telephone-based technologies have helped to create entirely new approaches to a firm's relationship with its customers. Telecommunications has also stimulated creativity elsewhere. Facsimile has revolutionized delivery of newspapers and advertising, and services have been delivered completely through electronic means.

This chapter reviews a few examples of how telecommunications can help accomplish these objectives.

Sanwa Bank's Global Strategy of Innovation

Although not a well-known name in the United States or Europe, the Sanwa Bank, Limited, ranks among the top five international banks, as measured by assets. The scope and volume of its services make it one of the world's leading financial institutions. The bank recorded gains in size, breadth, and profitability of its activities during 1987. Its total assets expanded 17.6 percent to the equivalent of $327 billion. Net income increased 48.2 percent to $991 million, placing Sanwa among the top ranks of international banking institutions. In 1987, its before-tax income was $1,951 million.

Sanwa Bank's strength lies in its operation throughout a large international network, which includes more than 100 offices outside Japan in 23 countries. It has a particularly strong presence in the Pacific Basin.

By the late 1980s, Sanwa was facing a growing trend toward financial globalization. The financial sector was undergoing liberalization within Japan, and a global trend toward more liberalization was under way, bringing more business opportunities (at least at first). Financial globalization changes the financial system in each country in response to development of universal standards or rules. This results in an integrated global financial system. Logically, business in the Tokyo market will be conducted in the same way as in the New York or London markets. The only difference between the world financial centers will be their time zones. The result has been a growing 24-hour-per-day financial market. Constant trading greatly increases the amount of information that must be handled by financial institutions on a global basis.

In addition, the intense competition among deregulated financial intermediaries has lowered the price of financial services, resulting in a situation in which firms must reduce expenses. Traditional financial restrictions had the political objective of protecting depositors, at least in Japan. The popular perception in Japan was that the banks could not go bankrupt. Two major pillars supported this network of restrictions. One was the deposit interest rate restrictions. The other, the division of commercial banking and investment banking, eliminated the risk that a bank would go bankrupt resulting from a bad position in equities. However, Sanwa faced the liberalization of large-scale deposit interest rates, and the narrowing gap between types of business. Both of these trends increased competition. In addition, due to the easing and then abolition of Euro-yen controls by the Japanese Ministry of Finance, the participation in the market by non-Japanese rival institutions resulted in a cutthroat market. The total result was a strong pressure to cut costs by all means possible.

Sanwa was also facing the expanded transaction volumes in the markets around the world made possible by innovations in information technology and the high computerization of financial services. As a result, Sanwa was finding that its present systems were inadequate to handle the present and expected growth. The growing transaction volume strained the system since practically every financial transaction creates one or more information transactions within the host financial institution. High computerization of financial services in Japan had made it possible to create new financial products such as the *Kokusai Teiki Kouza* (International Certificate of Deposit Account) and *Waido* (a banking package), which proved to be very popular with Japanese customers.

Based on an awareness of the changing business conditions and challenges facing the bank, the management of Sanwa embarked on a new strategy to enhance the information and telecommunications system of the bank. The project, the Universal Project 90, was started in April 1987. The principal goal under Universal Project 90 was expressed in the phrase, "Most innovative and strongest aim to be a global, universal bank." As part of this strategy, Sanwa recognized the importance of computer communication systems. In order to accomplish this, the bank established the Business Information Development Division to provide integrated domestic and international banking capabilities.

The strategy was aimed at achieving economies of scale through information and telecommunications technologies. The bank realized that the combination of computer and communications technologies had made it possible to transmit and process a wide range of information almost instantaneously. In addition, the bank was trying to build a telecommunications infrastructure that would support increased diversification of the business. This would enable Sanwa to develop diversified businesses such as credit cards, sales of national loans, and pension funds to complement its core deposit business. Importantly, it wanted to be able to add new products without increasing inflation in information technology and telecommunications infrastructure costs.

For telecommunications, Sanwa wished to cap the growth in its yearly costs, which had been increasing year by year. Sanwa management decided to accomplish this by building a private telecommunications network. By doing some research on its competition, the Sanwa Business Information Development Division determined that one of its competitors had reduced its telecommunications costs by 3,360 million yen ($26 million) per year.

The strategy Sanwa adopted was to link all the financial centers of the bank for all types of telecommunications transmissions. Voice, data, image, facsimile, and other applications would be linked together on a worldwide basis. Sanwa's plan called for linking the offices around a backbone network with the use of both digital and analog lines. (See Exhibit 15.1.)

Within each office, Sanwa installed a Digital Multimedia Information Multiplexer (DMIX) supplied by Fujitsu. The DMIX is able to transmit and route all types of transmissions through the backbone network. Transmissions of the various host computers, PBXs, digital facsimile machines, and other pieces of equipment are sent over the same network, which is leased at a constant price from international record carriers.

Sanwa's use of a Corporate Information Network System (COINS) supplied by Fujitsu allowed it to integrate all of its media, which it had previously operated as separate networks. Sanwa has been able to reduce telecommunications costs by one-third in comparison to traditional means of telecommunications. The Fujitsu system makes efficient use of the backbone telecommunications circuits by taking advantage of time differences and configuring the three relay points as a triangle, as Exhibit 15.2 shows.

This solution used three techniques to reduce telecommunications costs:

- *Integration of Multiple Media*
  Instead of separate networks for voice, data, and facsimile, the Sanwa Bank now has only one network. All of the transmissions in

Exhibit 15.1

Sanwa Bank Network of Offices and Communications

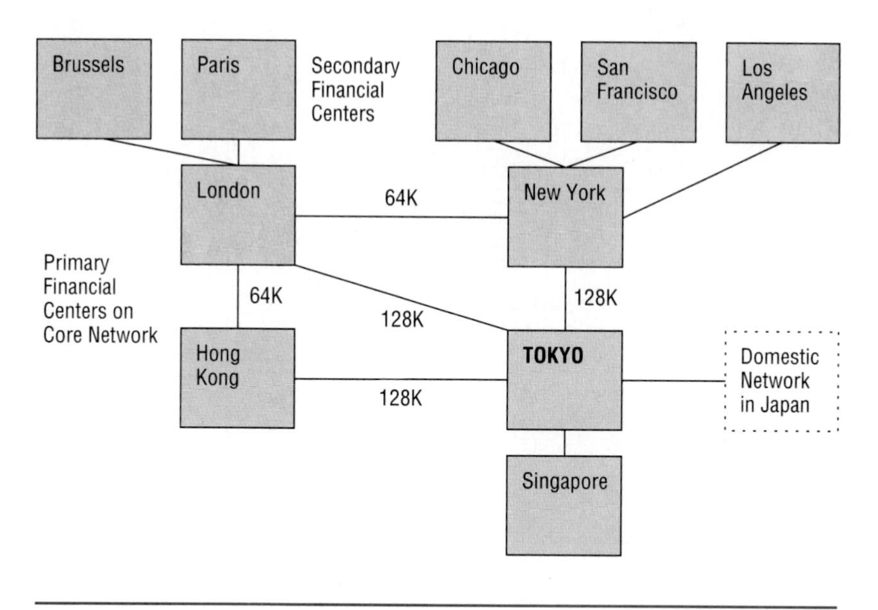

each of the different media travel through the same backbone network.

- *Cost-Efficient Use of Alternative Routing*
  Automatic rerouting of traffic through the telecommunications network reduces the costs that might be incurred through off-network additional line charges.

- *Efficient Multiplexing of Data Using Packet Switching*
  Data is efficiently multiplexed throughout the network to allow for load leveling and most efficient line utilization.

Sanwa has also improved its position through integration of information resources throughout its organization. Through the telecommunications system, all employees are able to access information resources from any location in the company. This has helped make the bank operate as a more cohesive unit.

Exhibit 15.2

Technology for Sanwa Network

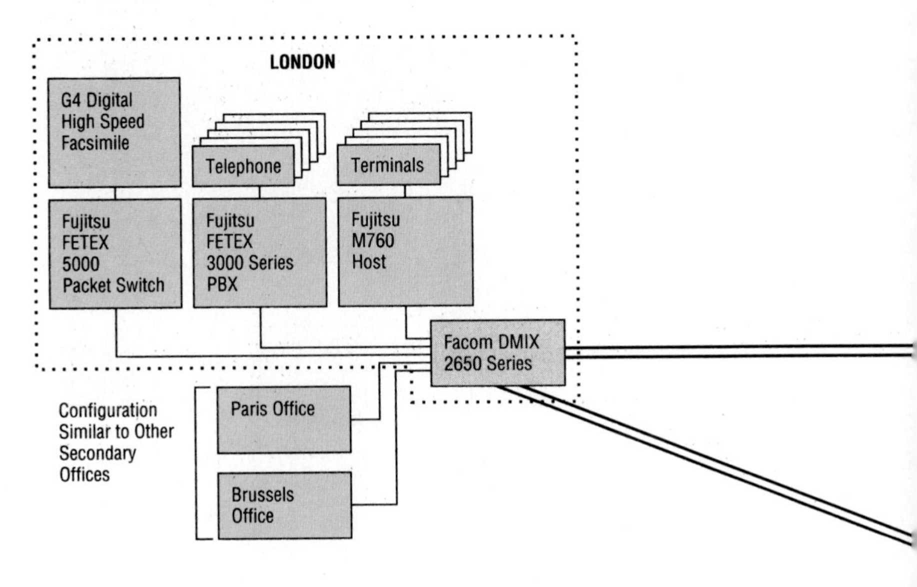

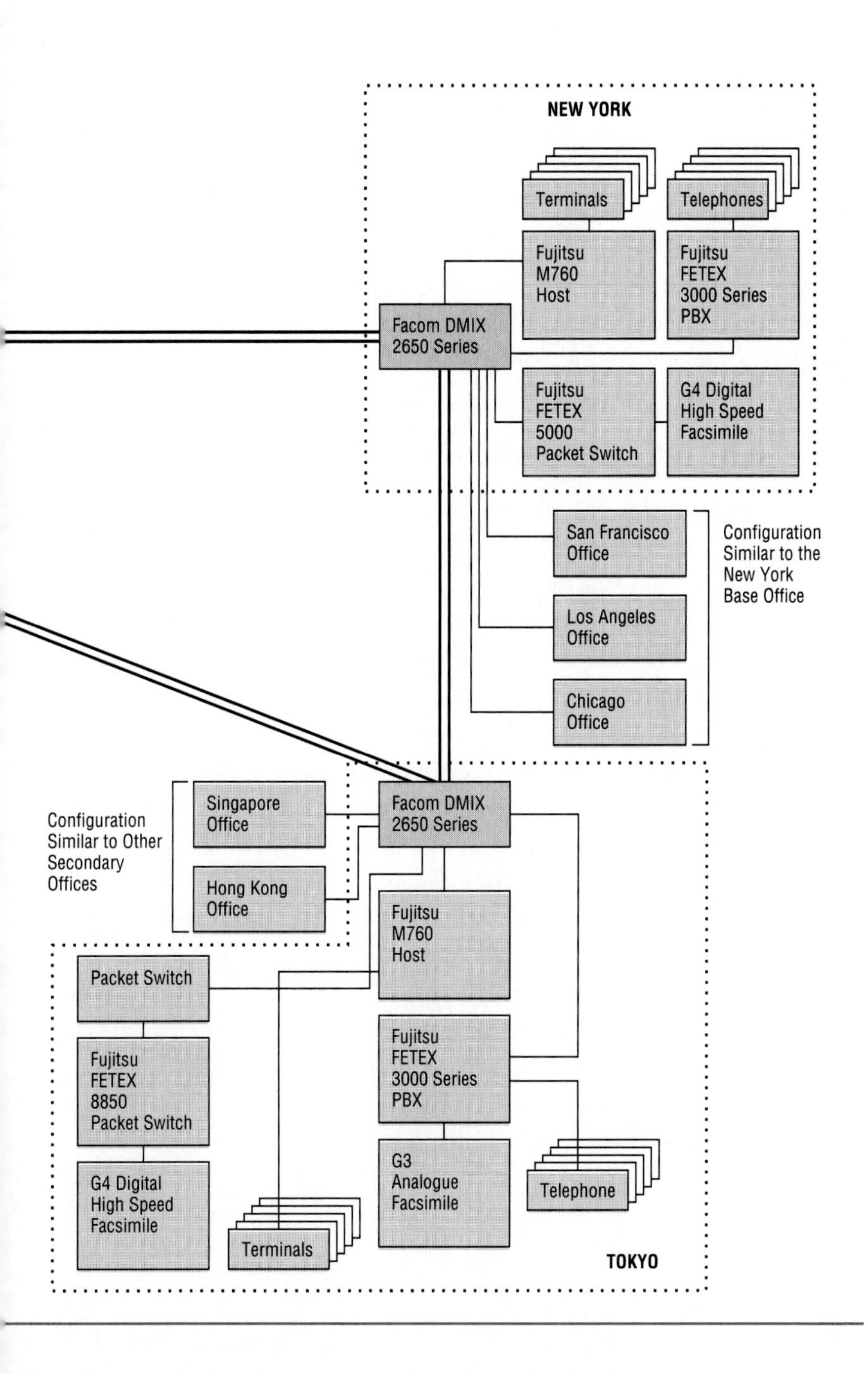
NEW YORK
Terminals
Telephones
Fujitsu M760 Host
Fujitsu FETEX 3000 Series PBX
Facom DMIX 2650 Series
Fujitsu FETEX 5000 Packet Switch
G4 Digital High Speed Facsimile
San Francisco Office
Los Angeles Office
Chicago Office
Configuration Similar to the New York Base Office
Configuration Similar to Other Secondary Offices
Singapore Office
Hong Kong Office
Facom DMIX 2650 Series
Fujitsu M760 Host
Packet Switch
Fujitsu FETEX 8850 Packet Switch
Fujitsu FETEX 3000 Series PBX
G3 Analogue Facsimile
Telephone
G4 Digital High Speed Facsimile
Terminals
TOKYO

DEC's Financial Information Service

In the United Kingdom, the insurance sector is emerging as one of the largest users of telecommunications systems for business operations. The Digital Equipment Corporation offers a Financial Information Service network that allows insurance companies and their representatives to access information on services available. After pricing services, the representative can place an order for insurance directly through the network. The DEC system is an improvement over other competitor systems, such as British Telecom Insurance Services (BTIS) and ISTEL, because the FIS system is oriented much more toward processing transactions, in addition to simply providing information for use in completing paper-based transactions.

Tsukuba Medical Diagnosis

Japan has been in the forefront of developing information and telecommunications technologies for the consumer. In the mid 1980s, Japan began to develop the idea of experimental science cities or science parks to foster advanced research on the new information society. As a by-product of that plan, Tsukuba was created. Tsukuba is a small village approximately 1 hour northeast of Ueno Station in Tokyo. Designed to be a combination of research centers and institutes, a university, and living quarters for government workers, Tsukuba was created to be a prototype city. One of the more successful experiments in the field of telecommunications involved linking doctors with patients through a teleconferencing system.

The high degree of privacy needed in patient–doctor relations was a drawback, but the use of a teleconferencing system directly connecting the home to the doctor's office appeared to point to success. It was found that 70 to 80 percent or more of consultations could be completed properly over the telecommunications system. Patients would call the doctor during "electronic office hours" and answer the doctor's questions, then receive a diagnosis.

Datum Company in Tokyo: Daily News Digests via Facsimile

Datum Company in Tokyo introduced a new service to the busy Japanese *sarariman,* who has little time to read and collect the critical information that seems permanently in short supply in the most literate, information hungry country in the world. Japan publishes and consumes more newspapers and mass media than any other country in the world on a per capita basis and also has the highest literacy and education rate.

To serve this need, Datum Company, located in the Shibuya district, sends daily digests of approximately 29 magazines to the personal fax machines of individuals as they leave for work, and also to those on the office desks of various executives. At the beginning of 1990, Datum was growing to serve several hundred clients. Having a digest of the information from different magazines delivered directly to one's desk saves those few minutes at the newsstand and, in Japan, every second counts.

Connecticut News

The *Connecticut News* started giving its customers the option of receiving information via facsimile, rather than through the traditional medium of newsprint. Instead of getting all of the newspaper every day, customers can place orders for only the type of information they read on a regular basis. Only the information a customer preselects is provided.

In addition, the delivery of the information can take place without the usual hindrance of physical distribution. The customer saves a great deal of time and is given a prescreened selection of information. This tends to simplify reading activities. On the other hand, the *Connecticut News* obtains enhanced revenue because it raises the price of the delivery of information via facsimile. Because of this difference and the speed at which it supplies information, it is able even to make a profit.

The Information Store and Dynamic Information: Facsimile Research Delivery

In the world of information research, the library is quickly becoming an anachronism. The delivery of information from library to recipient or researcher is coming to depend on telecommunications. This trend has two technological forks: advancements in delivery of information through on-line databases, and the delivery of information through facsimile systems.

During important projects such as an acquisition or consulting study, time is much more important than the cost of delivering information. Many times, even overnight delivery of information is too slow. In the past, specialized organizations such as Dynamic Information and The Information Store have developed businesses based on filling requests for information. In response to a request for an article, for example, these companies photocopy the article using nearby or in-house library sources, then send the copy to the client. This function in effect serves the purpose of sending an assistant directly to a public or corporate library to fill the same request.

The businesses have increased in volume as the demand for immediate delivery of information has proliferated. These companies have started to use facsimile systems to send information quickly to clients. When a photocopy of a document is received, the information provider sends the article to the client in the form of a facsimile. This type of service is possible only through the use of telecommunications, and it has changed the economics and customer service requirements of an entire industry. One advantage of facsimile or optical delivery of information is that both text and photographs, figures, diagrams, or tables can be transmitted through the information channel.

Minitel System in France

In the late 1970s, and probably even before, the French government and leading industrial circles and academics realized that information and telecommunications technologies were going to be one of the most important infrastructure issues in the coming years. The French had seen a few failures in their national effort to create a computer industry, and were hoping that in the area of telecommunications they would be able to achieve a technological advantage over others.

One of the strategies developed was to create a national on-line telephone directory through which any citizen equipped with a terminal could look up a telephone number. The next step was to start to mass produce the terminals, hoping that events would take place fast enough to help the French industry reach economies of scale in manufacturing, thus allowing the price of the terminal to drop to a world competitive level.

The result was slightly different from plans because the French encountered a few problems:

- Problems due to the price performance trends in the world market for terminal equipment were much more severe than originally thought.
- It took longer than planned to agree on design specifications for the terminals.
- Government reaction time was so slow that the manufacturing window closed.

The technical solution adopted was to give all telephone subscribers Minitel terminals to connect to their telephones. Using the terminal, the subscriber could make inquiries regarding telephone numbers. In addition, other important services were available through the system. Many advertisements appeared in good color on the Minitel terminals, and firms found that advertising on television while providing added information through Minitel was a good combination to deliver messages. Travel firms and other firms offering goods and services have found the system to be useful.

As the next generation of Minitel users comes of age, the provision of electronic information will be seen as a very basic part of life. The addition of electronic mail and messaging services, combined with electronic bulletin board services has also greatly increased the utilization rate for the system. These types of services and applications show that it is possible to use telecommunications to expand the influence of a few service offerings.

Home Banking—The Next Generation at Citibank

Although Citibank had achieved good economies of scale in credit card processing and delivery of ATM services by the late 1980s, it realized that technological progress would not stop. Firms that innovate must continually keep pace with competitors, and the price for failing to keep up can be heavy.

Even before it introduced its innovative second generation of teller machines, which used touch-sensitive screens and could operate in several languages, Citibank was already working on its plan to extend its influence out to the end user via computer and information networks.

In the mid 1980s, several banks around the country had experimented with home banking. Believing that the information revolution and home banking were part of the wave of the future, bankers depended on market research that indicated that by 1990 many of the households in the United States would have personal computers available. The logic was simple: if the customers have computers, then they should be able to dial up and get information from the bank, or even to receive a great number of services offered at that time either through the ATMs or through the old system of person-to-person contact.

The problem with this strategy, as many banks found out, was that the penetration of personal computers with modems was much smaller than expected. By the late 1980s, after much marketing effort, Chemical Bank, for example, had managed to attract only a few thousand users in the metropolitan New York area. It was realized that the growth of home banking was being held back despite the proliferation of personal computers in the home. Since the typical personal computer user was relatively wealthy, well-educated, and prone to experimentation, it was becoming clear that home banking should not be developed for the masses. Unfortunately, only mass programs would enable the banks to reach economies of scale to lower the cost per transaction and thus generate profits from the service.

Citibank's solution was to develop an inexpensive telephone that would also serve as a data terminal for accessing home banking services. The terminal/phone operated over the regular telephone lines, and provided a display and a keypad that also dialed telephone numbers when the customer was using it as a regular telephone. From the customers' house, the terminals would link directly to the bank's computers to enable the customer to accomplish a variety of banking services including:

- Reviewing account and balance information and electronic statements
- Making payments to creditors
- Utilizing a variety of information resources available through the Citibank network
- Transferring money between accounts
- Many other electronic services

Citibank spent much money and time on research and development to design the terminal/phone so that it would be easy to use for the average person. It also invested in a manufacturing company so as to greatly reduce the actual cost of the terminal.

Citibank's strategy rested upon providing customers with cheap terminals, changing the economics of banking so that customers would be encouraged to use the system, then marketing heavily. Again Citibank's next generation of home banking had placed it ahead of its competitors, and again this jump was made possible through the strategic application of telecommunications.

Summary

Telecommunications has played a strong role in helping financial institutions deliver electronic technology-based innovative services. Home banking, automated teller machines, and other systems are examples of this. In addition, telecommunications has provided a conduit for other companies to link themselves in with their customers. At the same time, telecommunications provides a strong current of access so that regular consumers can feel that what they are getting is targeted at them individually.

Questions

15.1 Discuss the role of telecommunications in the financial services sector.

15.2 What was the basic strategy behind home banking?

15.3 Why is it advantageous to a financial institution to give access to its computers to outsiders through schemes such as home banking?

15.4 a. How many businesses can you name which actually consist of nothing but information or flows of information and data?

b. What is the significance of telecommunications for information-intensive enterprises?

Chapter 16

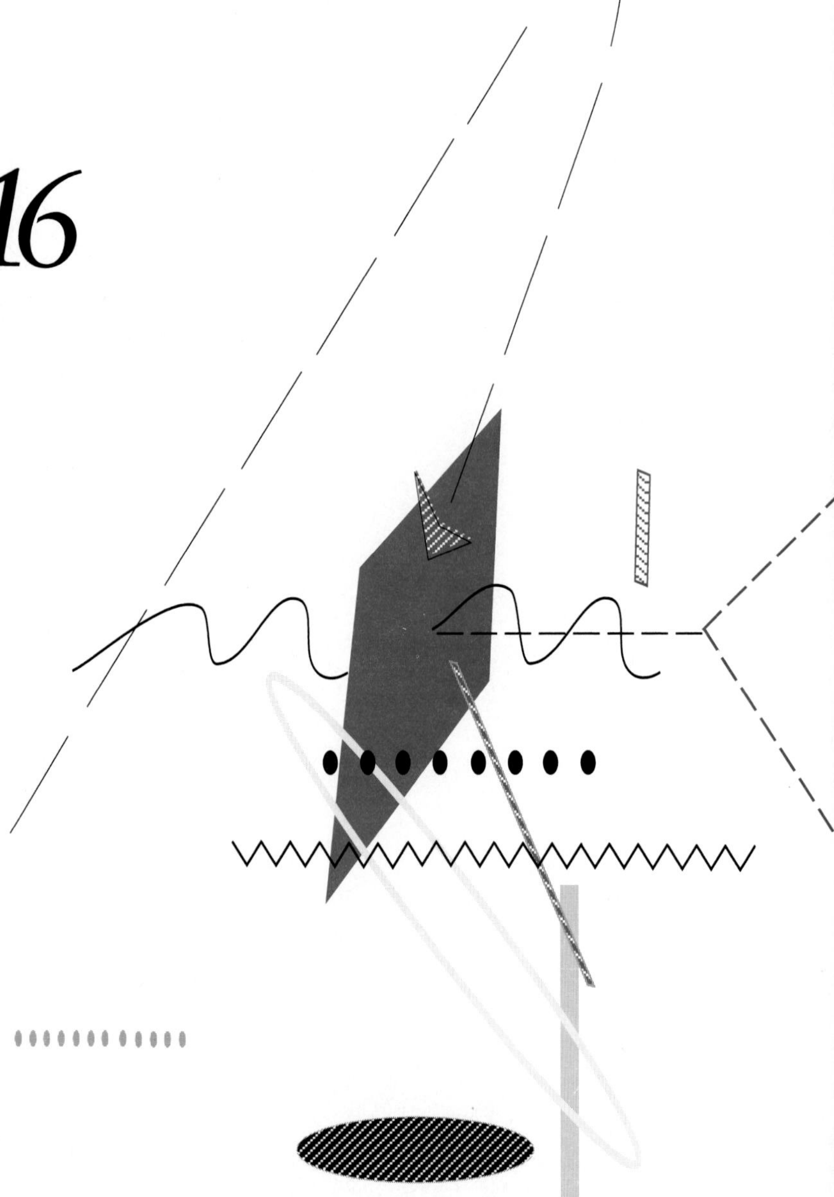

Link Strategic Alliances Together

Telecommunications can help to link together organizations, even, on occasion, competitors. Within organizations, such as Japanese gurupu, ***telecommunications can provide critical linkages between different sections. These connections can bring enormous improvements in operational efficiency. Also, between different organizations, telecommunications can provide a linkage that effectively creates a strategic alliance. Within sectors as a whole, entire groups of companies, even competitors, may use telecommunications-based utility functions to conduct their business.***

EFT/POS Systems in United Kingdom

Interorganizational systems are composed of links between the computers of different organizations through telecommunications channels. During the middle to late 1980s, electronic funds transfer (EFT) systems using point-of-sale (POS) terminals were developed and started to penetrate the market. The leadership for this technology came from Japan, which by 1990 had produced the greatest number of systems with the most customers and greatest number of stores linked together. Experiments were also going on in the United States, continental Europe, and the United Kingdom.

In electronic funds transfer POS systems, the customer pays for goods and services with a magnetic card. At the cashier's station, the customer passes the card through a reader and then enters a personal identification code number to validate the transaction. From each of the POS terminals in a store, transaction information is transferred to a concentrator hooked by leased line to a packet switched X.25 network. The information is then transferred to a switching computer which passes the message to the bank from which the debit card was issued. After verifying the transaction by comparing the PIN against the card number and verifying the availability of funds in the account, the bank notifies the POS terminal back through the network. (See Exhibit 16.1.) When the sale is confirmed, the bank first transfers the required money to the bank of the retailer, then debits the account of the customer, then includes the record of the purchase in the monthly statement to the customer.

There are several advantages to this system:

- The customer is able to make purchases without having to carry cash, and this is an advantage from the security point of view.
- If the card is stolen or lost, the customer still retains possession of the PIN, so the card is useless.
- The retailer is able to collect its funds immediately when the purchase is made, and yet avoids the time and expense of handling physical cash.
- The bank receives fees from both the retailer and the customer for use of the system.

Exhibit 16.1

EFT/POS System

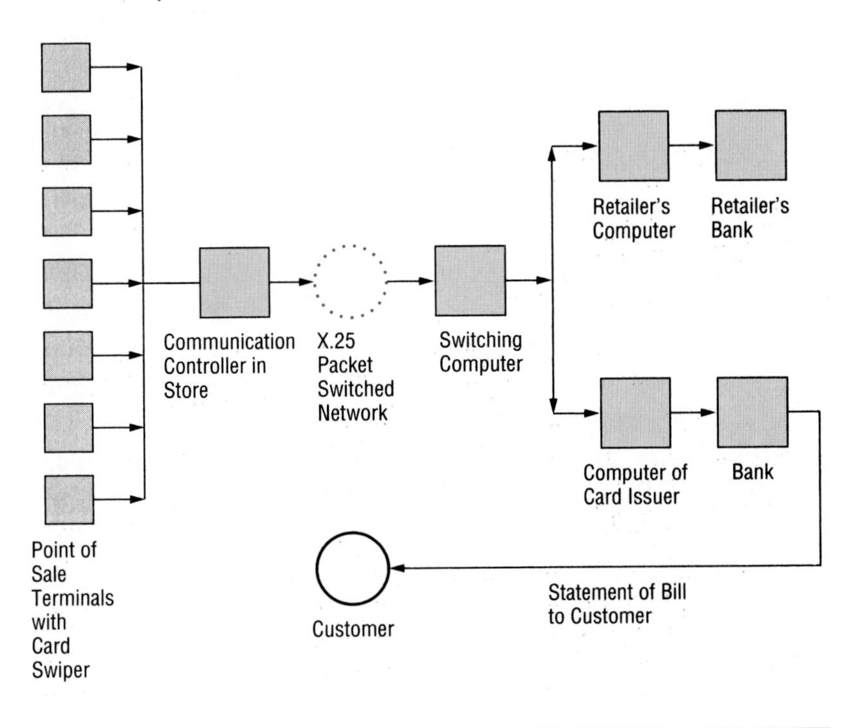

- The retailer is able to offer faster checkout than standard check writing in most instances.

EFT/POS technology is highly dependent upon telecommunications systems. The POS terminals are linked to the concentrator through a local area network in many cases. In addition, a packet switched network transmits the heavy traffic generated by short, individual transactions. The switching computer is able to use the heads and tails of the packet to switch the telecommunications traffic to the correct computers without having to reestablish connection each time. The electronic funds transfer from bank to bank is carried by interbank networks.

EFT/POS technology is one of the important developments that could lead to a cashless society in the future, although most expect that there will always be cash.

EDI and GEISCO

The General Electric Information Services Company (GEISCO) is one of the largest time-sharing and remote computing services in the world. GEISCO offers a large variety of products delivered over a packet switched network accessible from most of the developed countries of the OECD area. In time-sharing, a central computer facility offers a service to users through telecommunications channels. Time-sharing is a way to offer computer services to customers who, for either financial or other reasons, are unwilling or unable to build their own computing center. Time-sharing emerges as a cheaper option, particularly if workloads are not severe. Like any economic function, users suffer from the need to meet economies of scale in their own data processing and distribution of information. If their usage volume is not high enough, it is more economical for them to go to an outside service for a small fee based on usage. An additional advantage to using time-sharing is that it eliminates the necessity of maintaining a staff trained in computer science.

Although GEISCO started out as a basic time-sharing system, it suffered from the changing economics of time-sharing in the late 1970s. It had a profitable business for a few years, but with the introduction of, first, minicomputers and then desktop (personal) computers, it became less and less necessary for users to access remote time-sharing services. GEISCO faced a loss of customer base.

After trying a variety of types of services, GEISCO decided to stake a major investment in the provision of EDI services. Exhibit 16.2 diagrams the system it developed. **Electronic Document Interchange (EDI)** is the transmission of standard business documents through telecommunications channels typically linking different businesses. EDI systems are by their very nature interorganizational systems. EDI rests on the fact that much of business is conducted through the use of standard forms. Examples include purchase orders, invoices, shipping confirmations, etc. Even when business is conducted internationally, there is still a great deal of standardization in documentation (although, of course, some differences occur from one country to the next).

Agreement on a standard electronic format for an invoice, for example, allows the critical information to be sent through an electronic system. If the information is received at the other end of the channel in standard form, then the data entry function no longer seems needed, since the information can be put directly into the accounting or inventory control systems of the receiving firm. Exhibit 16.3 compares EDI and manual, paper-based transaction systems.

By the late 1980s, GEISCO was concentrating on building a worldwide EDI network linking together all of the major trading partners of the world into one standardized system. It was facing a few problems in building acceptance of EDI, and in getting agreement between different legal and regulatory environments on the standardization needed for EDI. On the whole, however, EDI was beginning to show the rapid penetration characteristic of the early stages of many new technologies.

The benefits to GEISCO of this approach are many:

- Once its customers start using EDI, GEISCO is virtually assured of a long-term contract.
- EDI is targeted at trade and commerce-related transactions, which generate a healthy volume by their nature.
- Linking users into an EDI system means that GEISCO can open the door to introducing more services, such as data communications or other time-sharing services.
- EDI becomes more and more cost effective as more firms join the system.

GEISCO has managed to do well with EDI, but its position is by no means assured. Other competitive EDI systems are emerging, and these are a direct threat to GEISCO. As the 1990s progress, we will expect to see the emergence of competing EDI systems, and the eventual building of gateways between the different systems based on standardization of the various popular EDI documents, listed in Exhibit 16.4 on page 389.

Exhibit 16.2

GEISCO EDI System

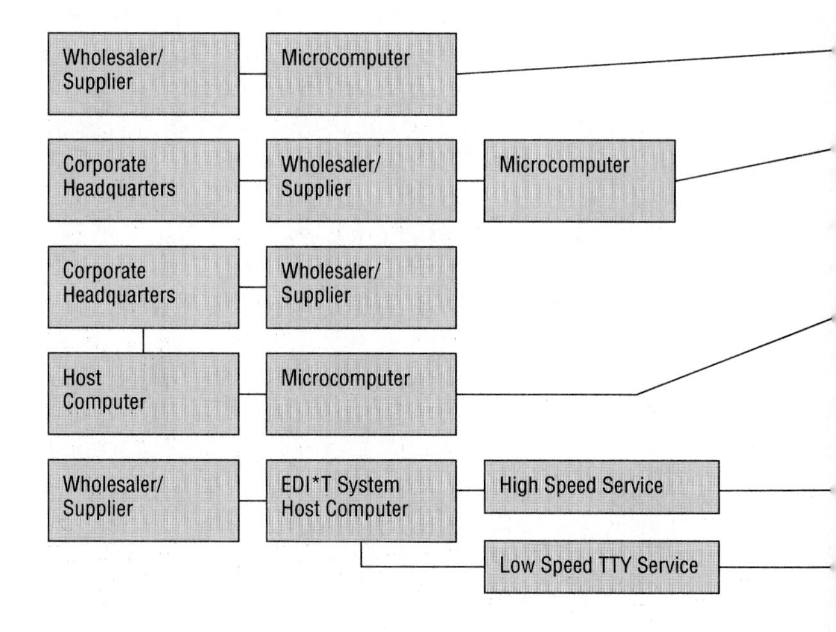

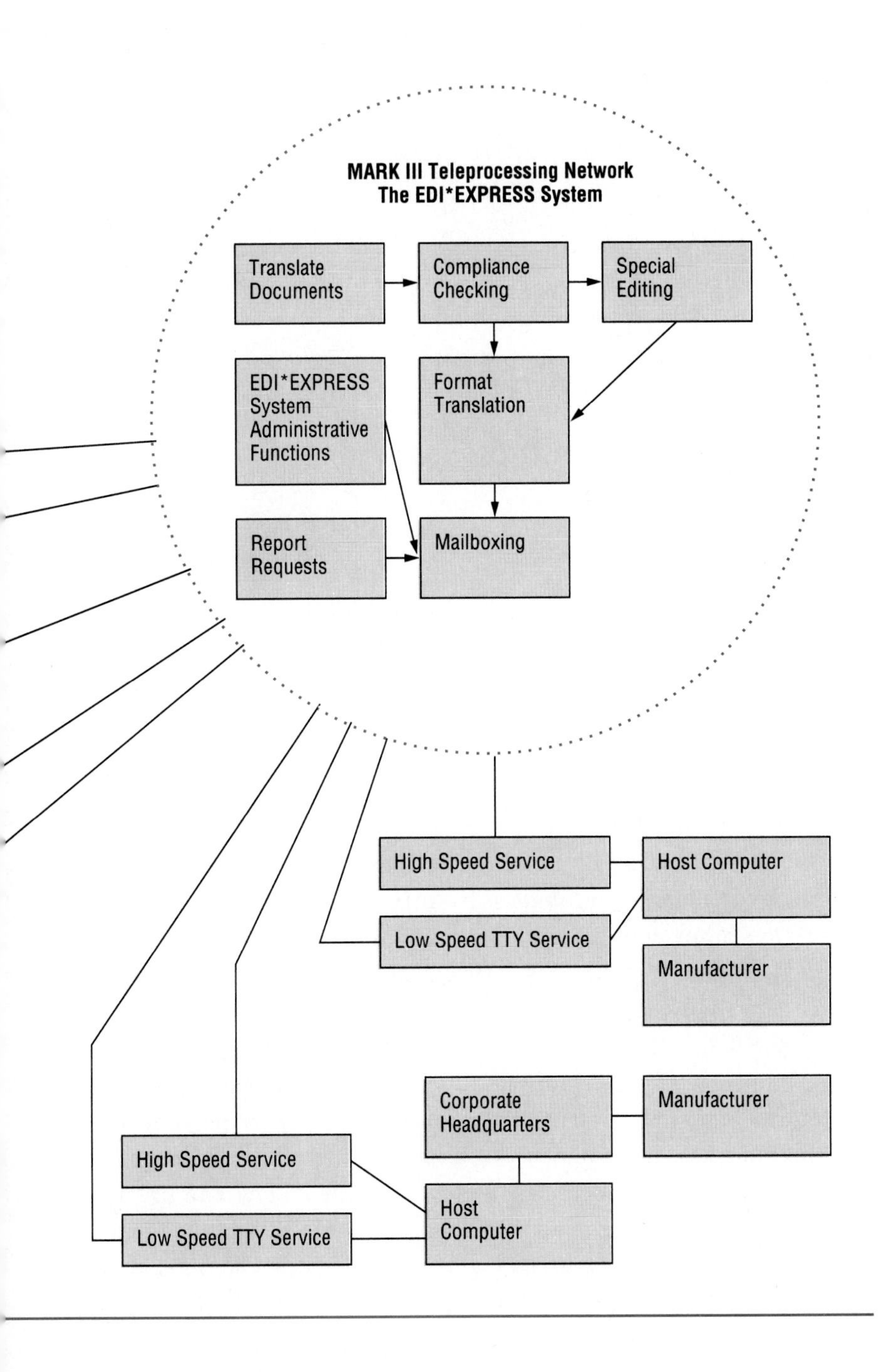
MARK III Teleprocessing Network
The EDI*EXPRESS System
Translate Documents
Compliance Checking
Special Editing
EDI*EXPRESS System Administrative Functions
Format Translation
Report Requests
Mailboxing
High Speed Service
Host Computer
Low Speed TTY Service
Manufacturer
Corporate Headquarters
Manufacturer
High Speed Service
Host Computer
Low Speed TTY Service

Exhibit 16.3

Electronic Document Interchange versus Manual Document Interchange

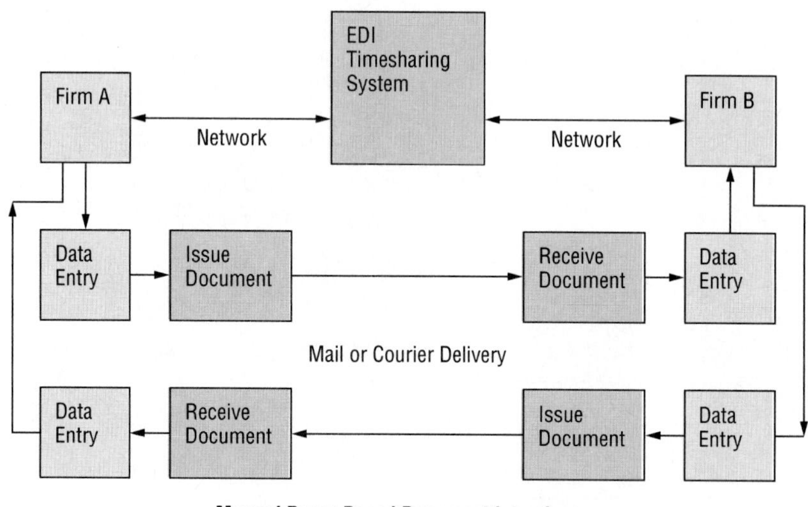

Legal Problems with Electronic Document Interchange

It is often said that legality and regulations follow technology. First technology changes, then solutions to emerging legal problems appear. Later, after a few court cases, a body of case law and practice grows up. (At least in the United States this sequence is common; in Japan there are virtually no legal suits.)

In the late 1980s, as EDI was gradually gaining acceptance, corporations were worrying about the unproven legal status of documents that were sent electronically. When a document is transmitted from one company to another, particularly a document that replaces a paper document with a set body of law and practice around it, changes might occur. Technically, it is possible to change a document stored in an electronic system such as a computer database leaving little if any auditing trail to show the source of the

Exhibit 16.4

Trading Documents Sent through EDI

Purchase Orders	Plan Schedules with Releases
Purchase Order Acknowledgements	Status Details Reply
Invoices	Shipment Information
Functional Acknowledgements	Acceptance/Rejection Advice
Freight Details, Invoices	Commercial Invoices

Source: Adapted from GEISCO literature.

changes. A problem could occur if two companies or organizations disagree about such a change. Consider the following sequence of events, for example:

- Company A sends an EDI order to Company B for 1,000 items.
- Company B ships 10,000 items to Company A (9,000 more than the 1,000 that were ordered).
- Company A rejects the shipment, claiming it ordered only 1,000 items.
- Company B sues Company A for nonpayment for the extra 9,000 items.
- The court examines the electronic documentation from both Company A and Company B, and finds that Company A has a copy of an order for 1,000 items, but Company B has an order for 10,000 items.
- Each company claims that the other has electronically altered its copy of the original order.

The court must make a decision for a settlement, but does not have any case law or practice to determine which document is valid. What should the court do? How can it decide the case?

Given all of these factors, it is easy to see why there were concerns in the 1980s that a set of legal practices would have to emerge. This type of situation, however, is not unique to EDI. Other examples arise in telecommunications, as well. The use of facsimile systems to send copies of bills of lading from the United States to Australia required an international agreement.

Mobil Corporation's Temporary Hold on EDI

As late as the end of the 1980s, Mobil Corporation was reluctant to install an optical storage system linked into a local area network because of legal concerns. "We do not wish to be one of the companies which has to go to court to justify the use of optical images," said one of its representatives at a cocktail party. "We would rather wait for other companies to prove the legal validity and sustainability of the technology before we commit ourselves." This type of prudent attitude is found in many organizations that want to exploit the best and newest information and telecommunications technologies, but are also cognizant of the legal risks involved.

Mitsubishi Heavy Industries' Links with Suppliers through Value Added Network

How can addition of a value added network virtually eliminate paperwork in an office, speed up communication with suppliers, and bring about approximately 4,500,000 yen in annual savings to a small work group with only eight members? Mitsubishi Heavy Industries, Ltd. has shown the answer. It is the largest heavy industry company in Japan, and one of the largest and most powerful companies in the world.

MHI's main products are large ships, large boilers and turbines, power plants, steel plants, cement plants, chemical plants, water treatment plants, paper machinery, printing machinery, space systems, turbochargers, diesel engines, various industrial machines, construction machinery, aircraft, etc. MHI has grown steadily since its founding in 1917, and by the end of the 1980s it had 54,500 employees and 14 major production centers located throughout Japan. Its annual sales were about 2 trillion yen (approximately $14 billion) and its capital equity was about 127 billion yen.

Within the vast enterprise known as MHI, the Nagasaki Shipyard and Machinery Works has the largest single plant. The Nagasaki Shipyard and Machinery Works produces about 20 percent of the total sales of Mitsubishi Heavy Industries. Nagasaki has approximately 8,000 employees producing its main products, large ships, power plants, and boilers and turbines. The plant has its own material purchasing department which handles inventory controlling and shipping. In the materials department, each of the several purchasing sections buys different materials. The machinery materials purchasing section, one of the purchasing sections, is mainly in charge of processed materials used for construction of thermal power generator plants.

Nagasaki purchases valves from a limited number of suppliers, with only a few exceptions. The suppliers, as is typical in many Japanese industrial organizations, are strongly related to Mitsubishi in terms of capital, personnel, and manufacturing technology. The valves used in power plants are standardized and specifications do not fluctuate very much. Valves play a very important role in a thermal power generation plant. They are critical enough that their failure can stop an entire plant. Mitsubishi developed the designs of the valves, then licensed the technology to its suppliers. The bodies of large valves are die cast, thus making delivery control difficult if standardized quality is to be maintained throughout the product line.

In order to get better control over the suppliers, Mitsubishi participated in the equity capital of the valve makers. Additionally, in order to strengthen the relationship between Mitsubishi and the suppliers, Mitsubishi sometimes sends important engineers and administrative personnel to the suppliers. Nagasaki works with a small number of suppliers in a tight relationship in which it can exert influence on the managers of the suppliers. However, the suppliers can expect constant orders from Mitsubishi, and therefore they avoid having to compete against each other.

There is a relationship between the high quality of valves needed and the type of information that must be sent between Nagasaki and its suppliers. Since the customers for power generation plants normally specify relatively strict quality standards such as those of the American National Standards Institute (ANSI) or the American Society of Mechanical Engineers (ASME), the valve specifications usually do not vary much in details. They must all meet the approved standards. As a result, the specifications and information sent to the suppliers is relatively simple. Usually the inquiry specification states only size, applied standard, capacity, and quantity.

Eight members in the valve purchasing group are in charge of purchasing the many valves used for constructing thermal power generation plants. The group has four purchasers, who are men, and four assistant purchasers, who are women. The four purchasers are in charge of deciding inquiry policy, negotiating with suppliers, and deciding purchase prices. The four assistant purchasers are in charge of sending inquiry documents and order sheets and making the filings.

At the Nagasaki Works, the purchasing work flow, including valve purchasing, was computerized as far as internal information is concerned. All external communication, however, had to be done by paper, telephone, and facsimile. Processing an order for a set of valves from a supplier was fairly complex, involving many steps (see Exhibit 16.5):

- The engineers entered detailed specifications of the valves on an inquiry sheet, then sent it to the valve group.
- The purchasers decided the purchasing policy and chose the supplier to which the inquiry would go, then instructed the assistant purchasers.
- The assistant purchasers added more information, such as inquiry number, to the inquiry sheet, then sent one copy to the supplier and another copy to the data entry clerks. They sent the inquiry to the supplier by mail.
- The data entry clerks inputted the information from the inquiry sheet into the mainframe. Data included the inquiry number, size, quantity, delivery time, etc. This data went into the inquiry data base.
- Upon receiving the inquiry, the valve supplier estimated the price and sent a quotation sheet by mail back to the valve purchasing group.
- When the valve purchasing group received the quotation, the assistant purchasers retrieved the inquiry data in question, then inputted the quotation information into the mainframe.
- After this, the purchasers negotiated with the suppliers by phone and settled on an agreeable price, payment terms, and delivery

Exhibit 16.5

Ordering Process before VAN

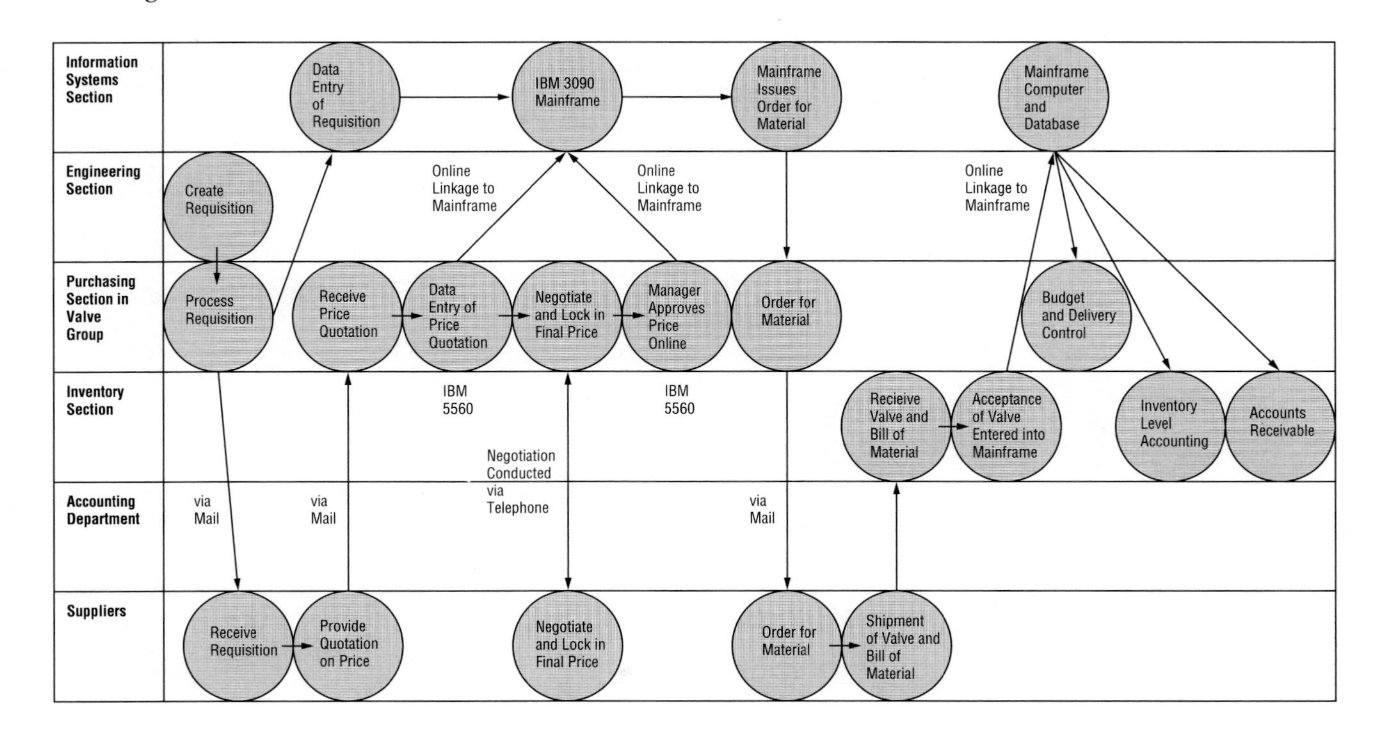

period. The purchasers asked for an approval of the order from the section manager through their on-line terminals. The manager approved the order using his own password. This process also added data in an order database.

- The next day the information systems section outputted all the approved orders as formal order sheets and then sent them to the valve group.
- The purchasers checked the order sheets for errors, then the manager of the machinery materials purchasing section stamped the order sheets.
- The assistant purchasers sent the order sheets to the suppliers by mail.
- The purchasers, at the end of the delivery period, delivered the valves with a bill of materials to a specified warehouse in the factory.
- The factory personnel checked the delivered valves and inputted acceptance information into the mainframe. This input increased accounts receivable in the accounting system and inventory in the inventory system.

This process generated very heavy paperwork for the assistant purchasers. More than 20 percent of the time of the four full-time assistant purchasers was spent in sending documents by either mail or facsimile. Approximately 100 sheets of inquiry data were sent from the engineering sections *daily*. The four assistant purchasers entered several instructions and inquiry numbers on the inquiry sheets, and wrote up the envelopes to send the inquiry sheets and formal order sheets to the suppliers.

For urgent inquiries, approximately 30 percent of all requisitions, it was not possible to use the regular mail service. Instead, the facsimile was used. As a result, occasionally the assistant purchaser would stand in front of a facsimile machine for a few hours sending inquiry documents.

Another problem was that handling the data for the quotations was very heavy work for the assistant purchasers. Approximately 100 sheets of quotations were sent from the suppliers each day. The number of items to input into the computer system for each quotation varied from 1 to 250, with an average being approximately 20 items.

The four assistant purchasers entered the data using their terminals. In the early stages of a plant construction project, the number of inquiries for valves might soar to 200 a day. When this happened, the valve purchasing group had to work overtime (without extra pay) to handle each day's paper.

For another problem, the use of the mail system could hurt when the critical delivery period was very short for a project. It was considered that any delay would do harm to the company reputation. (Contrast this to the attitudes found in many countries!) When all of the documentation was sent by mail to the suppliers, it normally took an entire working day for a document to reach the valve group from the suppliers and vice versa.

When a customer required a valve in an emergency situation, the delay caused by mail became critical. In order to continue to meet agreed delivery deadlines, Mitsubishi would order the valves without settlement of the agreed price. This would eliminate one of the delays in the ordering process and enable the company to cut delivery time. The consequence of this was that later, when it came time to pay, Mitsubishi would have to pay whatever price the supplier demanded. On the other hand, if Mitsubishi tried to minimize the purchasing costs, it would have to spend the time to settle the price, thus potentially resulting in a delay in the delivery schedule. Urgent orders might give only 1 or 2 weeks for delivery, and under these circumstances it was difficult for Mitsubishi to get an attractive price. In critical situations, the purchasers used facsimile, but this sometimes caused further delay, and sometimes caused miscommunication because of missing documents.

A further negative factor in the situation was the psychological dimension. The great amount of papers piled up in the purchasing office, and the resulting disorganization, depressed the purchasers, who always felt they were behind schedule. Most of the papers were quotations from suppliers. Although the purchasers managed to keep up with the work, occasionally a quotation went missing, causing an additional communication task with the supplier. When the papers for urgent orders went missing, the situation could become desperate. Each day when the purchasers went into the office, they would face a huge, depressing pile of pending papers.

At the end of the month, things usually got even worse. Suppliers would call into the office to inquire regarding the payment schedules for valves they had already delivered. Such calls interrupted the assistant purchasers even more. Once the valves were accepted by the inventory control section, personnel there inputted the acceptance information into the mainframe. Since the suppliers were never sure the input was made correctly and that the payment was on schedule, they would ask for confirmation by calling the valve purchasing department. At the end of the month, the assistant purchasers would be interrupted every few minutes to check the acceptance information on their terminals. Time devoted to this work would take up to 10 percent of total daily working time for an assistant purchaser, depressing productivity on the on-going work.

To solve these problems, the valve group adopted a value added network system to set up electronic telecommunications linkages between themselves and their suppliers. Mitsubishi adopted IBM Japan's VAN which is named NMS. NMS is a nationwide telecommunications network with a center in Mitaka, Japan. It is a computer-controlled, multimedia/multipath, private, packet-switched, data-only telecommunications system. At first, some of the managers were dubious about the benefits of using the value added network, but approval was obtained from higher up in the organization. The VAN was set up to link the valve purchasing group with eight key valve suppliers of most of the normal and simple automatic control valves, approximately 60 percent of the total valves purchased by Mitsubishi (see Exhibit 16.6).

The total cost for utilization of the value added network system is 650,000 yen per month, which includes the network use fee and additional computer operations cost. This figure is calculated by prorating the shared cost for the valve group's number of employees.

The processing of orders between the valve purchasing group and the suppliers became completely automated as a result of using the value added network. (See Exhibit 16.7.) The suppliers were linked into the mainframe supplied by the Choryo subsidiary of Mitsubishi, thus making it possible to obtain needed information. Exhibit 16.8 diagrams the new, automated purchasing process.

Exhibit 16.6

Total Annual Valve Orders by Type

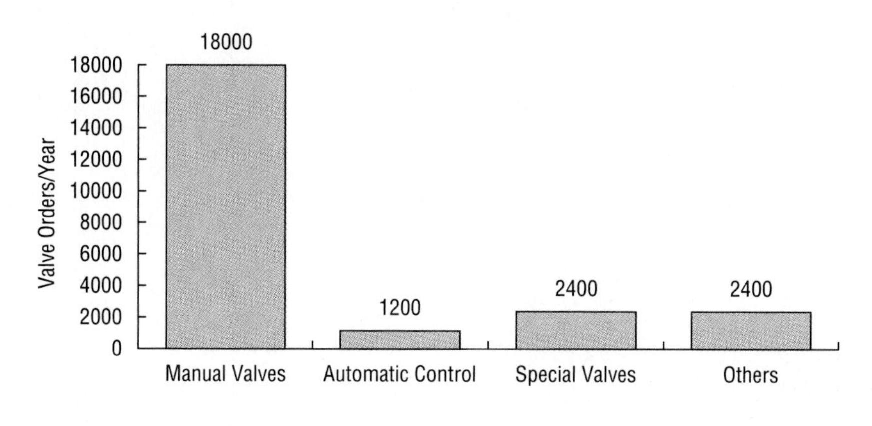

Exhibit 16.7

Nagasaki Works Value-Added Network

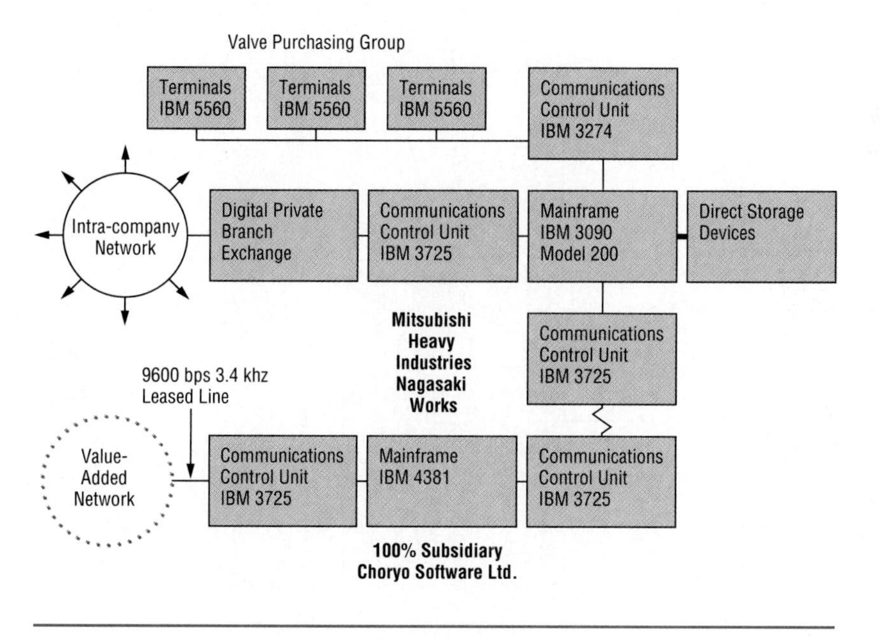

Exhibit 16.8

Ordering Process after VAN

The value-added network tied Mitsubishi's suppliers directly into its information system.

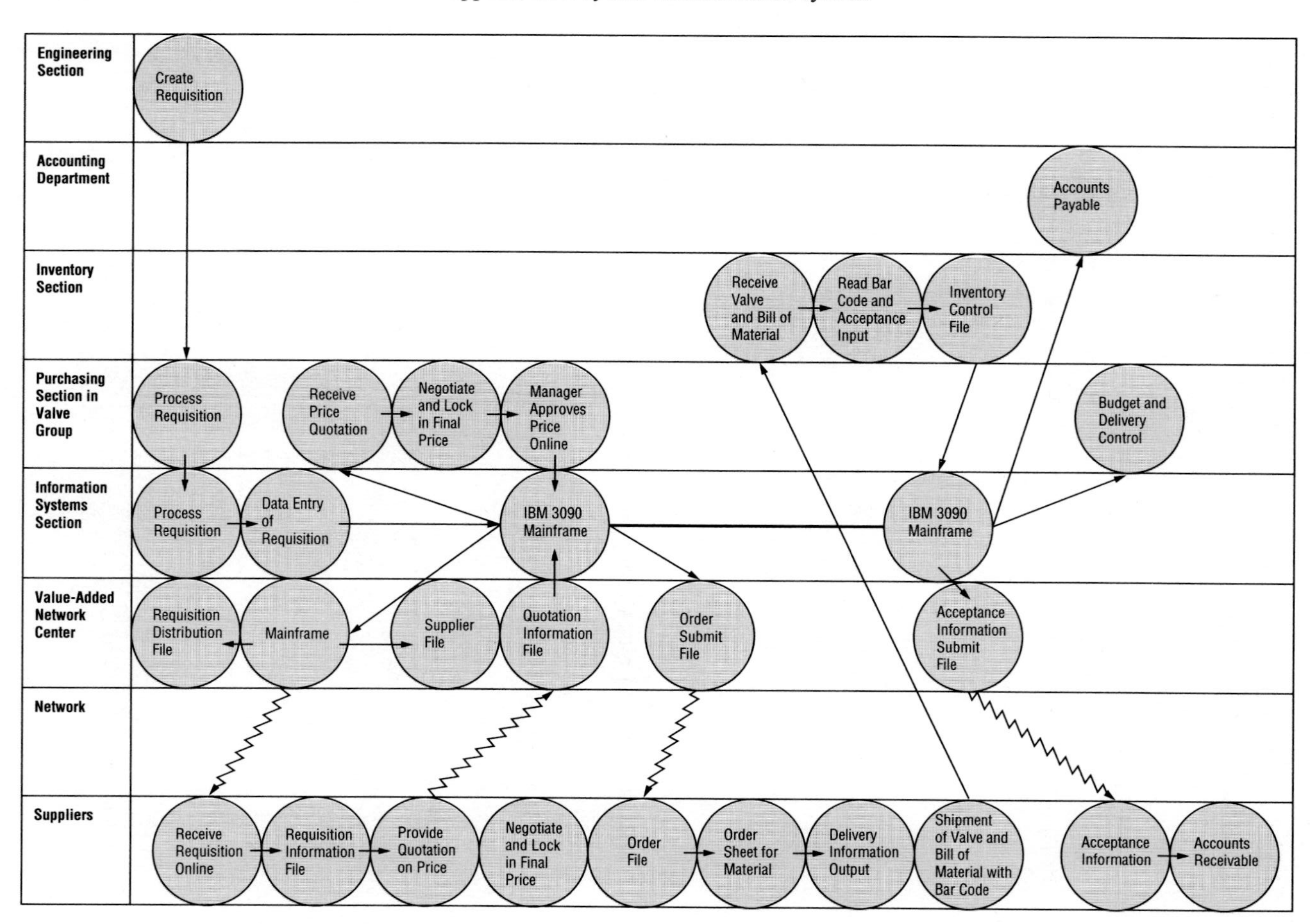

The benefits of this linkage between Mitsubishi and its suppliers can be measured in several different dimensions. All of the inquiries, quotations, and other information are sent back and forth through computer facilities instead of through the mail and via facsimile. As a result, there is no need for assistant purchasers to prepare and fill out envelopes, or to try repeatedly to send the documents through facsimile. The approximate saving was 12 work hours or 36,000 yen per month.

As all of the information input by data entry clerks into the database is simultaneously sent to the suppliers through the computer telecommunications facilities, most of the mail-related paperwork is eliminated. This saves many clerical steps, including:

- Copying numbers on inquiry sheets for multiple inquiries
- Putting address stamps on inquiry sheets
- Checking addresses of companies
- Stuffing envelopes with inquiry sheets
- Addressing envelopes
- Sending urgent inquiries by facsimile

This saves approximately 10 percent of the time of the assistant purchasers. This improvement increases time spent in creative work and betters the overall situation in the office. The assistant purchasers no longer have to enter a great number of quotations into the mainframe, because they are directly inputted through computer communications. The suppliers input the quotations to the system, not to the valve purchasing department's mail basket. This saved about 50 percent of all data entry work of the assistant purchasers. Particularly during the peak times for purchasing, when a plant is in the early stages of construction, the system has eliminated the need for working beyond normal hours. Also, the purchasers do not have to wait for the assistant purchasers to input the quotation to the mainframe during peak periods. This saves 32 work hours per month, equivalent to 96,000 yen.

For another saving, the communication through computer facilities instead of through the mail makes it possible to save time that had been needed for dispatching. The quickness of the communication prevents the company from delay in the delivery of spare parts to customers. This ultimately improves the reputation of Mitsubishi, while saving approximately 225,000 yen per month. The additional value of the better reputation with customers is impossible to quantify.

The inquiry sheets and quotations which in the past had been set by mail came to be telecommunicated. This fact alone saved at least 2 working days simply in issuing one order. Normally, if a purchaser accepts an inquiry in the morning of a particular day, it is now possible to look at the related quotation the next afternoon.

Furthermore, the purchasers no longer have to worry about having to specify the short-time delivery orders and emergency orders. Costly purchases of emergency orders have been eliminated because there is now enough time to agree on a price before shipment. Exhibit 16.9 contrasts the number of orders overall and the number over budget with and without the VAN. There is no longer any delay in delivery of emergency orders.

Additionally, many piles of papers that cluttered the office in the past have disappeared. As a result, other valuable files of information can be kept in better order. The psychological productivity of the workers in the valve purchasing section has improved as a result of the clean and neatly arranged office. In general, the consensus at the valve purchasing office was the following:

- Paperwork was eliminated and thinking time increased.
- Time in front of terminals was reduced, further increasing thinking time.
- The office became cleaner and more organized than before, thus enhancing productivity.
- Structured, routine work was eliminated, and the work is now more interesting.
- Communication with suppliers is smoother than before and friction has been reduced.

The most important result of the value added network was that the overall relationship between Mitsubishi and its suppliers was highly strengthened. Upon receiving valves at the warehouse, the inventory control section checks the quantity, quality, etc., and then inputs the

Exhibit 16.9

Total Orders and Over-Budget Orders

Mitsubishi increased the number of orders while decreasing the number of over-budget orders.

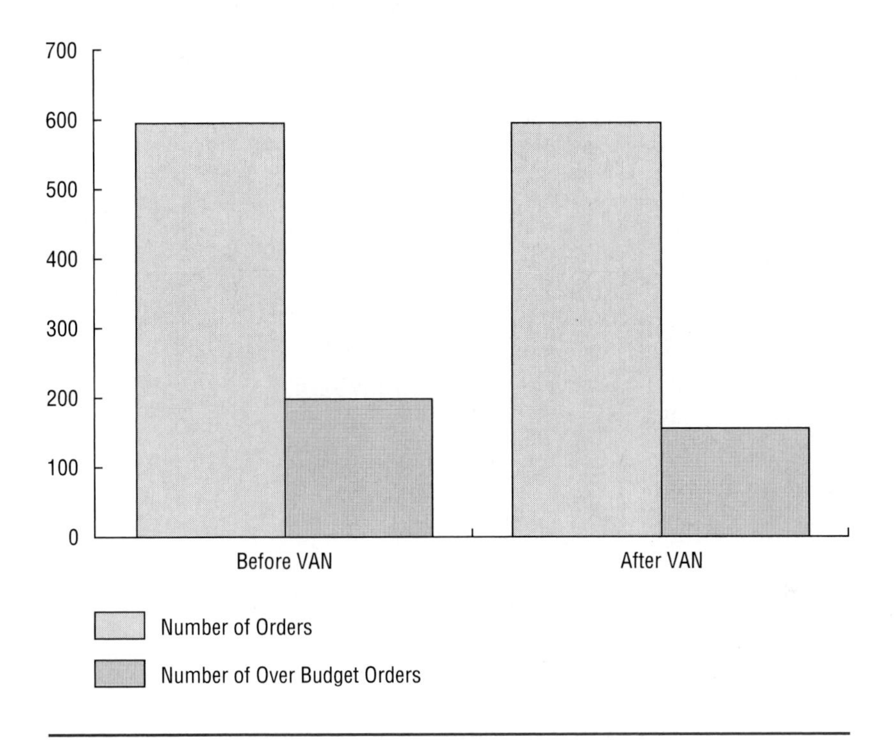

acceptance information to the mainframe for inventory control purposes, as before. At the same time, the information is relayed through the value added network to the supplier. As a result, at the end of every month, the suppliers can output the acceptance list at their offices and have no need to call the valve purchasing department to make inquiries. The assistant purchasers are therefore not interrupted by many telephone calls at the end of every month. This alone saves eight working hours, equivalent to 24,000 yen per month. Exhibit 16.10 shows the change in volume and overall costs of valve purchasing.

Exhibit 16.10

Cost per Order before and after VAN

Mitsubishi increased the orders and lowered the cost.

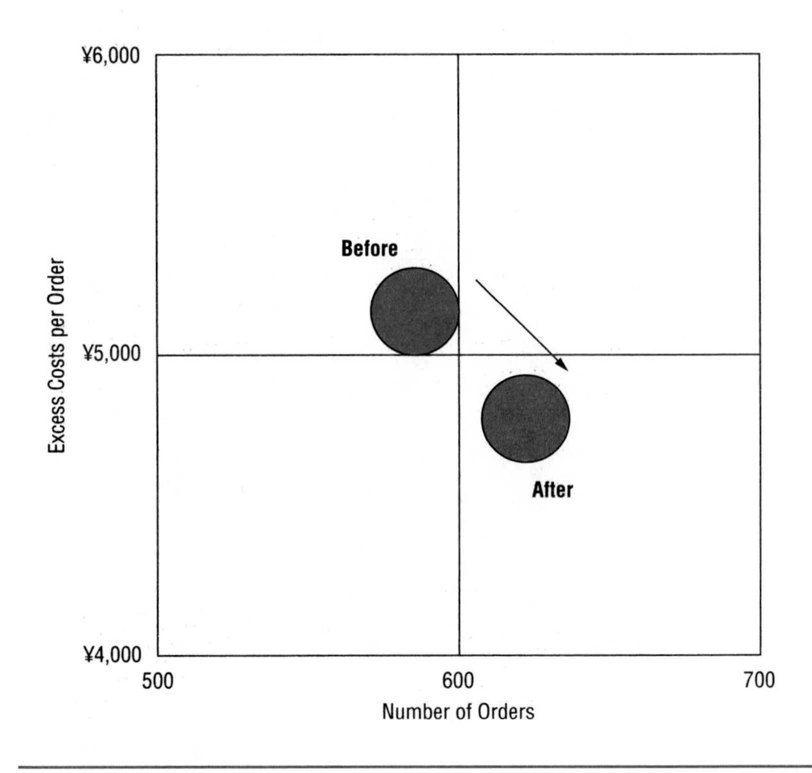

Ford's FordNet EDI System in Europe

The automotive industry has been one of the first to use Electronic Document Interchange (EDI) on a large scale. For example, Ford Motor Company is typical of automotive manufacturing operations in Europe. It has a multitude of suppliers from which it buys large amounts of goods with a frequency and pattern determined by sales and production requirements. In order to support this type of operation, millions of documents—bills, remittance advice notes, invoices, etc.—had to flow between Ford and its suppliers. The amount of time and money to process this mountain of paperwork was considerable, and errors inevitably crept in when something went slightly askew. For example, in any dispute about a receipt or bill, the Ford accounting department had to step in and investigate on a case-by-case basis. This resulted in holding up the payment until the dispute was resolved.

Realizing that it was swimming through an ocean of paper, Ford sought a telecommunications-based solution to automate its relationship with its suppliers. It had several goals for the system:

- Increase the satisfaction of suppliers and further link them into the Ford manufacturing system as reliable partners
- Decrease the mountain of paperwork handled on a daily basis to work with Ford suppliers
- Increase the velocity of information throughout Ford's system
- Decrease the drag from errors, corrections, and disputes between Ford and the suppliers

To accomplish these goals, Ford created FordNet, an electronic data interchange system for its European operations. As FordNet became operational, the old paper-based system was replaced by an Evaluated Receipts Settlement system in which invoices and other information are sent to the suppliers over the network, instead of through paper transactions. The information is coded into a standardized format to eliminate problems with definitions of documents. In addition, actual payments are now being made electronically through electronic funds transfer directly to the accounts of the suppliers.

The result has been a significant improvement in efficiency of the interface between Ford and its suppliers. The velocity of information has been greatly improved, and the dispute settlement procedures are under control.

SWIFT

One of the greatest interorganizational systems in the world is run by the Society for Worldwide Interbank Financial Transfers (SWIFT). SWIFT was set up by a group of banks as a telecommunications-based system to transfer money from one bank to another in the settlement process. Linking together hundreds of banks around the world,

SWIFT is the most important standardized conduit for electronic funds transfers.

EFTs were increasing at a very rapid rate at the end of the 1980s. It is expected that more and more additional information and services will be added in the future, with the introduction of SWIFT II, a new upgrade of the system. In spite of the name, SWIFT II has been virtually redesigned from scratch to take advantage of new capabilities in information and telecommunications technologies.

When a bank in one country wishes to send funds to a bank in another, or to acknowledge a change in the status of an account, it transmits the information to the other involved banks over the SWIFT network. SWIFT has a special encrypting and coding feature to provide the type of security needed for transferring large amounts of funds.

The SWIFT network has many features that make the transmissions efficient, secure, and auditable. If some problem occurs with using the network, SWIFT has quick recovery features. Most importantly, it provides an audit trail that leaves traces for each transmission of information and transfer of funds.

Summary

Links between either related companies, competitors, or others into electronic networks greatly improve economic efficiency and make possible many activities that were impossible in the past. In banking and finance, for example, organizations such as SWIFT show how competitors can get together and build a telecommunications-based system to provide infrastructure for conducting business. Electronic Document Interchange is another example of how companies can use telecommunications as an intermediary between different organizations. Technologies such as EFT/POS also provide critical linkages between totally different sectors, i.e., retailers and banking institutions.

Questions

16.1 a. Explain how EFT/POS systems work and discuss the role of the telecommunications infrastructure.
b. Why are such systems worth setting up?

16.2 a. What are some of the problems in implementing EDI systems?
b. What are the advantages and disadvantages of EDI systems?

16.3 Give some examples of how various sectors have set up telecommunications-based cooperatives for business purposes.

Chapter 17

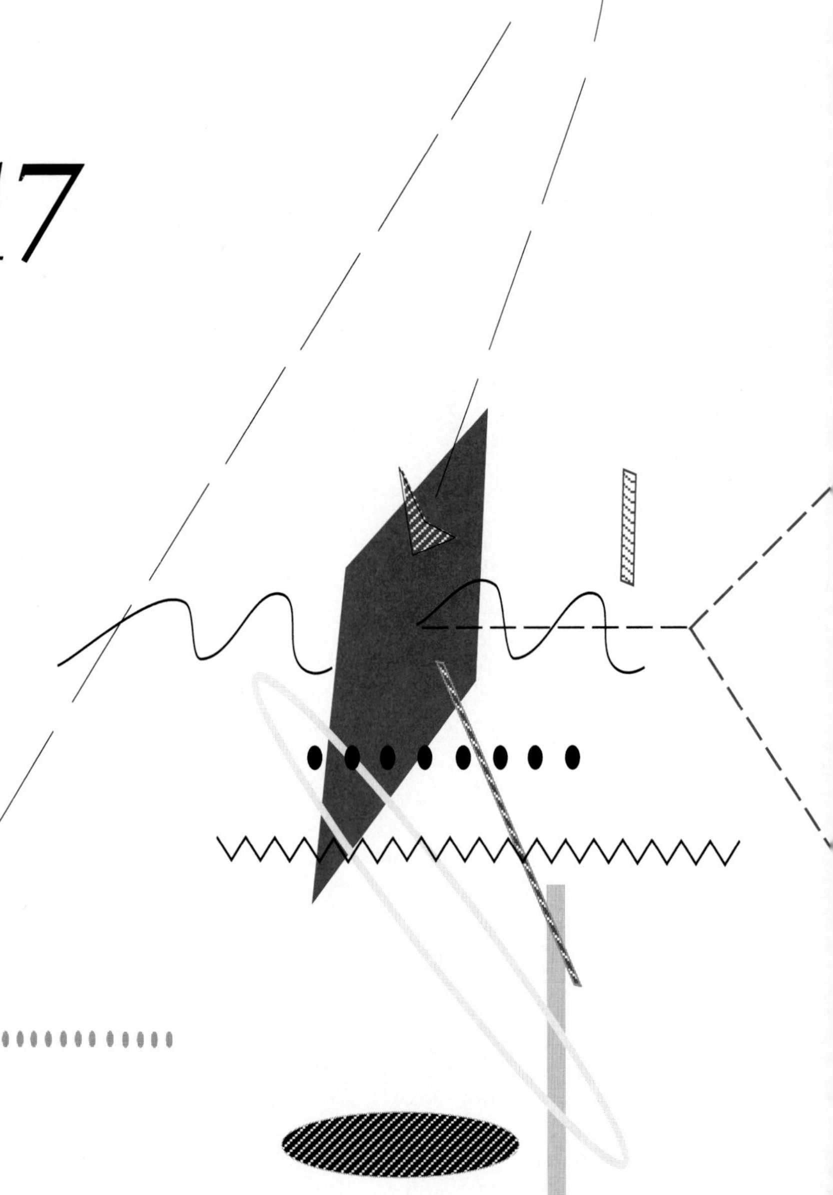

Key Challenges for the 1990s

This chapter summarizes the challenges and problems facing management in the 1990s. It also discusses some key opportunities presented by the current uncertain, fluid situation, then it discusses what the future will hold and how business operations will change.

Business in the 1990s

Business in the 1990s will face many problems arising from developments in telecommunications technologies. The most significant will revolve around two basic issues: how to maximize benefits from the investment already made by the corporation in its telecommunications resources, and how to respond to rapid advances made by competitors in their use of information and telecommunications technologies.

As we have seen, the investments made in telecommunications are a significant part of a company's MIS budget. The amount spent on telecommunications as a percentage of overall expenditures has been steadily rising, as well. Telecommunications expenditures have increased from about 5 percent of the huge MIS budget to more than 10 percent, depending on the economic sector to which the business belongs. With this type of rise, the economic and managerial risk of error increases. In many industries, such as financial services, we have shown that there is virtually no room for error whatsoever. With the eroding position of any fixed technology, given rapid technological change and resulting accelerated obsolescence, management will face a shifting set of equations defining the economic payoff from investment. This shift will mean essentially that payoff must come faster and faster.

Firms have always faced a major problem in that competitive advantages gained through information technology may quickly erode under the pressure of competitors. Technological changes increase the ability of competitors to equalize the playing field. Firms can leapfrog the technology of competitors more often, since new technologies that can give advantage arise more rapidly. It is no longer possible to plan using a fixed mode of analysis, basing assumptions on stability in the environment. Both the technical and business environments are changing rapidly, and will continue to do so. The effect of this is that a business must plan its utilization of telecommunications technologies under pressure of *permanent uncertainty.* The question becomes not so much what the best type of fixed planning methodology and form of analysis might be, but rather, how does one change methods of thinking in response to permanent, on-going change. This is one of the great management challenges of the 1990s.

Is Telecommunications the Only Source of Competitive Advantage?

You have seen how telecommunications has been used successfully to radically shorten response intervals and thereby change customers' expectations of service. This change in expectations on the part of the consumer sets up a barrier to entry against other competitors who must, as a result, offer the same level of service. Those that can offer the same level of service can stay in the game. However, if they must pay a relatively higher cost to move to the new level of service, then they face an even longer-term problem in economic survival.

It is critical that the reader understand, however, that many other factors in addition to telecommunications may be responsible for the increase in competitive advantage. Reward and training programs for employees, basic changes in internal operations, adoption of new technologies or processes, changes in quality or pricing of services, adoption of lean manufacturing techniques, and many more factors may be critical ingredients in reducing the response interval and providing superior customer service. The reader should always think of telecommunications and information technology in their proper places. Firms use them in collaboration with many other management actions to improve the operations and competitive advantage of the business. There is little room in this world for monodimensional thinkers in business strategy. There is no single answer or strategy or factor on which one can depend for business success.

After the Millennium

The 1990s will see continued rapid advances in all telecommunications technologies. Different cycles of competition and innovation will continue to sweep through the business world. Successful competitive strategies of today will crumble, to be replaced by better ideas and implementations. Business will be faced with a series of changed or intensified conditions, which will come in some sectors to dominate the shape of competitive strategy.

Time and interval reduction will continue to be more important in both internal operating efficiency, as well as providing the type of instant customer service required by the new generation of consumers—and the new generation of time-based competitors. Many homes in the developed countries of the world will have access to advanced services such as ISDN through their regular telephone lines, thus opening up entirely new markets for information-based goods and services.

In business establishments, as well, the coming of ISDN will open up very great opportunities for linking together networks of advanced, high-quality workstations. This makes end-users unaware of where their future comes from.

As international business moves into the next millennium, telecommunications will be a seamless web linking together every imaginable type of equipment in an information system of a complexity never before seen in human history. In the past, when it took days or even weeks for messages to travel internationally, and when overseas operations of organizations were run as feudal kingdoms exempt from headquarters control, telecommunications consisted of little more than international telex. Later, as we have seen, telex gave way to international computing and facsimile systems. The late 1990s should see the final extension of control to reach decentralized operations and a strong tendency toward centralization of power and control in transnational corporations.

But in spite of the negative consequences that may result from heavy centralization, the linking together of large resource control and distribution networks internationally through telecommunications channels and networks will help bring about much greater efficiency and much less waste of resources. It is almost as though the entire world, with the bulk of its resources, will be linked together into a giant point-of-sale system in which inventory control will be highly efficient and waste will be reduced.

What we saw in the 1980s with just-in-time inventory systems, point-of-sale systems, and lean manufacturing will hopefully spread throughout the world, particularly the developing countries, in the 1990s. This will spread the great strides in efficiency to all sectors of the world.

Ultimately, telecommunications brings together the human community and makes the world seem smaller. Geographical and time differences have been erased, and this may in the future lead to much greater understanding on the part of all humans as to their common destiny and common responsibility to promote peace and ecological and economic rationality. As with all such great social and political transformations, there is a profit to be made somewhere by the clever corporation!

Part ■ *III*

Minicases with Critiques

Case studies provide one of the best ways for the student to learn about the practical difficulties encountered by a telecommunications team working on project development or problem solving. Under these circumstances, the problem solving requires a combination of vision, methodologies, specific information on technical details, and ability to present a coherent argument and reasoned idea to top management.

This chapter presents several detailed problem solving exercises meant to be realistic for the environment faced by today's telecommunications manager. These cases present background, problems, solutions, and results in different telecommunications applications. At the end of each solution and case, a critique alerts the reader to any possible errors. The reader should first absorb the case to learn how the solution was developed, then study the critique carefully.

Each of the minicases presents some specific problem. It is the job of the reader to examine each of the cases and take a critical view. When reviewing the critique, see if there are any additional problems you may have overlooked.

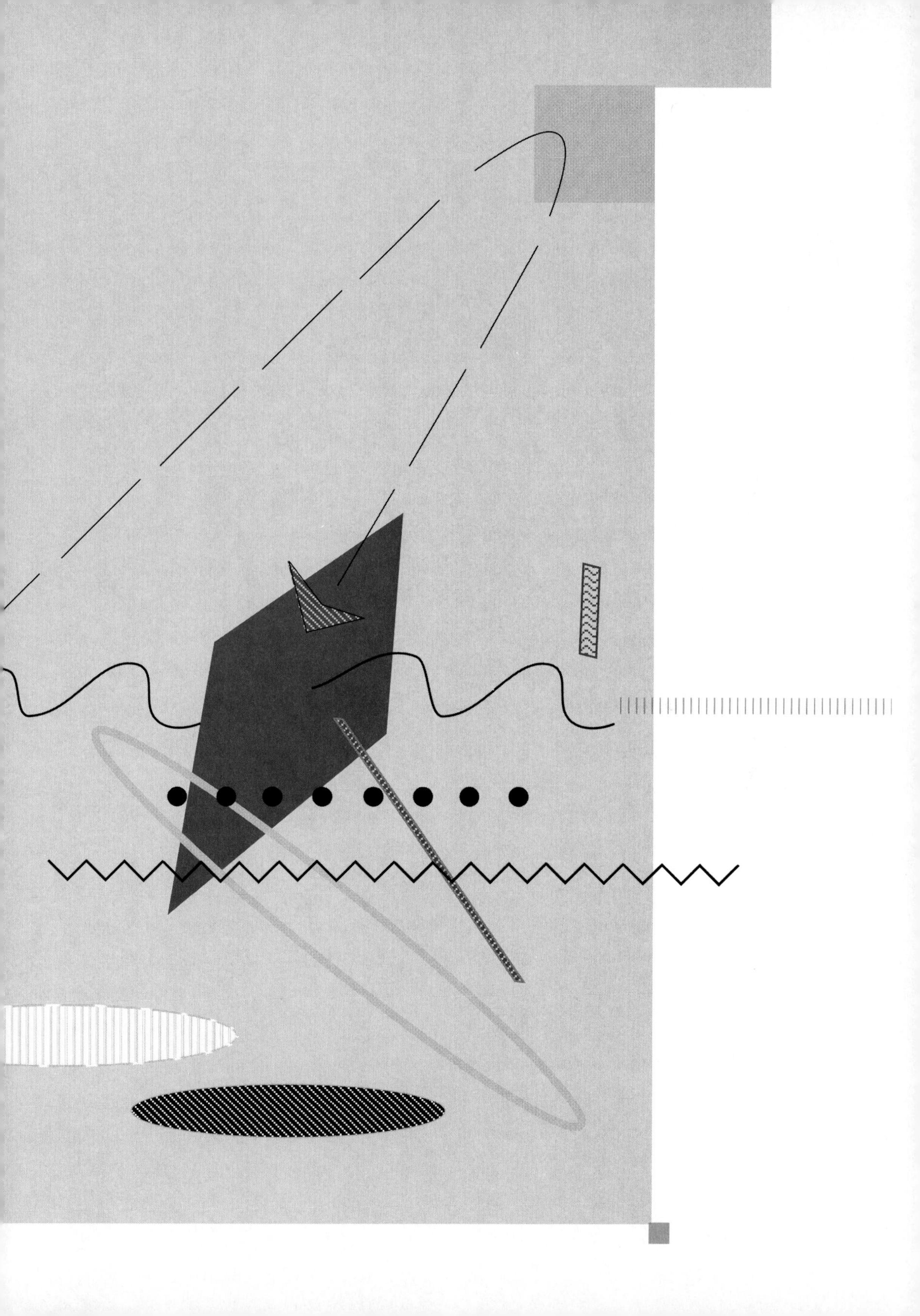

Minicase *1*

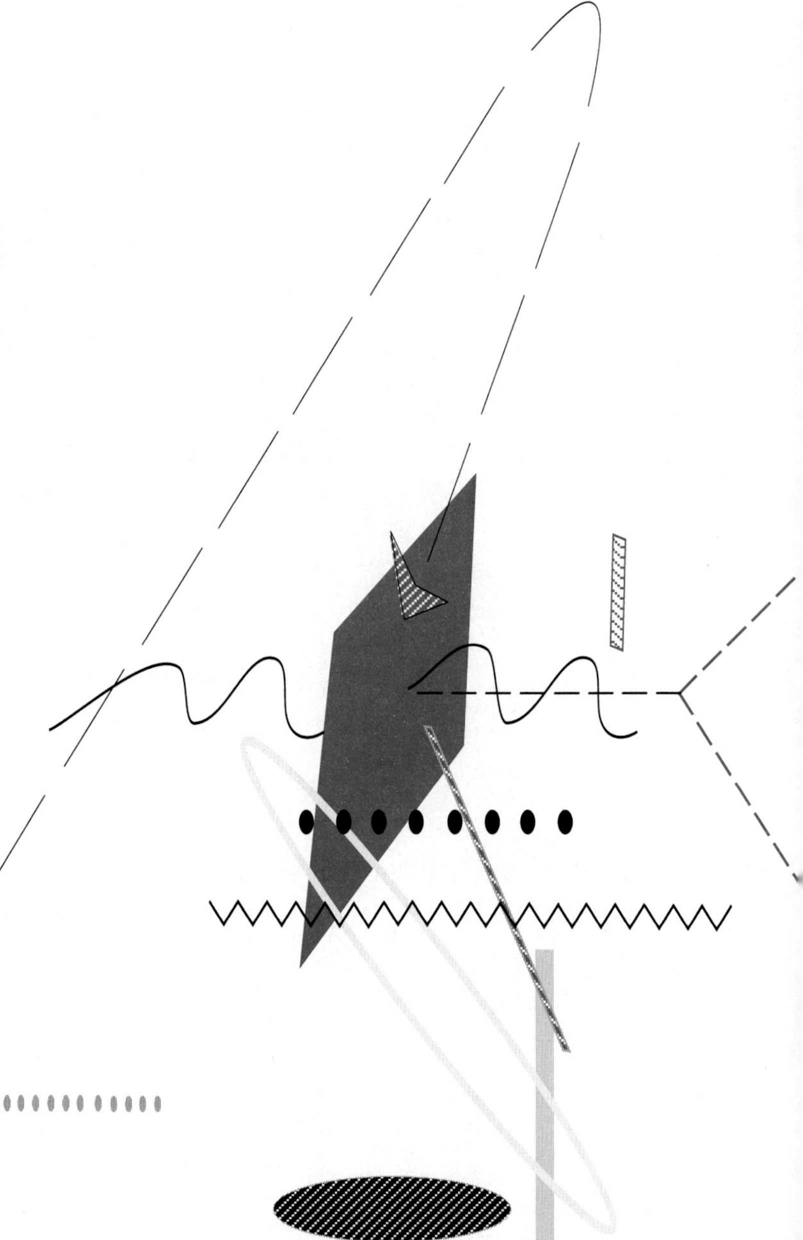

The Edo Company

Teleconferencing between Tokyo, Milan, and Silicon Valley in a High-Technology Firm

The Challenge

The Edo Company is a Japanese multinational company. Its headquarters office in Tokyo employs 3,000 engineers and 30,000 people overall. Edo Company produces computers, semiconductors, and software. Its annual sales are approximately $1.5 billion.

Over the past few decades, Edo Company has concentrated on heavy industry. Now, however, top management has recognized the opportunities in high technology, particularly advanced semiconductor manufacturing, systems integration, and software design.

In order to move more vigorously into high technology, Edo Company has taken over a Silicon Valley high-technology company and formed a joint venture with a very large Italian computer and telecommunications company. Edo took over the Silicon Valley startup company because it was composed of very young, brilliant software and systems engineers who did not face the same creativity problems as are rumored to exist in Japan.

The general plan of Edo Company is to use the creative technology from the Silicon Valley company, manufacture it in Japan, where manufacturing efficiencies are much greater than in the United States, then place the components into the body of Italian-made equipment so as to guarantee entry into the protected European Community market after 1992.

In order to accomplish this global feat, Edo Company is forced to coordinate very highly technical activities, including the product design in Silicon Valley, the manufacturing in Japan, and the specifications of the systems integrator in Milan. This coordination process requires frequent meetings between different teams of engineers. As the project starts to enter active manufacturing, these meetings become much more frequent, and coordination becomes more difficult.

As a result, Edo Company's Research and Development department is considering setting up a three-way teleconferencing system between Tokyo, Milan, and Silicon Valley. At each location, four top engineers will attend an approximate annual average of 48 conferences so that the different design teams can communicate with each other without travel. Edo Company has learned that one of its chief competitors has adopted the same strategy for placing manufactured components into the European market, and Edo is looking toward teleconferencing for the edge over its competitors to get to the market first.

A team in the telecommunications area met and worked out the analysis of the problem excerpted in the following text.

The Response

The frequency of project meetings between Tokyo, Silicon Valley, and Milan has caused problems of increasing time loss during trips, traveling cost, and project delays. Since Edo Company began active manufacturing, the project meetings have become so frequent as to cause several troublesome situations. As an important decision has to be made, top engineers and technical managers should meet in one place because three remote companies are involved in their own tasks within the entire project; Silicon Valley is involved in designing the advanced semiconductor, manufacturing and fabrication are being set up in Japan, and the systems integration is being done in Milan. Therefore the project meetings require two-way, or even three-way communication, between those geographically dispersed areas of the world.

The major problems incurred in this situation are as follows:

- Increasing travel costs for air fares and accommodations
- Increasing time lost during business trips for top engineers
- Delay of the project

Next, the telecommunications team discusses the travel and time costs being incurred by Edo Company.

Travel costs, which include air fare and accommodation fees, amount to approximately $50,560 per month. As the project meetings become more frequent, the cost factor becomes a major issue for the company.

Current Cost of Travel for One Month

Round-Trip Air Fare	$1,350
Accommodation Fee (1-2 days)	170
Car Rental and Miscellaneous	60
Number of Trips	4
Number of Engineers	4
Two such parties travel, one from the United States and one from Japan, to Italy.	
Approximate Total	$50,560

After pointing out the basic costs involved in the current arrangement for coordination of the engineers, the telecommunications team then estimates that approximately $24,000 worth of time per month is being lost.

The time of top engineers lost during business trips has increased to about 800 hours per month. Usually, a project meeting needs the presence of four top engineers from each of the three areas. However, as the project advances and gets more complicated, more people who work on specific tasks in the project should be involved in the project meetings. Furthermore, although the meetings are currently held four times per month, it is anticipated that the number of meetings will increase.

Our calculation estimates that currently 100 hours, or 4.2 days, per month are lost on average per engineer. During the time spent for travel, engineers cannot work on their jobs, thus the project is likely to be slowed down. The aggregate time loss for trips amounts to 800 work hours for the company. Converting the time loss factor into a dollar value based on an average engineer's salary, time equivalent to $24,000 per month is being lost by the company as an opportunity cost.

The telecommunications team then discusses the overall danger to the entire enterprise caused by these delays:

The project is likely to be slowed down because the frequent project meetings restrict the schedules of the engineers. A number of factors, illustrated in Exhibit M1.1, delay the project:

- It is difficult to coordinate the timing of meetings of more than 10 engineers from three remote areas. A couple of days, sometimes a

Exhibit M1.1

Sources of Project Delay

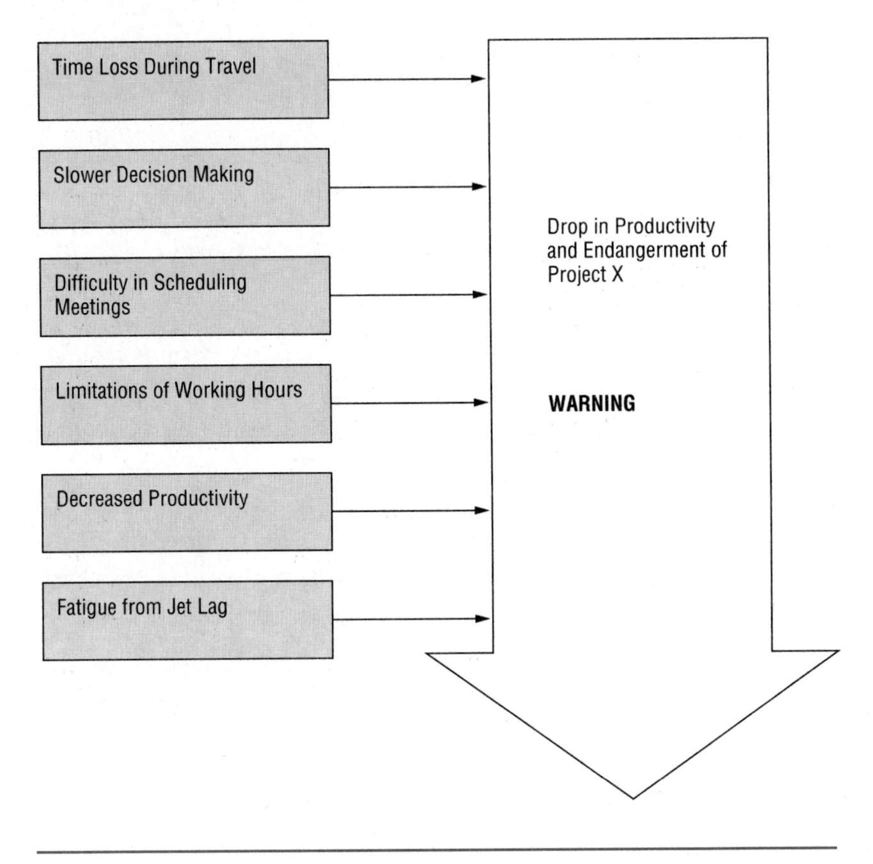

week or more, have to be spent until the meeting is scheduled. Therefore, quick decision making is almost impossible under the current situation.

- Time loss during the trip also delays the project. Even though the daily work is quite hectic, the engineers have to spend five days monthly for traveling between their laboratories and the meeting destinations.
- Although it is hard to estimate quantitatively, productivity of the engineers appears to be decreasing. The frequency of the 10- to 13-

hour flights makes them tired and less productive because of jet lag and time differences.

These factors clearly reduce the efficiency of the project. Nevertheless, the project has to be completed as early as possible. Since our rival company has adopted a similar strategy, there is tremendous pressure to bring the system to market.

Observing these problems, we conclude that the company should develop a telecommunications network to link together the different engineering centers. The problems of skyrocketing travel expenses and time losses will likely become even worse. More importantly, the success of the company's new project depends heavily on efficient communication between the design teams.

This system will be based on teleconferencing. Teleconferencing will enable the engineers to meet with one another through an international linkage of video cameras, facsimile equipment, and write-boards. They will be able to meet, essentially, without having to travel, thus saving a great deal of time and speeding up the completion of the project.

The telecommunications team then discusses how teleconferencing works and some of the advantages it will bring to Edo Company.

Some information can be fully communicated only by the speaker in person. Communication between two distant points must always overcome the obstacles of time and space. Modern technology has made enormous progress in overcoming these obstacles—the telephone and facsimile. However, some information can only be fully communicated in person. This factor is even more important when several people are meeting together.

Teleconferencing can bring some great advantages to the Edo Company:

- The need for travel will be greatly reduced.
- The habit of face-to-face meetings will be broken, thus accomplishing much more in less time.
- Decision making speed will be increased, thus helping organizational efficiency.
- More people can attend meetings (since they need not travel), thus increasing organizational loyalty and enhancing job satisfaction.

The system for teleconferencing uses international telecommunications services to link customers in one country to overseas destinations. The image and audio information from the conference is translated into digital form, then transmitted to a central station by the domestic telecommunications network. At the central station, the digital signal is switched and relayed to a satellite communications earth station, which transmits it through satellite to another country. Submarine cables are not used for teleconferencing, as they are reserved primarily for telephony.

Satellite communications are performed by placing a satellite in orbit about 36,000 kilometers above the equator. Radio waves transmitted from earth stations to a satellite are amplified by the satellite's relay and then transmitted back to receiving earth stations. The radio waves broadcast from a single satellite in geostationary orbit cover about one-third of the earth's surface. This means that all of the earth stations in the covered area can simultaneously receive and transmit the same signal.

The International Telecommunications Satellite Organization (INTELSAT) has constructed a network that covers the entire world with satellites located in geostationary orbits over the Atlantic, Pacific, and Indian Oceans. INTELSAT is an international organization that provides space links necessary for international telecommunications to governments or their designated telecommunications entities regardless of membership or lack thereof in an INTELSAT organization. As of May 1988, INTELSAT had 115 member nations. INTELSAT was provisionally established in 1964 and became a permanent organization in 1973 with the establishment of an operating agreement between government-designated telecommunications entities.

The telecommunications team then begins an analysis of three options: leaving the status quo in place, building a proprietary teleconferencing system for Edo Company (which the team defines as teleconferencing rooms), and renting the teleconferencing facilities as needed. The telecommunications team goes through a complex set of variables, to which they assign different weights, and then comes up with the recommendation to rent the teleconferencing facilities as needed from international telecommunications providers. The analysis rests finally on a cost basis, since the amount of time required for the option of building a proprietary system to pay back the time and money it costs is far too long. The report continues:

There are three options to be considered for Edo Company in responding to this problem. Each has advantages and disadvantages.

Option 1—Status Quo

The advantages of continuing with the present system and doing nothing are that Edo Company would not have to make any effort to introduce a new system of working into an already crowded schedule. The face-to-face meetings would continue to maintain close relationships between the employees in the different countries. Also, continuing to provide international travel to the top-level engineers acts as a type of job perk that keeps up morale. This is important in the final stages of a project.

The disadvantages are that the existing problems will more than likely not be solved. It is hard to underestimate the potential advantage to our competitor of its rumored teleconferencing system.

Option 2—On-Site Teleconferencing Facilities

The advantages of installing teleconferencing rooms on site at each office and utilizing the teleconferencing network of a provider are that most of the existing problems can be solved as well as present technology allows. Also, it would be possible for Edo Company engineering teams to open meetings at any time of the day and respond quickly to urgent situations.

The disadvantage is that the very significant time difference between the three countries creates managerial problems in scheduling teleconferences. (See Exhibit M1.2.) Also, it appears to be somewhat expensive to install and build a private teleconferencing room. Edo Company has no practical experience with teleconferencing, thus introducing an element of risk. The technology is still improving and changing, therefore we would eventually be left with outdated technology. Further, it takes a significant amount of time to install the equipment. Finally, technology for three-way meetings is not widely available.

Exhibit M1.2

Time Differences for Global Teleconferencing

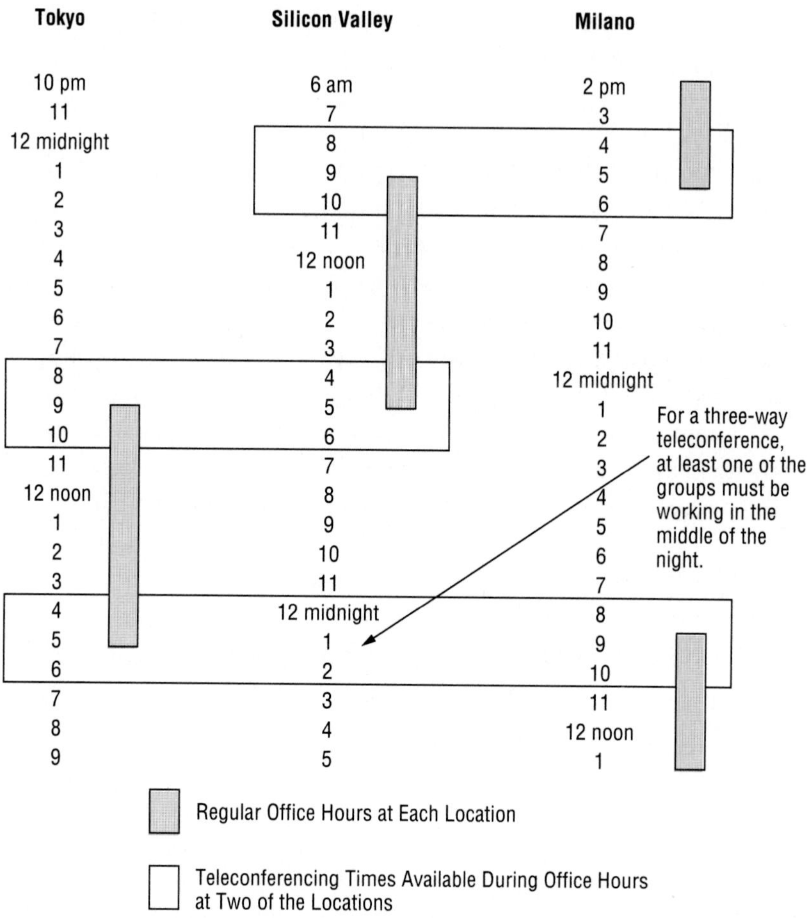

Tokyo	Silicon Valley	Milano
10 pm	6 am	2 pm
11	7	3
12 midnight	8	4
1	9	5
2	10	6
3	11	7
4	12 noon	8
5	1	9
6	2	10
7	3	11
8	4	12 midnight
9	5	1
10	6	2
11	7	3
12 noon	8	4
1	9	5
2	10	6
3	11	7
4	12 midnight	8
5	1	9
6	2	10
7	3	11
8	4	12 noon
9	5	1

For a three-way teleconference, at least one of the groups must be working in the middle of the night.

Regular Office Hours at Each Location

Teleconferencing Times Available During Office Hours at Two of the Locations

Option 3—Public Facilities

The advantages of utilizing currently available public teleconferencing rooms are that Edo Company does not have to spend money for either teleconferencing equipment and installation or on-going maintenance. The public teleconferencing rooms are available in many countries of the world. People could open a teleconference immediately, if necessary.

The disadvantages are that the engineers must leave their work places and travel to the public teleconferencing facilities, which are some distance away. Also, Edo Company will have to pay rental money. Finally, there is a significant chance that the rooms will not always be available at the exact times they are needed.

Our conclusion is that the Edo Company should select Option 3 and rent teleconferencing facilities as needed in order to coordinate activities worldwide. Kokusai Denshin Denwa (KDD) of Japan, American Telephone & Telegraph (AT&T) of the United States, and Italcable of Italy are capable of providing good services for the teleconference requirements of Edo Company. KDD is providing advanced color video communications between Japan and overseas sites for immediate, high-impact, total communications. Videoteleconferences and visual telephones supplied by KDD are highly cost effective means to enhance professional productivity, and have institutional and educational applications.

The teleconferencing rooms supplied by these vendors have many features which will aid the engineers in their decision making:

- *Video Images*
  The equipment has the ability to transmit and receive split screen images, closeups of people or write-boards, still images such as designs and drawings, and VCR images.
- *Audio*
  The equipment gives participants the ability to mute audio output for private consultations at any point during the meeting. Echo cancellers make voice and other audio transmissions very clear.
- *Miscellaneous Capabilities*
  The equipment can transmit VCR recordings and project VCR images into each of the conference rooms. It can produce hard copies of the contents of the write-boards at all locations. It also allows simultaneous G3 facsimile transmissions and separate telephone lines parallel to regular transmissions.

Next, the telecommunications team makes a preliminary calculation for the cost of videoteleconferencing, using published rates available from the various carriers.

The fee for the teleconferencing service is very reasonable compared with the expense of air fares, as the following table shows:

Cost of Teleconferencing

Rental Fee for Room (30 minutes)	$150
Fees for Cables (30 Minutes)	
First time Segment	570
Second time Segment	530
For a two-way teleconference between Tokyo and Silicon Valley for 2 hours, the costs would be as follows:	
Room ($150 x 4)	$ 600
Cables ($570) + ($530 x 3)	2,790
Total (¥781,200)	5,580

At this point, the telecommunications team is still unable to make a clear recommendation regarding its three options. It then presents a detailed cost/benefit analysis based upon comparing the scores of each of the three options against the costs and the cash flow of the operation. The team recommends that the rental option be used, even though the scoring reveals that building a proprietary teleconferencing system is superior from the point of view of operations.

There are several factors to consider when evaluating the costs and benefits of the introduction of teleconferencing to the company:

- The significance of the costs and benefits themselves have to be identified.
- A value system has to be created in order to rank the teleconferencing options.
- Certain variables that can affect valuation have to be identified. For example, the number of engineers per conference and number of conferences per year might be considered.
- Once each option has been evaluated for each cost/benefit factor, quantifiable dollar factors have to be weighed against unquantifiable factors.

We have decided to begin to evaluate our three options by evaluating our three quantifiable cost/benefit factors to derive dollar amounts for the sum of those factors for each decision option.

Our seven unquantifiable factors are weighted from 7 (most important) down to 1 (least important). For each factor we give an option a score, either 2 (best), 1, or 0 (worst), then we multiply this score by the factor weight. We add the results for each option to arrive at a total score. We then make an intuitive judgment based on both the quantifiable dollar totals and the unquantifiable score totals. (Ranges such as 17g to 67g show the cost/benefit, had certain variables been halved or doubled.)

At this point, the telecommunications team begins the detailed analysis point by point.

Equipment costs for any system normally include the initial cost to purchase hardware plus the cost of its installation, and the per-month costs of rental of equipment and communications services. For the Option 3 system, this would be $0 initial + $33,480 per month (assuming four two-hour conferences that take place approximately once per week).

Equipment +	Initial Installation		Usage Rental per Month
Option 1	$0		$0 per month
Option 2	$1,050,000	13g to 53g	$6,570 x 4=$26,280 per month
Option 3	$0	17g to 67g	$8,370 x 4=$33,480 per month

The savings on travel for the new system would be $50,560 per month. Travel costs cover sending engineers from two locations (Silicon Valley, Tokyo, and Milan) to one other location, including air fare, hotels, and car rental. Based on current international estimates for these costs, our new system would save $43,200 in air fare + $5,440 in hotel fees + $1,920 in car fees = $50,560 per month. This assumes the need for eight engineers to travel from two locations to a third for four conferences per month.

Travel Savings per Month

Option 1	$ 0
Option 2	50,560
Option 3	50,560

The savings on employee time through teleconferencing would be $24,000 per month. A trip half-way around the world normally wastes many useful working hours for the engineers involved. We assume that the typical employee spends 25 hours away from useful work when required to go on a long distance business trip, that eight engineers have to travel to four meetings per month, and that the typical engineer's salary is $60,000 per year. If the employees are paid for travel time (which they certainly are) then the cost equals hours wasted times their salary. The time saved equals 25 hours times 4 or 100 hours per month. The average salary of the eight employees of $60,000 each per year gives $30 per hour ($60,000/2,000 hours; 8 x 250 hours = 2,000 hours). Therefore, the amount of savings = (100 hrs./month)($30/hr.)(8 employees).

Employee Time Savings per Month

Option 1	$ 0
Option 2	24,000
Option 3	24,000

Having completed a simplified financial analysis of the cost savings available through teleconferencing, the telecommunications team analyzes the unquantifiable costs and benefits of adopting teleconferencing. The team accomplishes this by multiplying a ranking score for each item by a weight signifying overall importance attached to each variable.

The new teleconferencing system will bring a vital improvement in quicker decision making and faster feedback between engineers. The quickening of the decision making process is very important to our company since it speeds up the whole R&D process. This can enable the rapid introduction of new, profitable products. The ability to quicken the pace of new product introduction is vital to the company since it gives us a competitive advantage over other companies in the industry. Our reputation would grow stronger among our customers. The amount of competitive advantage that our new teleconferencing system would contribute is very difficult to estimate since one cannot predict the future. However, we rank it as most important among our unquantifiable cost/benefit factors, giving it a weight of 7.

Quicker Decision Making

Option 1		0.0
Option 2	1.5 x 7 =	10.5
Option 3	1.5 x 7 =	10.5

The new teleconferencing system will eliminate the need to travel away from home and therefore it will increase job satisfaction. Since travel will be eliminated, our engineers will be closer to their families and they will be less weary, in general. This could dramatically increase their productivity. Since much of this benefit depends on the psychological makeup of the employees, it is very hard to quantify. We still believe it is an important factor and are giving it a weight of 6.

Job Satisfaction

Option 1		0
Option 2	2 x 6 =	12
Option 3	1 x 6 =	6

The implementation of the system will greatly increase the convenience of arranging meetings when necessary. Since a meeting simply involves reserving time at a teleconferencing facility, it gives our company much greater flexibility for scheduling meetings. In the past, engineers would have to coordinate their schedules to meet at the same time. This would often cause much inconvenience. This is another difficult factor to quantify. We give it a weight of 5.

Convenience

Option 1		0
Option 2	2 x 5 =	10
Option 3	1 x 5 =	5

The new teleconferencing system will cause some engineers to stay up late at night for teleconferences due to time differences between countries. There is an 8-hour time difference between California and Italy, and another 8 hours between California and Japan since each is located approximately one-third distance around the globe from each other. This makes scheduling the time for a three-way meeting a difficult problem. To schedule a meeting with the least intrusion on anyone's normal sleep time would require a starting time of 8 a.m. at one location, 4 p.m. at the second location, and 12 midnight at the third location. This would cause an inconvenience for at least one group of engineers. As mentioned before, convenience is difficult to quantify. We assign this factor a weight of 4.

Time Zone Problem

Option 1	2.0 x 4 =	8
Option 2	0.5 x 4 =	2
Option 3	0.5 x 4 =	2

The new teleconferencing system will involve the cost of learning how to maximize utility. At the beginning of our use of teleconferencing, our engineers will be unfamiliar with it. There will be a period during which the system's utility will not be fully realized while the engineers become comfortable with the protocols and behavior of teleconferencing. It is difficult to quantify how long this learning period will last. Since it is only a one-time cost, we give it a weight of only 3. Permanent costs would be valued higher.

Learning Costs

Option 1	2.0 x 3 =	6.0
Option 2	0.5 x 3 =	1.5
Option 3	0.5 x 3 =	1.5

Since meetings will not be face-to-face, they will leave less opportunity for wasting time. Meetings will be more to the point. Often during face-to-face meetings, people feel no time pressure and will tend to concentrate on one point for an improper amount of time. In a teleconference, people will feel more restricted about wasting time. This benefit is weighted at 2.

Time Wastage Factor

Option 1	0.0	0
Option 2	1.5 x 2 =	3
Option 3	1.5 x 2 =	3

Now the telecommunications team, after much analysis, is prepared to compare the quantifiable and the nonquantifiable costs and benefits of adopting the teleconferencing system. The team experiences a conflict between the solution indicated by the cost analysis and the solution indicated by the benefits analysis.

The total quantifiable dollar gains from the new teleconferencing system would be $41,080 per month. To obtain this figure for dollar gains from the new system, the travel savings ($50,560) are added to the employee time savings ($24,000), and then the equipment costs ($33,480) are subtracted. The total is $41,080 per month.

Total Quantifiable Gains

Option 1	$0
Option 2	-$1,050,000 + $48,280

Note: The worst case is $6,000 per month; the best case is $243,000 per month. Comparing this option with option 1 shows purchasing has a 2-year breakeven period. However, compared with option 3, purchasing our own teleconferencing system has a 13-year breakeven period.

Option 3	+$41,080

Note: The worst case is $2,000 per month; the best case is $229,000 per month.

The calculations for the unquantifiable scores are as follows:

Unquantifiable Scores

Factor	1	2	3	4	5	6	7	Total
Option 1	0.0	0	0	8	6.0	0	0.0	14.0
Option 2	10.5	12	10	2	1.5	3	0.0	39.0
Option 3	10.5	6	5	2	1.5	3	1.5	29.5

The telecommunications team now presents its decision and recommendation:

Our decision is that Edo Company should rent the teleconferencing facilities from the dominant providers as needed. Our recommendation is based on an estimation that this option will have superior cash flows.

It can be clearly seen why we did not choose the status quo. It scores last in both quantifiable and unquantifiable factors.

We decided to choose the renting option over the purchase option even though the latter came out ahead in monthly savings and in the unquantifiable score. The reason is that, based on the expected cash flows, it would take the purchase option 13 years to overcome the rent option. This is too long a period, especially since newer and better teleconferencing technologies might become available in the interim. In order to make up for this deficiency, the purchase option would have had to overwhelm the rent option in the unquantifiable factor score. Although it has a higher score, it is not overwhelmingly higher.

The Critique

In considering the problem, the telecommunications team has applied a cost/benefit analysis. They have compared the benefits of having a teleconferencing system and shown that some type of system was clearly advantageous.

Their next problem was a lease-or-buy decision. After careful analysis of the costs involved in time and salary savings, the telecommunications team was able to show that both options were superior in savings to doing nothing and leaving the working situation as the status quo. They calculated the breakeven point for the lease-or-buy option and demonstrated that the breakeven time for building their own system would exceed the expected life of the generation of technology they would be using.

Based on this analysis, they recommended the lease option. The engineers of the Edo Company would cut their travel, but would hold teleconferences from public facilities as needed, providing the facilities were available.

This type of methodology has both strengths and weaknesses for consideration of this type of problem. Some of the factors to consider are the following:

- Strengths
  The methodology allows a comparison of both quantifiable and nonquantifiable data or measurements. The methodology creates a way to evaluate many different "soft" factors. The methodology is able to account for both real cost and expected cost savings from improved human productivity.
- Weaknesses
  The telecommunications team failed to consider the additional benefits to other sections of the company, including different teams working on different products, when they used the in-house teleconferencing system during off-hours when the project engineers were not using it. The telecommunications team may have underestimated the amount of teleconferencing time that would be required in the final stages of the project. The telecommunications team failed to take into consideration the great amount of time to get from Silicon Valley to the nearest teleconferencing facility in San Francisco, and the associated costs. The telecommunications team

> *built its analysis upon old technologies, assuming a need to build a million-dollar plus teleconferencing room at each location. In fact, small portable desktop units for full-motion videoteleconferencing cost less than $30,000 for each unit. Slow scan black-and-white units cost even less.*

In general, although their methodology was sound, as far as it went, the telecommunications team failed to adequately investigate the different types of teleconferencing options available.

Minicase 2

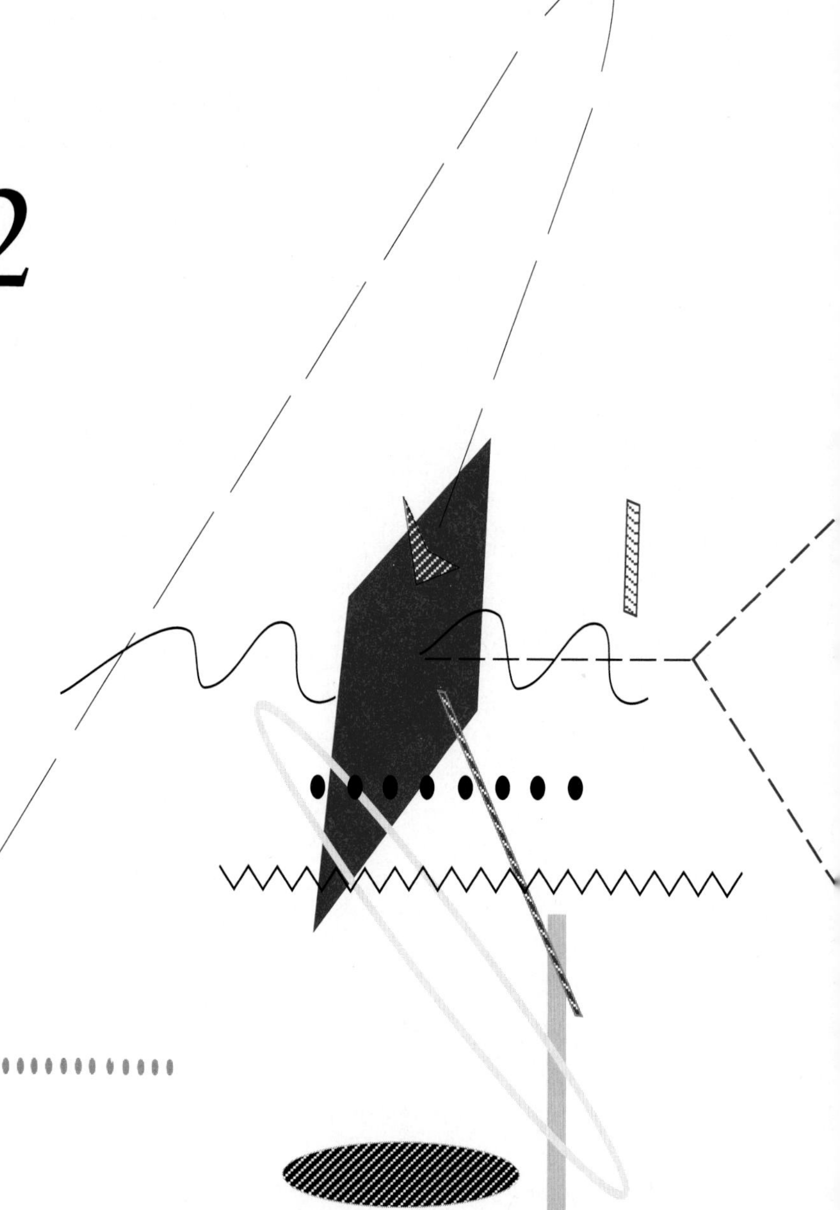

The Sports-R-Us Company

Telecommunications Implications of a Hostile Takeover of a Sydney, Australia Subsidiary

The Challenge

You are the head of telecommunications for a multinational corporation headquartered in New York. You are strictly an IBM shop, because the head of MIS has managed to control the proliferation of all non-IBM equipment throughout the company. Unfortunately, top management just completed a hostile takeover of a company in Sydney, Australia and announced that the company would be "totally integrated" into your global operations. As the head of telecommunications within the MIS department, your duty is to integrate the Sydney operation with the company branch in Hong Kong and the headquarters in New York.

On your recent trip, you found that the entire operation in Sydney was based on Digital Equipment Corporation's VAX equipment. This very large installed base of equipment could not be thrown away. You must help integrate the DEC equipment into the IBM network. Linkages between Sydney, Hong Kong, and New York must have a subsecond response time.

You have found that it is difficult to get a good circuit from Sydney to New York directly, but your organization already has a very powerful circuit between New York and Hong Kong. You have been told that if you cannot get the system working properly, you will be terminated.

After having received the challenge described above from top management, the telecommunications team presented the following report.

The Response

Our company, Sports-R-Us, runs IBM equipment exclusively. We have just completed a hostile takeover of a company with only DEC equipment. Our task was to completely integrate our information system with that of the new company to achieve a subsecond response time.

Sports-R-Us produces sports memorabilia. It manufactures and sells sports accessories such as T-shirts, bags, knit hats, baseball caps, scarves, banners, and socks. The company was founded in 1975. Since then, the company has grown from a small storefront to a large, multi-outlet corporation.

The company is headquartered in New York. It maintains retail stores in 20 cities in the United States, and also sells wholesale. The New York headquarters maintains a staff of 750 and coordinates the operation worldwide. The stores account for 50 percent of total sales with the remaining 50 percent coming from direct sales to sports teams and department stores.

All manufacturing occurs in Hong Kong because of low labor costs. The Hong Kong office has a staff of 40 persons. When orders are placed for merchandise over the communication network shown in Exhibit M2.1, the Production Manager contacts small, independent groups of people to produce the goods.

To diversify the current product line and increase worldwide market share, Sports-R-Us has just completed a hostile takeover of Hats-R-Us in Sydney, Australia. Hats-R-Us has a diverse product line of hats that should sell well in the United States. Additionally, the company was acquired as a backup against the 1997 inclusion of Hong Kong within China. The company is based in Sydney with stores in Sydney and Melbourne. Company offices are linked together by a telecommunications network which must be joined to the parent company's network, as illustrated in Exhibit M2.2.

After discussing some of the large-scale dimensions of the problem, the telecommunications team then discusses the types of information the telecommunications network must carry. After discussing this, the team turns to a first-order estimation of the size of the telecommunications circuit that will be needed. They decide on a 56 kilobit per second line to tie in the Australian operations.

Exhibit M2.1

Sports-R-Us Communication Network before Acquisition

The Sports-R-Us telecommunications network links all locations plus the Hong Kong office and corporate headquarters.

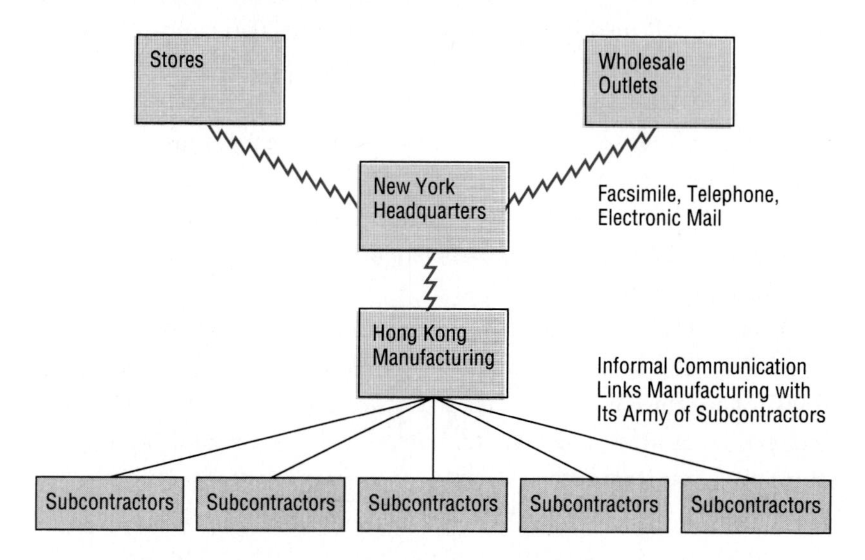

Exhibit M2.2

Sports-R-Us Communication Network Including Sydney Office

The newly acquired Australian operation must be linked into the network.

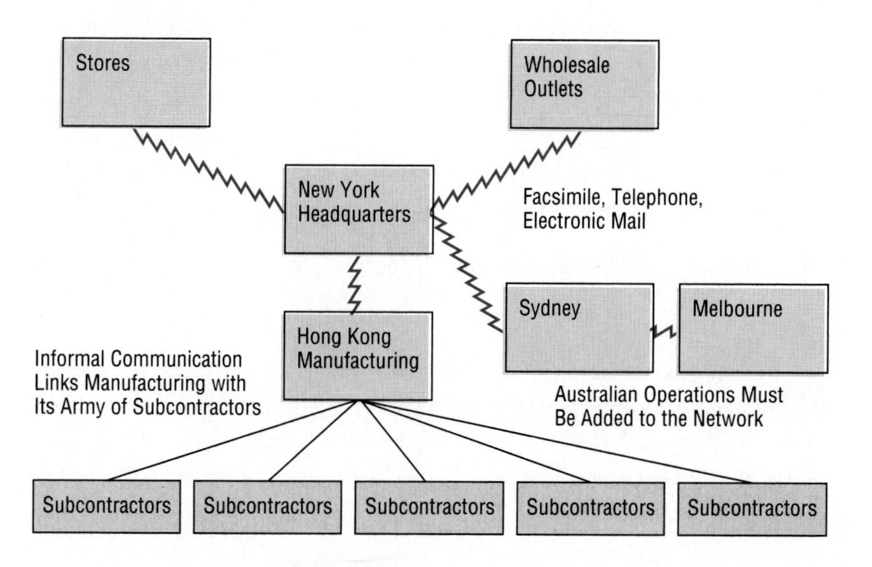

Six types of information will be transmitted between New York, Sydney, and Hong Kong. (See Exhibit M2.3.) The primary data transmission need is for inventory maintenance. Order processing, sales information from stores, marketing and design information, and financial information will also be sent. The system must also support queries between the different locations on the network.

Sports-R-Us has acquired only IBM equipment, at the insistence of the Vice President for Management Information Systems. The New York headquarters utilizes an IBM 3090/XA mainframe computer, a 4341 mainframe, a System 38 minicomputer, and multiple personal computers. Each U.S. store has a personal computer linked to the New York headquarters.

Hats-R-Us owns only Digital Equipment Corporation (DEC) equipment. The Sydney headquarters, with a staff of 300, uses a VAX 8800 mainframe, a VAX 8700 mainframe as a backup, and many personal computers and minicomputers. The Sydney and Melbourne stores have minicomputers.

In order to link the two companies together, a 56 kilobit per second telecommunications channel will be needed. The largest data flow, inventory information, will be sent every two hours. At most, it contains 120 bytes (equivalent to 960 bits) per record and approximately 60 records may be sent per transmission. Therefore, a 56 kbps system is necessary to link the conglomerate together.

Now that the telecommunications team has established the speed of line it needs, it must choose the basic route the line is going to take. Should it link to Sydney through Hong Kong, or directly from New York?

The question now arises, what route should we choose to connect New York and Sydney? There are two possibilities:

- Connect New York to Sydney directly
- Connect New York to Sydney through Hong Kong

We have chosen the first option, to connect New York to Sydney directly, mainly because of the political situation in Hong Kong. Hong

Exhibit M2.3

Types of Information To Be Telecommunicated

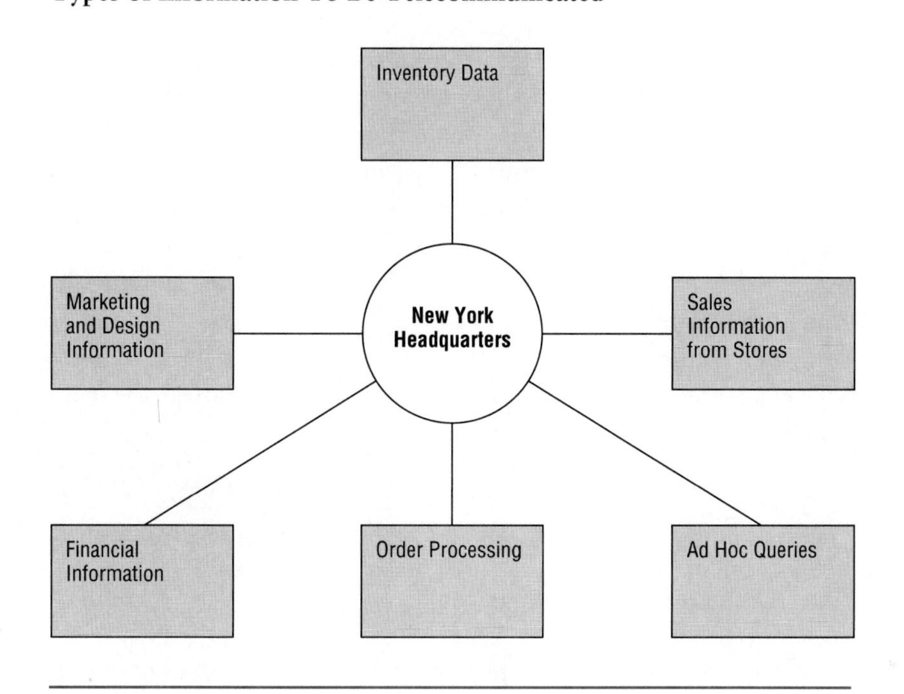

Kong is going to be returned to The People's Republic of China in 1997. China has been in political turmoil. Since telecommunications may be more severely restricted by the Chinese government, it may not be possible for Hong Kong to function as an information center in the way it does now. Furthermore, for security we need to decide what types of backup are needed for the expanded network.

The telecommunications group then discusses the different types of international linkages available that can provide the quick response times the company needs. They point out that no undersea fiber optic cables link the U.S. mainland with Australia, and this fact limits their options. They appear unable to reach the subsecond response time requested by management.

We have determined that three groups of overseas network linkages appear to provide the speed requirements we are seeking. Our mandate from management states that "the linkages between Sydney, Hong Kong, and New York must have a subsecond response time." However, it is physically impossible to send information that quickly. Instead, we have investigated high-speed overseas links. Fiber optics, a growing part of international telecommunications, is not available between the United States and Australia at this time.

After investigating the various available options, we chose the following three categories, all of which appear to come close to the speed requirement:

- Leased Lines (available from AT&T and GTE Sprint)
- Telecommunications Satellite Service (available from AT&T, Contel, and MCI)
- Value-Added Network (available from IBM)

The telecommunications team then discusses the different criteria and considerations that influence the choice of one vendor over another.

We have established several criteria which we have taken into consideration in making our recommendations:

- *Cost*
  The total cost includes the fixed rental fee, variable costs, and installation costs for each installation, including the New York and Sydney offices.
- *Security*
  We need to ensure that the information and data we transmit are not intercepted by our competitors.
- *Installation Time*
  We need fairly rapid installation times so as to begin the process of integration as quickly as possible.
- *Quality of Transmission*
  The probability of data transmission errors, of data being sent to the wrong place, or of missing data must be taken into consideration.

- *Maintenance of the System*
  We must consider what backup systems are available from each of the vendors in case of system failure.

- *Service Quality from Vendor*
  We must evaluate how quickly the vendor responds, and the general cost and efficiency of the service.

- *Ease of Use*
  We must gauge how much training would be required to operate the new telecommunications linkages.

In addition, it is necessary to choose the best telecommunications software capable of linking the IBM system in New York with the DEC system in Sydney. The software factor can be divided into two categories from the point of view of functions: (a) packet switching interface software, and (b) multivendor communication software.

The telecommunications team now turns to technical analysis of the various factors which must be considered: the line speeds, the medium of telecommunications, the hardware required, and the interface and multivendor communications software. Each of these factors is considered in substantial detail.

The major concerns in networking the DEC system in Sydney with our IBM system in New York were transmission speed, the media over which data were to be transmitted, and the software necessary to allow the systems to communicate with one another.

For transmission, a high-speed 56 kilobit per second (kbps) line was found necessary due to our interactive communications, which require a short response time, and to the volume of data to be transmitted, primarily from the inventory management system. Though a subsecond response time is reported to be an electronic impossibility, we are able to keep the response time within the 2- to 3- second range.

A personal computer screen of 25 by 80 characters holds 1,920 bytes (15,360 bits) of information. At line speeds of 9.6 kbps, 1.6 seconds are needed for transmission. A speed of 56 kbps would need 0.27 seconds to transmit the same data. While it may be rare that a full screen would be transmitted at any one time, the possibility should not be discounted. It must be noted that actual transmission time is not the only factor affecting the response time. The time necessary to poll the data from the terminal, CPU time, packet switching, etc., add further delays. On a multidrop line such as ours, polling alone may add as much as 0.2 seconds per segment (at 9.6 kbps). Even at speeds of 56 kbps, a subsecond response time between Sydney and New York may not be possible, though to keep response time to a minimum, a 56 kbps line is clearly preferable.

The telecommunications team then presents the detailed calculations and estimates that relate the expected volume of traffic and the various transmission speeds available through the network.

Transmission volume is affected by our newly installed just-in-time inventory and manufacturing system. This means large amounts of data are to be transmitted between New York and Sydney at 2-hour intervals. With approximately 1,000 products offered by the firm, considering the various models, sizes, and colors produced, there is a potential of 1,000 records with 100 bytes per record, or 800,000 bits per transmission. Over traditional voice line speeds of 9.6 kbps, this transmission would require 1.74 minutes. This includes a 25-percent penalty overhead for terminal polling, error checking, etc. This is equivalent to 21 minutes per day, assuming 12 transmissions per day. At a speed of 56 kbps, the same data are transmitted in 17.85 seconds per transmission, or 3.6 minutes per day.

Calculation

At 9.6 kbps

(800,000/9,600) x 1.25 = 104.16 seconds = 1.74 minutes

(9,600,000/9,600) x 1.25 = 1,250 seconds = 20.83 minutes

At 56 kbps

(800,000/56,000) x 1.25 = 17.85 seconds

(9,600,000/56,000) x 1.25 = 214.28 seconds = 3.57 minutes

The weekly, monthly, and quarterly consolidation figures for sales, manufacturing, and finance require an additional 500 records each, at 100 bytes per record, equaling 50,000 bytes, or 400,000 bits. Transmission of these data at 56 kbps requires 8.9 seconds versus 52 seconds at 9.6 kbps.

Next, the telecommunications team discusses the cost differences between leased lines, satellite services, and value added networks. All of these technologies might be used for this particular application.

We have concluded that, for our purposes, the MCI Satellite system is the cheapest option for our company. The price of the satellite service is higher than that of the value added network, however, when the company is sending smaller amounts of data.

Cost Analysis of Different Carriers

	Leased Lines	
Type of Charge	**Sprint**	**AT&T**
Charge per Month (United States)	$5,000.00	$10,250.00
Charge per Month (Australia)	5,000.00	10,056.80
Installation (United States)	2,000.00	2,000.00
Installation (Australia)	2,000.00	1,934.00
Variable Costs	4,200.00	0.00
Yearly Expenses	$174,400.00	$247,615.60

	Satellite Service		
Type of Charge	**AT&T Satellite**	**AT&T Skynet**	**MCI**
Charge per Month (United States)	$12,000.00	$7,500.00	$5,500.00
Charge per Month (Australia)	10,604.00	9,283.20	7,736.00
Installation (United States)	2,000.00	4,000.00	850.00
Installation (Australia)	1,934.00	1,934.00	850.00
Variable Costs	0.00	0.00	0.00
Yearly Expenses	$275,182.00	$207,332.40	$160,532.00

	Value Added Networks	
Type of Charge	**IBM**	**Tymnet**
Charge per Month (United States)	$3,969.00	$3,858.00
Charge per Month (Australia)	3,969.00	4,150.00
Installation (United States)	1,500.00	1,600.00
Installation (Australia)	1,500.00	1,550.00
Variable Costs	7,257.60	7,000.00
Yearly Expenses	$185,347.20	$183,246.00

Assumptions include the following:

- All transmissions are at the speed of 56 kbps.
- The cost in Australia is converted into U.S. currency at the exchange rate in effect on June 22, 1989.

- The variable costs are based on the following formula:

  8 records x 20 bytes x 10 hours x 60 minutes x 60 seconds x 21 days x $0.03/1,000 bytes x 2 offices

- The variable costs figure for Tymnet (3 cents per 1,000 characters) is approximated based on the standard amount of data transmitted.

Taking these figures into consideration, the amount of data the company is sending will give a price for the leased line starting at $174,400. This option is not under consideration.

The satellite services charge a fixed monthly rate for service no matter how much communication time the company uses. The value added networks charge for transmissions by the character in addition to their monthly charges.

The company is currently updating its data every 2 hours. The querying function is also used very often. Data transmission for our 1,000 products requires the transmission of more than 8 million characters per day. This obviously makes it more reasonable to choose the satellite service instead of the value added network.

Finally, since the company is expecting a big boom in its business in the Pacific Rim area, production and sales are expected to grow 200 percent in the next 3 years. The satellite system can handle the increased amount of data transmission without extra cost.

The cost graph presented by the telecommunications team (see Exhibit M2.4) shows clearly how the problem was solved: the costs for each carrier were plotted on the graph as a function of the amount of data communications traffic being telecommunicated. At the expected traffic level of 8 to 10 million bytes per day, the least expensive option is MCI Satellite. This is what was recommended.

After considering leased lines, value added networks, and satellite communications, it was concluded that the MCI Satellite service provided a reliable and affordable alternative while satisfying our line speed demands. MCI entered the realm of satellite communications in March 1986 by purchasing Satellite Business Systems. MCI is both a carrier and a lessor of satellite equipment, i.e., it both owns its own satellites and leases transponders from third parties.

MCI will provide us with synchronous data transmission, in which data are sent in blocks or packets. The entire block is retransmitted in

Exhibit M2.4

Cost versus Telecommunications Traffic for Various Options

The MCI Satellite service appears to offer the best economy at the firm's expected usage level.

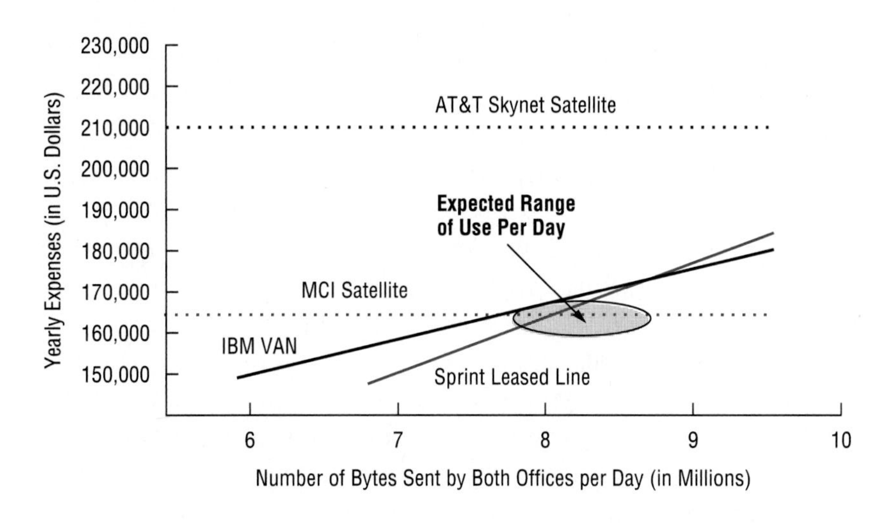

the event that the data have been corrupted or missed entirely. Traditionally, blocks range from 100 to 560 characters, though they may be greater than 1,000 characters in length. Often, this is a function of terminal characteristics or buffer size. While most satellites currently in use employ analog transmissions, most, if not all, proposed satellites will utilize digital transmission. The equipment can provide 56 kbps service in both analog and digital modes, as a function of available channels.

Since satellites attain geosynchronous orbit at approximately 23,300 miles above the earth's surface, there will be a lag time of 270 milliseconds between transmission from the earth station and the satellite. There is also a switching time resulting from the conversion of the signal from the uplink to the downlink frequency. (The frequencies of the uplink and the downlink signals differ to reduce signal interference.) Total transmission time from the New York earth station to that at Sydney is approximately 0.6 seconds. While this lag time is relatively inconsequential for our batch uploads, it does add over 1 second to the turnaround time of our query facilities. However, on a high-priority terminal, the response time is easily within the 2 to 3 second range.

The telecommunications team now turns its spotlight on the hardware and software of the IBM and DEC installations, focusing on how to connect them through the telecommunications system. As pointed out, there are two basic aspects of this problem. First, a question arises about interfaces, data communication protocols, speeds, and the like. Second is the question of the software that will allow the telecommunications to take place. Each of these topics is considered in turn by the telecommunications team.

Any option that keeps the DEC machine in the Sydney office will reduce the cost of setting up the whole system. The cost of the necessary hardware and software to connect the IBM and DEC systems is much lower than that of replacing the DEC systems in Sydney with new IBM mainframes. Since the company already has a communications network in place to connect New York to Hong Kong, it will not cost too much extra to purchase the software and hardware for the connection between New York and Sydney. The company does not have to purchase an extra Network Control Program (NCP) for its IBM system because it has already installed them for telecommunications with Hong Kong.

Excluding the leased circuits and the satellite system costs, the main expenditure will be for only two items. The first is an X.25 Portal 2000, a service package needed to make the VAX 8800 compatible with X.25 networks. We chose JNet by Joiner Associates instead of the SNA Gateway to accomplish network job entry, because it is cheaper and does not require additional hardware. The costs of the X.25 Portal 2000 and JNet are $16,500 and $12,800, respectively, plus an installation fee of $610 for the X.25 Portal 2000. This cost is significantly lower than installing a new IBM system in Sydney. This savings in switchover to the new type of operations will save the firm some of the cash it needs to pay down its debt.

We are currently running an MVS/XA operating system on our IBM 3090 in New York. The 3090 is our largest machine and is capable of running over a 16 mbps line. As such, it is perfectly suited for running the memory-intensive applications in New York, in addition to serving as the host machine for our international operations. We communicate with our IBM AS/400 minicomputer in Hong Kong via satellite, with an undersea cable backup system. In Hong Kong, we run X.25 protocol over INTELPAK, one of the two public Packet Switched Data Networks (PSDNs) available there. The X.25 protocol requires installation of the IBM X.25 Network Control Program (NCP) Network Packet Switching Interface (NPSI) running in the Communications Controller. The DEC VAX equipment based in Sydney consists of a VAX 8800 running the VMS operating system, a VAX 8700, and several MicroVAXs, all of which are networked through the DEC Ethernet system.

To allow the 8800 to communicate with the X.25 PSDN, we will employ DEC's X.25 Portal 2000, which can also run a number of different communications packages. To allow communications between the DEC and IBM systems, we propose using Joiner Associates' JNet NJE Version 3.3. JNet is a software package that runs on VAX systems running VMS 4.2 or later. It is a store-and-forward transport system that utilizes IBM's Network Job Entry (NJE) protocol. JNet provides full routing services, batch job submission and execution, query and management of the NJE network, and real-time communications among network users. JNet runs at 56 kbps when used with the JNet NJE BSC Assist Option between IBM and VAX systems. Therefore, the JNet software allows us to query both systems and to transmit our inventory and financial data. Also, due to its use of NJE, it allows a job in Sydney to call data from both Hong Kong and New York. The output could be printed at any of the three sites. (Exhibit M2.5 diagrams the full system recommended by the telecommunications team.)

The telecommunications team now summarizes the benefits it believes will accrue to Sports-R-Us as a result of adopting the MCI Satellite scheme.

The company will enjoy many benefits from choosing the MCI Satellite alternative. The operating efficiency of the company can be improved by the new network system. The new network system between New York and Sydney gives each office the ability to communicate with the other with a very short response time. Any information shorter than 5,200 characters may arrive at either destination within 1 second. The real response time will take only a little longer, including terminal and processing time.

This speed of data transmission will enhance the operating efficiency of the company. Both offices can get the updated information

Exhibit M2.5

Recommended Sports-R-Us Telecommunications System

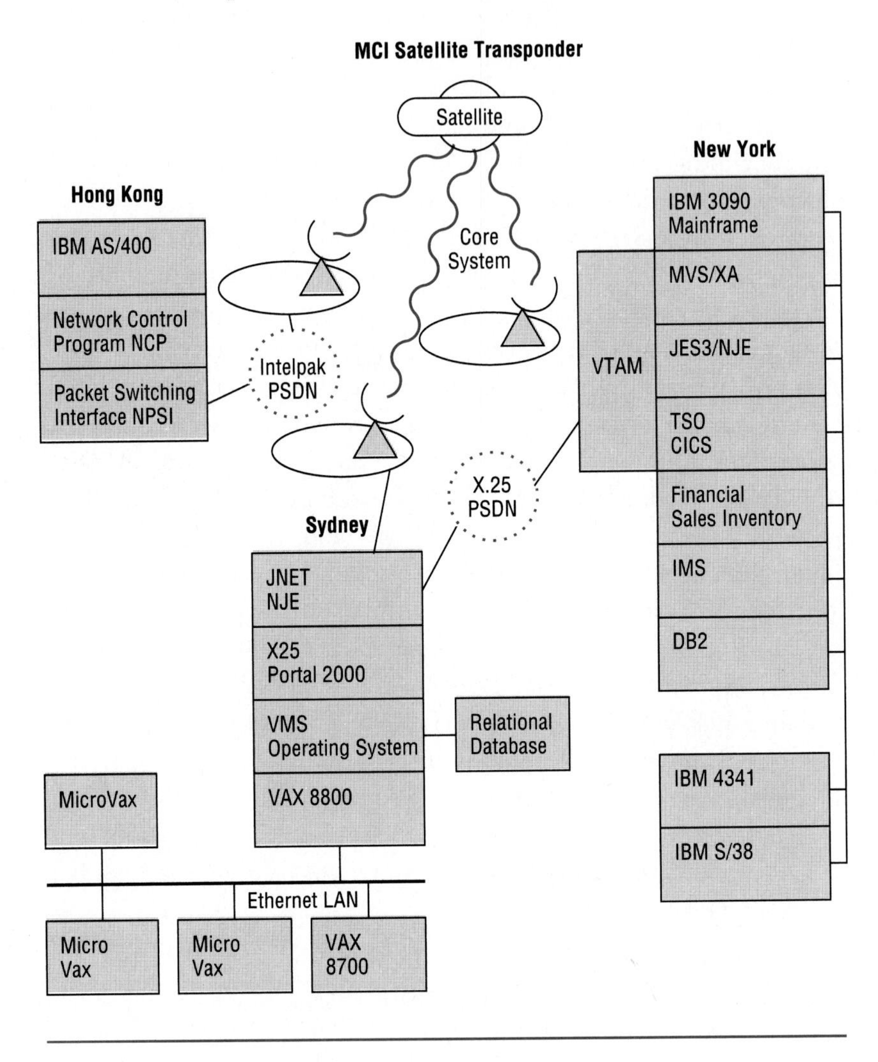

about production, delivery, and the newest market situation almost simultaneously from the other side of the world. The company can then adjust its production lines to the optimal production combinations and output levels. The levels of inventory can be kept at the minimum. The delivery schedule can be altered to inventory's needs. All the above operating efficiencies make the company more competitive. This will result in a higher profitability and also increase market share for the company.

The Critique

In this case, the telecommunications team appears to have done a very good job in making both cost estimates and comparisons of different carrier options, and in matching the telecommunications hardware and software to the task at hand. The steps and major calculations the team went through included the following:

- *Estimate the total amount of data to be transmitted per 24-hour period and derive the speed and size of the telecommunications linkage needed*
- *Identify the carriers that can provide the needed linkage and rigorously compare costs by matching expected volume requirements against costs*
- *Once the carrier is chosen, analyze the type of telecommunications interfaces to be needed (e.g., X.25 interfaces)*
- *After the interfaces are chosen, pick the telecommunications software to support the interfaces and work with the operating systems of the different machines*
- *Allow for some type of backup in the telecommunications circuits in case the primary system fails (e.g., the PSDN network)*

A few items are left unanswered in this report. What is the resolution of the long discussion regarding the many different factors they considered when choosing a telecommunications vendor? It appears as though they made the decision based almost exclusively on cost. Does this appear sufficient, or can we assume that the decision is in harmony with all of the other factors they mentioned?

Another question concerns the integration of voice with the satellite network. It is certain that the coordination of international operations requires a great deal of telephone traffic. If this is the case, then why is the extra capacity of the satellite network not exploited by adding the capability of carrying voice traffic? The failure to consider voice integration, even though it was not called for in the initial challenge from top management, means that a large well of potential cost savings is not being tapped.

On the whole, however, the telecommunications team appears to have done a reasonably convincing job for top management.

Minicase 3

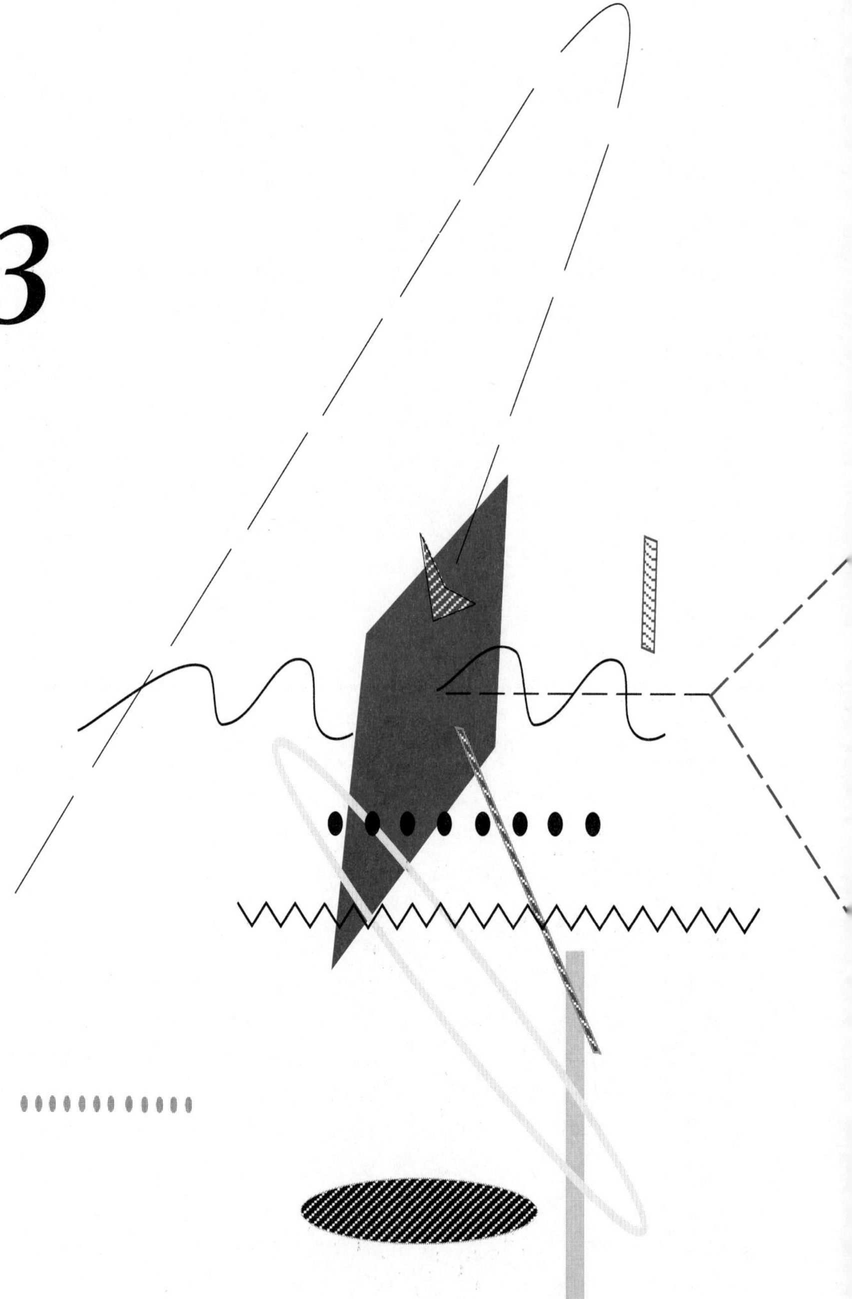

Hong Co. International

The Search for Subsecond Response Time for Traders

The Challenge

You are the general manager of telecommunications for a major Hong Kong-based foreign exchange trader with offices in Hong Kong, Tokyo, London, and New York. Over the past year or so, complaints have started to come in from traders that their equipment is outdated. Their trading stations are about 8 years old. They argue that they are losing money because the response time on their systems is not fast enough.

Sometimes it takes 5 to 10 seconds or even more to get a response from the system. The traders claim that in that period of time, deals can be lost because of quickly shifting prices. In addition, they complain that their telephone connections are terrible—they have trouble dialing numbers quickly enough, and many calls are lost.

With a general mood of belt-tightening in the company, a task force has been set up within the MIS department to choose a new trading workstation strategy and a new telecommunications strategy (for both telephone and data traffic). Your responsibility is telecommunications. You are new to your job because the previous telecommunications manager recently committed suicide due to work-induced stress.

Complaints have centered around the bad response time. You suspect this derives from satellite linkages, dropped calls, complex dialing procedures, and network congestion during peak traffic periods.

You must present a plan to the task force within about two weeks. However, you have noticed that a few new graduates from MIT have been hired into the MIS organization. You were able to get unofficial copies of their resumes and have noticed that they have strong telecommunications backgrounds. You suspect that if your presentation is bad, the director of MIS will replace you with one of the younger graduates. Unfortunately, you just took out a 30-year mortgage on a new house and moved in your parents, who are invalids.

After having received the above described challenge from top management, the telecommunications team considered the situation and then produced the following report, which summarizes their recommendation for the firm's telecommunications strategy. Before discussing the technical details, the team reviews the general background of the firm and then links this to the needed telecommunications strategy.

The Response

Hong Co. International was established in 1974 as a foreign exchange trading center. With headquarters in Hong Kong, Hong Co. International has additional offices in New York, London, and Tokyo. Hong Co. activity is aimed at generating profits from a large volume of low-margin transactions. These margins are based on the spread between the buy and sell prices of foreign currencies. On a daily basis, $50 trillion is exchanged through trading activity. Hong Co. trading represents 1 percent of this activity, or $500 billion daily.

Hong Co. employs 288 people. Each office consists of 50 traders, 2 managers, 10 researchers, and 10 back office clerks, for a total of 72 employees per office.

Hong Co. International's telecommunications system was established in 1981 on a satellite linkage. (See Exhibit M3.1.) Hardware, lines, and satellite receiving dishes have remained unchanged since 1981. Unlike the largest companies or money center banks, Hong Co. leases all of its equipment, based on a 5-year useful life before obsolescence.

After reviewing the background of Hong Co. International, the telecommunications team begins to discuss the problems that have emerged since the telecommunications system was established in 1981. The problems involve response time, reliability, disaster recovery capabilities, and basic security of transmissions.

The critical problem is that system response time is too slow, resulting in lost revenue. The trading stations and existing equipment are 8 years old. Transactions are delayed by the slow response time of

Exhibit M3.1

Hong Co.'s Outdated Telecommunications System

The system is based solely on satellite linkages.

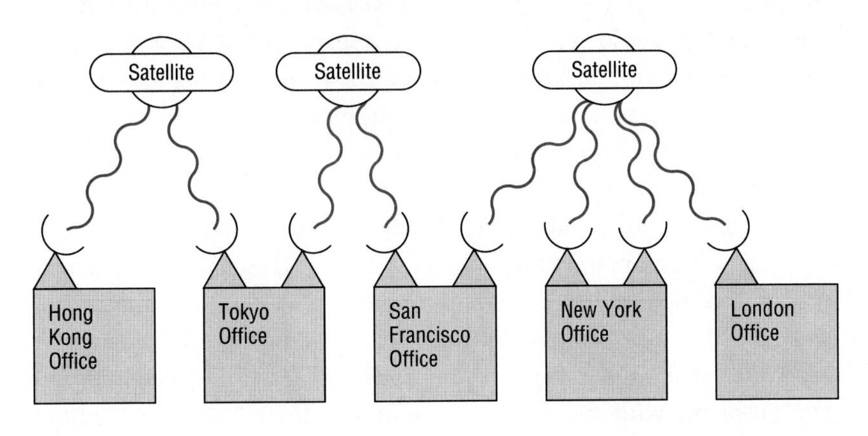

this dated system. Hardware capacity is limited by the memory and speed capacities available in the 308X series equipment available in 1981. Information transmission is limited by baud rates of 1,200 on local lines.

Another problem is that the transfer of information is unpredictable because the transmission process is unreliable. As the present telecommunications system is continually taxed by utilization at 100 percent of capacity and support by outdated telecommunication devices, calls are often dropped or lost. These telecommunication devices include a front-end processor, a bandwidth multiplexer, and a Private Branch Exchange (PBX) that are all slower than the devices presently available. Calls are usually lost at the PBX due to misdialing or misrouting.

In addition, no backup system exists. Hong Co. relies solely on the satellite-based system. Therefore, in the case of a system failure, all activity ceases and all data in transit are lost. Though Hong Co. has experienced a system failure only once, in 1985, senior management felt that the lost revenue plus the cost of restoring information justified the expense of leasing an alternative system through vendors of underground/underwater lines.

Another problem is that the data are transmitted across the phone lines unsecured. At present, all data transferred between traders across the link between the PBX and the traders' phones are unsecured. The information shared between employees can be of a sensitive or confidential nature and Hong Co. senior management suspects that a competitor has been trying to sabotage Hong Co. operations by tapping into its lines and gaining access to information on present and planned strategies.

Now, the telecommunications team discusses its suggested approach to solve the problem.

Beginning in January 1991, Hong Co. International will use an updated telecommunications system. In 1987, senior management recognized the need to improve upon Hong Co.'s telecommunication system. It committed itself to leasing state-of-the-art information technology for a superior telecommunications system. After identifying the problems with the existing system, conducting a systems study, designing an appropriate system, programming, installing the new equipment, conducting thorough testing, and providing training for end users, the system will be ready for use in 1991. The speed of transmission will be enhanced by the following:

- Upgrading local lines from 1,200 baud to 9,600 baud
- Upgrading intersatellite and satellite-to-mainframe lines to T1 lines with speeds of 1.544 megabytes
- Utilizing hardware with significantly higher processing speeds (higher mips values) and larger memories

In order to ensure the fastest possible response time, alternative routes will be established, as well. (See Exhibit M3.2.) A network will be established to allow calls to reach any of the four offices through the first route available. Previously, calls could only travel by one route to reach a destination point. Therefore, if the route were busy, the originator of the call would be forced to wait for a clear circuit.

Under the new system, calls will be automatically routed and rerouted across the first available path. For example, if the channel between Hong Kong and Tokyo is busy, a call from Hong Kong to Tokyo will be automatically rerouted to San Francisco, on to New York, and then from New York to Tokyo.

Exhibit M3.2

Satellite-Based System Supplemented with Fiber Optic Links

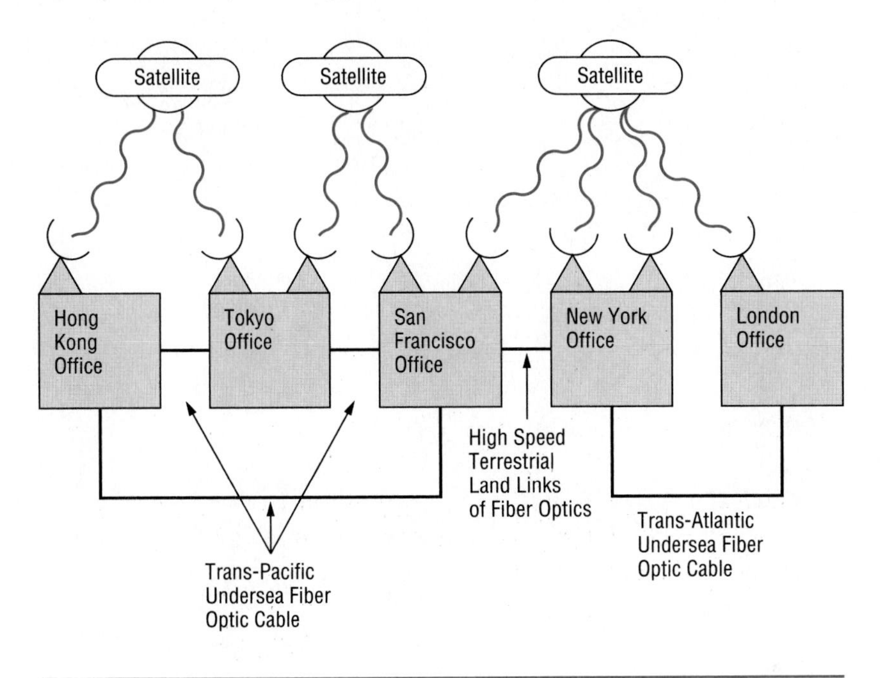

In addition to enhancing the speed of data transfer, the underwater/underground lines will also serve as a backup telecommunications network should the satellite network suffer from some form of failure or obstruction. In order to prevent the loss of data internally, Hong Co. will lease a Stratus XA 2000 Continuous Processing System. This fault-tolerance system is created by equipping all hardware with backup processors.

Data security will be handled by encryption. All voice data will be encrypted/decrypted by devices located between the PBX and the dial phones, the PBX and the direct line phones (for long distance traffic), and the bandwidth multiplexer and the direct line phone (for local calls). Exhibit M3.3 diagrams the new system.

Now, the telecommunications team is prepared to discuss the benefits to Hong Co. International of implementing the new telecom-

Exhibit M3.3

Recommended New Telecommunications System for Hong Co.

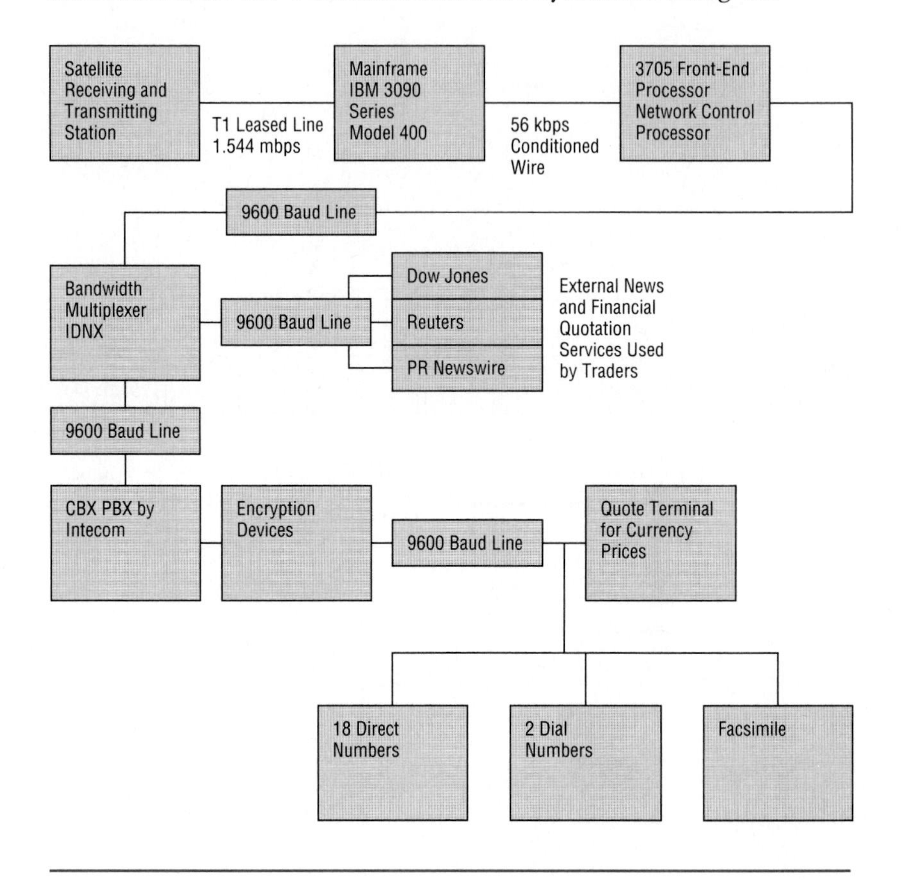

munications strategy. Its primary calculation of these benefits is based on a tradeoff between response time on the new system and the productivity gains from the time it saves.

We expect that the response time will be decreased by 50 percent, resulting in a productivity gain of 25 percent and cost savings of between \$20.8 million and \$44.5 million. The response time has slowed to the point where previous levels of productivity can be maintained only by hiring 50 additional traders for a total of 250 traders. The alternative telecommunication routes and the enhancement of equipment capacity will not only pre-empt the need for any additional traders, it will increase the productivity per trader.

Revenue and Expense Projections

	Worst Case	Best Case
Revenue		
Increment in Productivity per Trader	25%	37.5%
Average Revenue of Trader to Firm	$ 350,000	$ 350,000
Resultant Increment in Revenue/Trader	87,500	131,250
Increment in Revenue for 200 Traders	17,500,000	26,250,000
Expenses		
Resultant Increment in Salary/Trader	$25,000	$37,500
Increment in Salary for 200 Traders	5,000,000	7,500,000
Increment in Telecommunications Leasing Expense	5,000,000	5,000,000
Increment in Telecommunications Interest Expense	2,000,000	2,000,000
Total Increment in Expenses	$12,000,000	$14,500,000
Net Increment in Income	$5,500,000	$11,750,000
Net Present Value at 10 percent	$20,849,327	$44,541,745

Besides the financial benefits, the telecommunications team goes on to discuss other benefits, including better customer service, better recovery from a potential disaster, and more secure transmissions, guaranteed by encryption.

Dropped and missed calls will be reduced by 99 percent. The increased speed and capacity of the new PBX will result in a dramatic decrease in misdialing and misrouting. As a result, the majority of transactions will be completed. Traders will be encouraged by this reliability, and this encouragement will translate into increased profits for the company.

The backup systems will create a worry-free telecommunications network and strengthen the relationships between Hong Co. and its trading partners. The 1985 system failure resulted in a loss of $7.5 million in one day. Additionally, records were lost, invoices were delayed, and customers were lost. The Stratus XA 2000 Continuous Processing System will prevent such disasters in the future.

The secure system will maintain Hong Co.'s competitive advantage. By securing the data, competitors interested in obtaining Hong Co.'s detailed transaction data for purposes of self-advancement or sabotage would be unable to tap into Hong Co.'s data transfers. All data will be kept secret, satisfying corporate and employee political, financial, and personal needs. This new system will help Hong Co. keep competitors from obtaining confidential information and will possibly attract larger and more impressive clients.

The Critique

There is little evidence to suggest that this telecommunications team has solved the problem. In addition to the numerous technical errors (i.e., no undersea fiber optic link runs directly from Hong Kong to San Francisco, as called for by the telecommunications team), the following questions could be asked:

- *What is the basis for projecting such a large change in the system's response time after adoption of the newer telecommunications system?*
- *How is voice traffic handled? How is data traffic handled? What different technologies are involved in each?*
- *What is the cost of the system? What is its payoff? What other alternatives have been considered in the analysis? How do we know that the solution presented is the best?*

Some of the more obvious problems with analysis are indicated in annotations to the team's system diagram in Exhibit M3.4. There are so many errors, top management might be better advised to bring in an entirely new telecommunications group!

Exhibit M3.4

Hong Co. Recommendation with Annotations

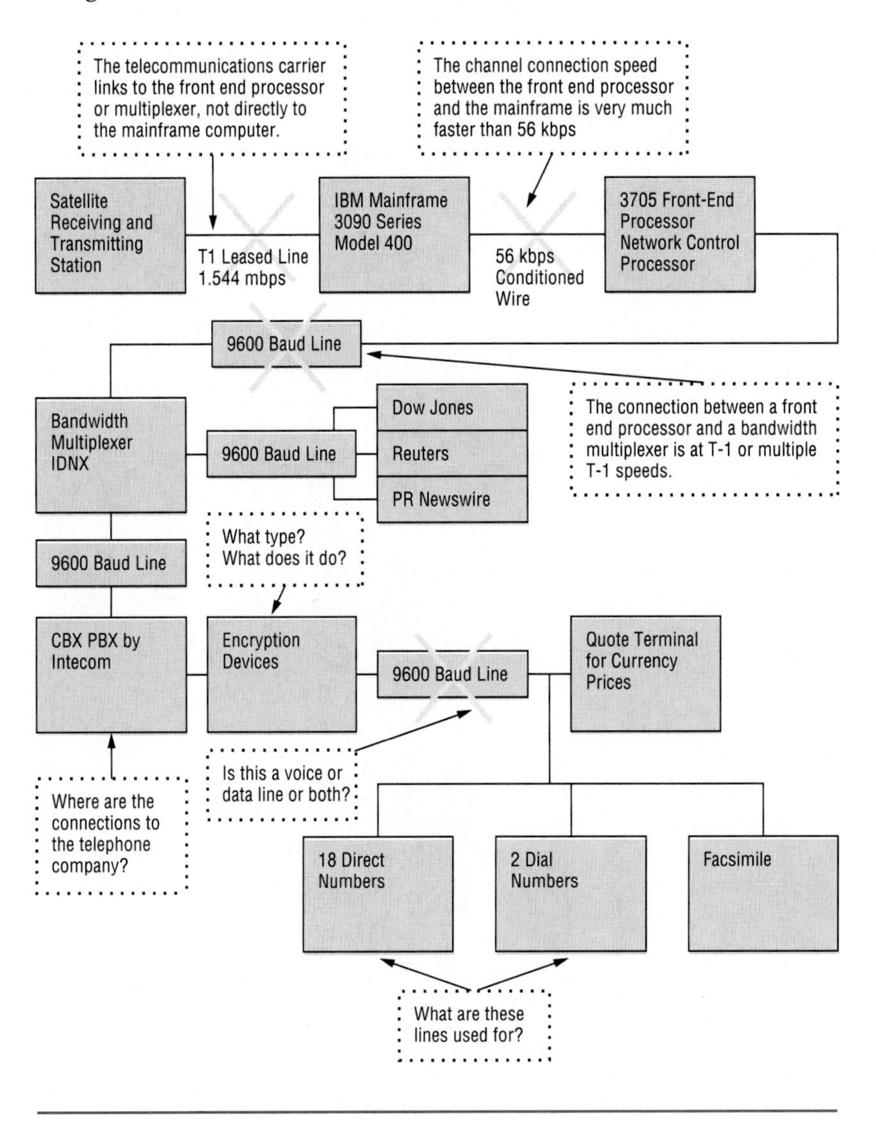

Minicase 4

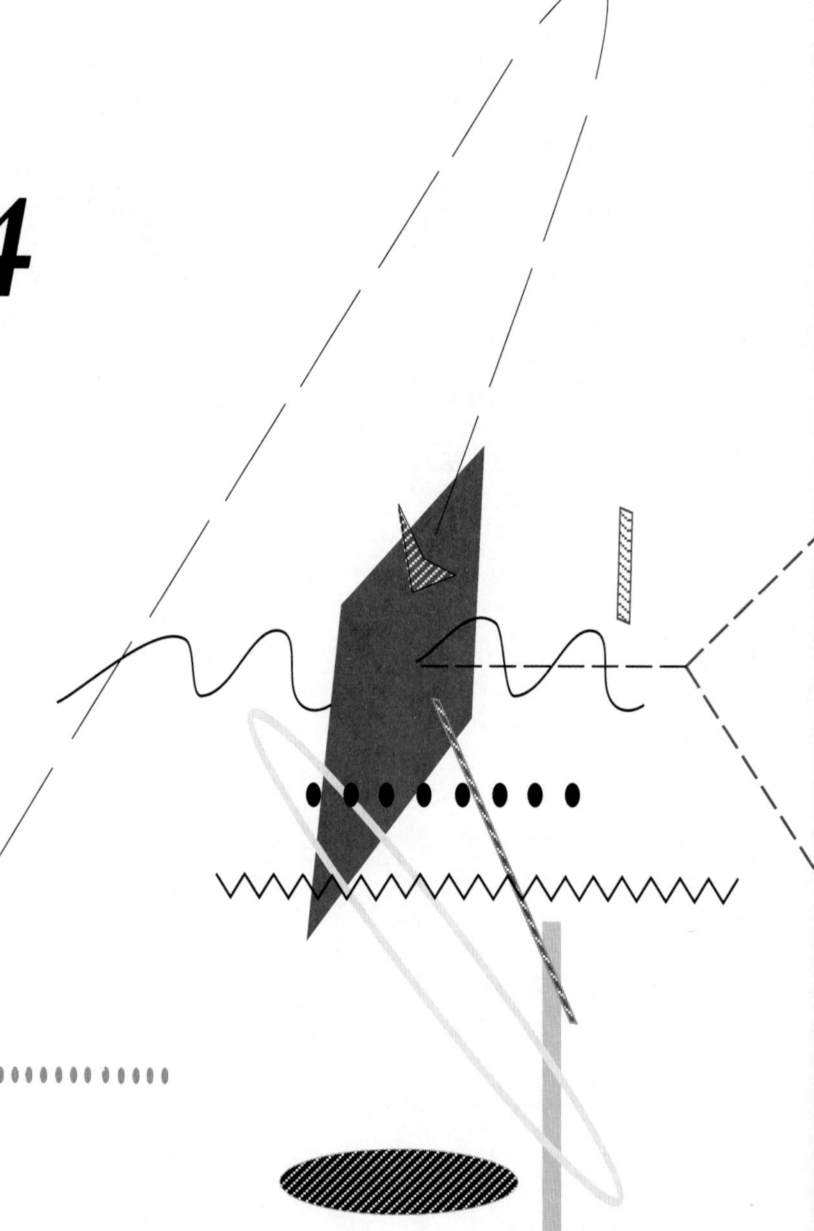

Mori Industries

Building a Global Inventory System

The Challenge

You are the telecommunications manager for a large company that has adopted a global strategy of being a major player in its industry in every market in the world. Although in the past your company produced different equipment for each national market, the trend in recent years has been for parts to be interchangeable internationally. You sell a very wide variety of industrial parts, particularly electrical supplies. Many of your customers are systems integrators and designers who rely on very quick access to spare parts.

Your competition is very aggressive, and top management suspects that other firms have been developing centralized inventory systems. Unfortunately, you are not sure of the technical solutions others have adopted because they are keeping everything secret. In spite of this, top management has decided to adopt a centralized strategy. In order to accomplish this, you must keep track of inventory on a global basis. At this time, there is no way to know what your hundreds of distributors around the world need. You are afraid that your competitors might be trying to place terminals in the offices of distributors.

You must now design a global network to link your distributors to both regional and centralized inventory control points. Since your business is extremely cost competitive, you must minimize the costs of the network. You have operations on all six continents.

Last year your company laid off about 15 percent of the work force, and there are rumors of more layoffs, particularly in middle and upper middle management areas, which includes you. You have been told that the entire company is looking to your team for an answer, and you must present it to the board within two weeks.

After an extensive review of the situation, the telecommunications team presented the following report.

The Response

Mori Industries manufactures industrial parts and electrical components. Annual sales top $1 billion. The main plant and corporate headquarters are located in Cleveland, Ohio. Three distribution centers warehouse and distribute the firm's products for the following regions:

- The New York Distribution Center covers North and South America.
- The Rotterdam Distribution Center covers Europe, Africa, and the Middle East.
- The Yokohama Distribution Center covers Asia and the Pacific Rim.

The firm's total inventory system controls over 20,000 different parts. Most parts can fit into an envelope or a small parcel. Bulk shipments of goods between the manufacturing plant in Cleveland, Ohio and the regional distribution centers are usually sent by water. All shipments from distribution centers to customers (independent distributors) are sent by air freight.

Retail sales are handled by a total of 300 independent distributors whose daily sales amount to 10 million units. These distributors are not exclusive, and may carry competitors' products. Currently, these distributors place all their orders for new products through our local sales offices, usually by phone or facsimile machine. (Exhibit M4.1 illustrates the communications network before the proposed new system.)

From locations in various countries throughout the three distribution regions, 30 sales/service offices serve the distributors in their areas. They place orders for the distributors, but do not maintain product inventories. Sales offices send daily orders to the regional distribution centers in batches over leased lines. Ordered products are shipped directly to the distributors from the distribution centers.

Products shipped from the main plant to the three regional warehouses remain there until required by orders placed by independent distributors through local sales/service offices. At that time, products are shipped directly to the distributors. Each sales office maintains its own sales records and customer database. Monthly reports are sent from the sales offices to the regional offices. Each regional office compiles information from its associated sales offices into a monthly

Exhibit M4.1

Mori Industries Communication Network before New System

Mori relied on telephone and facsimile to process orders for parts.

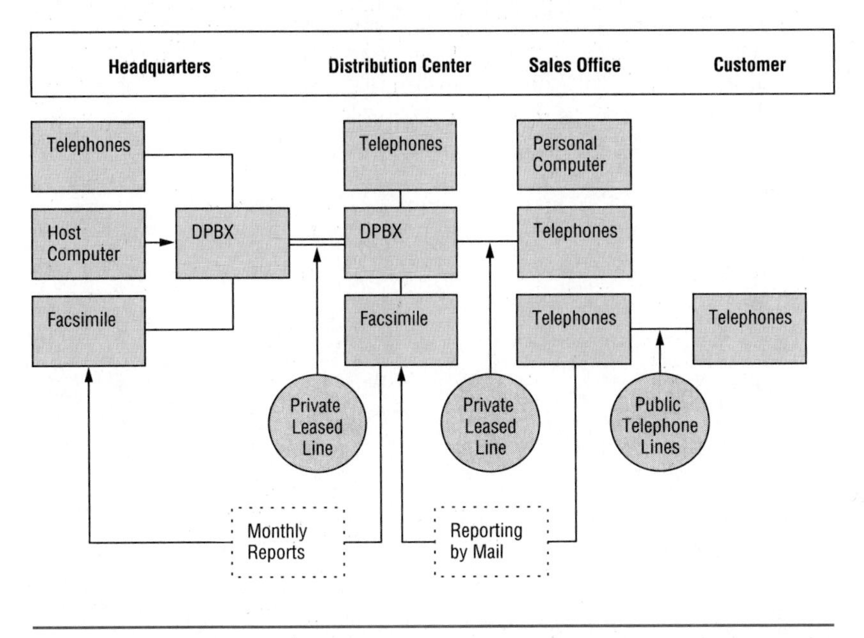

Note: FETEX 2050 DPBX has a standard capacity of 32 lines. Features include all key telephone features, plus display and one-touch feature buttons.

composite report and forwards this report to the main office in Ohio.

Currently, a mainframe computer is in place at the corporate headquarters. Microcomputers at the distribution centers process inventory reports, purchase orders, and billing. The microcomputers also handle all accounting and office automation functions. Weekly and monthly income statements and reports on inventory levels and sales office performance are transmitted to the central office over public telecommunications lines. Recently, an IBM 3083 was installed at the main plant in Cleveland to coordinate a new just-in-time inventory control program with suppliers. This mainframe also receives and processes all reports and data from the regional distribution centers.

Our team has been given the task of developing a communications network to make it easier for customers to access accurate, fresh information, place orders, and, if necessary, get support or service from a personal representative. Included in this task, we must propose a new inventory system. Upper management insists that we take a centralized approach. We are under pressure from competitors who are preparing to offer terminals and software to their customers, some of which are our customers.

Our objective is to gain a competitive advantage by offering service and information access at least as good as our competitors', then provide some added value so customers will be more inclined to work with Mori Industries in the future.

Now that the telecommunications team has introduced the problem in general, it goes into an analysis of the problems faced by Mori Industries with the present system.

Our analysis has revealed several major problems with the system as it currently operates. These problems occur in the corporate telecommunications and information system, in the product distribution system, at the customer interface, and in inventory control.

Corporate Telecommunications and Information System

The headquarters has no way to directly monitor local sales; this makes it difficult to project revenue and earnings. The present corporate reporting system requires each sales office to report to the distribution center on a monthly basis. The distribution center is supposed to compile information from all sales offices into a composite report which it forwards to the headquarters monthly. However, the composite report is usually delayed about one month; therefore, it has become necessary for the headquarters to reach local sales offices directly, without delay to generate accurate revenues forecasts.

Feedback to the headquarters on inventory levels is inadequate, causing shortages and surpluses at each distribution center. Without better information, Mori Industries will not be able to compete in the coming years. To solve this problem, top management has decided to develop a centralized inventory system. In order to accomplish this, Mori Industries must keep track of inventory on a global basis. At this time, however, there is no way to know what kind of information is needed by the hundreds of distributors around the world.

Company telecommunications costs are projected to rise when

management implements this centralized inventory system because feedback to the headquarters on inventory levels will have to come more often. Our team will design a global network for linking our distributors to both regional and centralized inventory control points. When this network is developed, telecommunications costs should decrease, and easier access to the information will allow frequent, large transmissions of critical data and information. In addition, the network itself will be able to reduce costs. Since our business is extremely cost competitive, we need a system that will provide us with the necessary information at a minimal long-run cost.

Product Distribution System

Production schedules often do not coincide with the movement of products at the regional warehouses. Customers have complained of a 2- to 4-day lag time between placement of an order and shipment from the regional warehouse. Customers have also complained of receiving orders that were incorrect either in type, characteristic, or amount of item ordered.

Customer Interface

Customers have complained of difficulty reaching local sales offices to place orders due to busy phones and unavailable salespeople. Customers are often unaware of new products or discounts and sales options offered by the firm.

Inventory Control

Under the manual inventory system, order processing at the regional distribution centers normally takes between 1 and 3 days from receipt of the order to the time of shipment. Orders are presently being received by telephone or facsimile machine.

From the incoming order, a pick list is generated manually and sent to the warehouse for shipping, where a bill of lading is generated manually. The operator locates the product by memory, pulls it from inventory, and brings it to the loading docks. Another employee records the part numbers and quantities in a log book as goods are loaded into trucks. The system is highly error prone and shipments are often delayed because products are out of stock. In 1988, 5 percent of all orders were shipped late or incomplete due to stock outages.

The company has no formal method of tracking or controlling the inventory. Inventory planning and control are based on sales forecasts and historical data which have been becoming less reliable as competitiveness in the industry has increased. Sales planning and logistics often use inaccurate data that are several days old to track specific product performance and inventory levels. Consequently, forecasts have been inaccurate.

Distribution centers incurred inventory holding and movement costs of nearly $12 million in 1988 alone. Inaccurate logistical planning and poor warehouse space utilization resulted in average inventory levels as high as 500,000 units, equivalent to 5 percent of annual sales at each individual warehouse. As the industry becomes more competitive, these costs will become more and more important in the growing battle to remain competitive. Inventory carrying costs generally run at around $4 million per year. In addition, direct labor costs with the manual system were at $4.2 million per year and increasing.

Estimating expected demand for each product and determining how much and when to ship is one of the most difficult tasks we face; the telecommunications system must play a critical role in this regard.

The company often cannot give customers quick access to spare parts. Many of its customers are systems integrators and designers who rely on very quick access to spare parts. Nevertheless, the distributors are reluctant to hold large inventories, and sometimes stockouts occur. In case of a stockout, the distributor (1) immediately places an order for additional parts to the local sales office or (2) substitutes our product with the product of a competitor. In the case of option 1, the local sales office places an order with the regional warehouse, which looks for the parts in stock, and immediately air ships them. If there is no stock in the warehouse, the regional center checks the availability of the part from the main factory in Cleveland, Ohio and the two other regional warehouses. It takes at least 2 weeks for the customer to obtain the parts, because it takes time to search and confirm their availability.

In the case of option 2, the firm bears an opportunity cost. Making matters worse, with present data collection methods, Mori Industries cannot estimate this opportunity cost. Firms do not suspect an opportunity cost until a distributor does not place an order for a considerable period of time.

It takes time for a distributor to place an order for parts. The distributors presently place orders over the phone or by mail. An order normally specifies 3 to 10 types each of 180 kinds of parts. As about 10 distributors call each local sales office every day, the line is frequently busy. Many distributors complain about this situation. Furthermore, the sales staff is so busy answering phones and processing order sheets that they sometimes make mistakes in shipping parts.

Now that the problems have been clearly analyzed, the telecommunications team turns to the details of the solution they propose. Their solution is a three-part strategy to expand and extend Mori Industries' information system directly to the customer. The result, it is argued, should be a user-friendly customer query, ordering, and information access system. The telecommunications team argues that the improved telecommunications network will back up the inventory control system resulting in superior service—the ultimate goal of the project.

It is important to notice the parallelism in the analysis. After discussing the problems, the telecommunications solutions are presented in exactly the same order: first the corporate telecommunications and information system, then the distribution system, then the customer interface, and finally the inventory control system.

Revamping the firm's telecommunications network and information processing capabilities will support changes in the customer interface and distribution system. (Exhibit M4.2 diagrams the proposed system.) At the corporate headquarters the existing IBM 3083 will help us build a worldwide customer database, track daily sales, and monitor inventory at the distribution centers. An X.25 packet switched network linkage will be set up through Telenet between headquarters and the distribution centers. The distribution centers will transmit reports of all orders and inventory transactions in batches every 12 hours. This information can be used to create models and make projections and to develop and improve the firm's marketing strategy.

Minicomputers will be installed at the regional distribution centers. These minicomputers will interface directly with distributors' personal computers to provide product information and process orders. Additionally, they will generate warehouse pick lists and bills of lading for

Exhibit M4.2

Mori Industries' Proposed New Telecommunications System

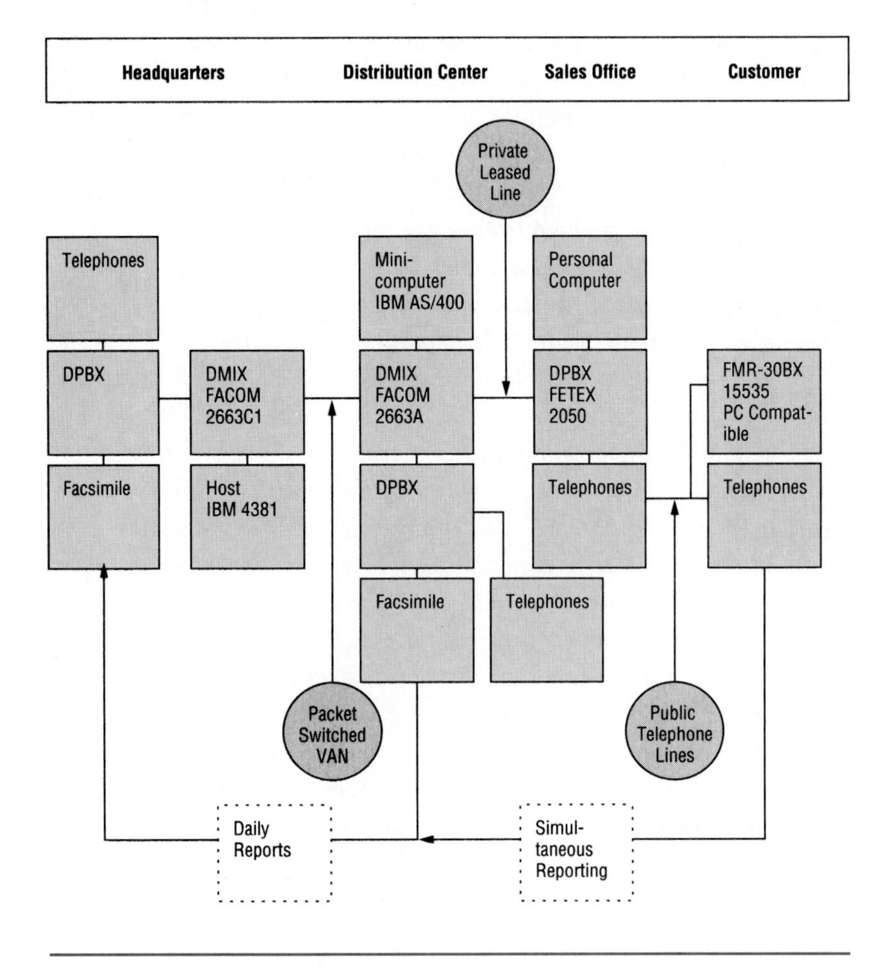

Note: FACOM 2663C1 DMIX has a date rate of 64 kbps to 6 mbps. It connects each multimedia system to a unified transmission network.

FACOM 2663A DMIX has a date rate of 64 kbps and all the features of a 2663C1.

FMR-30BX CPU 80C66 is a portable, high-quality personal computer.

shipment of orders. Through bar code scanning equipment, the minicomputers will monitor inventory. Their interface with the mainframe at corporate headquarters will help ensure adequate stock levels.

All 30 sales/service offices will be connected to this network through new IBM PS/2 machines. Through these workstations, the sales/service offices may access corporate information and monitor sales in their areas. Information on all orders placed through their office PBXs will automatically be captured on the sales office workstations. Billing will still be conducted through the sales/service offices. These workstations may also be tied into local banks electronically to expedite customer billing.

The only change in telecommunications produced by this strategy will be at the upper level, where an X.25 value added network will carry headquarters/distribution center telecommunications. The existing leased lines between the distribution centers and sales offices within their regions will both carry communications between sales offices and the distribution centers, and link customer personal computers to the distribution center minicomputers. To contact the local sales office, the customers will still use public telephone lines at their own expense.

Budget Estimate

IBM AS/400 minicomputers for regional distribution centers ($150,000 x 3)	$450,000
Private Branch Exchanges for sales offices ($10,000 x 30)	300,000
Multiplexers for regional distribution centers ($10,000 x 3)	30,000
Total	$780,000

The report now turns to the second problem, the distribution system.

A bar code inventory management system will be implemented to track products from the factory to the point of shipment from the regional warehouses. Each product will be labeled with a bar code sticker at the main plant. Bar code scanners at the main plant and the regional distribution centers will read these stickers and input data to the information system to record the following transactions:

- Shipment from the factory on a commercial carrier
- Receipt at regional distribution centers
- Placement on a warehouse shelf
- Retrieval from a warehouse shelf
- Shipment from a regional warehouse on a commercial carrier

When first applied to a product, the record accessed by the bar code will identify only the item's stock number, lot, quantity, and production date. However, as the item moves through the distribution system, the record will store more information related to the bar code ID, giving each item a history within the information system. Every time the item is moved, the date and other pertinent information will be recorded. When an item is loaded on a commercial carrier, the date of shipment, mode of transport (i.e., truck, ship, air, etc.), carrier name, and bill of lading number will be recorded. When the item is placed on a shelf, the location will be recorded. Hence, every item can be tracked as it moves through the distribution system. This system will allow real-time information management capabilities at the regional distribution centers. Also, inventory managers at the regional distribution centers, along with independent distributors, will be able to query the system to ascertain shipment dates of needed products.

The estimated cost for the scanners, related equipment, and controlling software is $679,000.

The telecommunications team then discusses the customer interface.

Each independent distributor will receive a computer hardware/software package that will allow direct access to the minicomputer at the appropriate regional office to place orders or obtain product information. Approximately 162 of our 300 distributors already use IBM-compatible microcomputers. For those distributors without IBM-compatible microcomputers, the firm will offer IBM-compatible Fujitsu personal computers at a 50 percent discount. We believe that this offer of a multiuse computer will prove an attractive selling point for this program.

All distributors will be provided software through which they will be able to operate interactively with the mainframe at the regional office. Through a modem, a distributor will be able to dial a local number, and be connected to a PBX at the nearest sales office. This DPBX will connect the distributor to the leased lines that link the sales office to the regional distribution center. Once connected to the minicomputer, the distributor can receive product reports, query the system for product information, and place orders. The system will not accept incorrect orders or orders for out-of-stock items. On placing an order, the customer will receive immediate confirmation of price (in the local currency), shipping costs, shipping date, and expected delivery date. The printed copy of this confirmation will serve as the customer's documentation of the order.

At any time after an order is placed, the customer will be able to query the system for order status. The system will maintain and provide information on actual shipping date, mode of shipment, and carrier. The local sales offices will be bypassed in this new ordering process. Although commissions will still be paid by the sales district, orders will no longer be placed through the sales offices. Therefore, the emphasis of sales office activities must change from handling sales transactions, to product promotion and market development. Routing orders directly to regional offices should free sales personnel to spend more time in the field visiting distributors. This will be especially important in the initial stages of this project to sell it to customers and explain the system.

A personal computer will be installed at each sales office to provide access to the minicomputer at the appropriate regional office. Daily feedback will be provided to each sales office on sales in its area. Sales offices will also be able to access the corporate database to obtain information on sales trends in other areas, which will help them plan strategy. All corporate memos, reports, and other pertinent information will be sent to the sales offices via computer.

Customer Interface/Sales Offices Costs

Personal computers and modems for distributors (148 x $2,000 x 50 percent)	$148,000
Software for distributors	15,000
Personal computers for sales offices (30 x $3,000)	90,000
Total	$253,000

Next, the telecommunications team discusses the final portion of the solution, the bar scanning equipment that feeds information into the system through a local area network.

A bar code inventory control system will be installed to streamline inventory tracking and control operations at all three distribution centers for a total cost of $700,000. The budget is detailed in the accompanying table.

Inventory Control System Budget

Four Matthews TMP 1900 Direct Therma Print and Dispense Printers (with RS-232-C ASCII serial asynchronous interfaces)	$18,760
Eight Matthews TMP 1904 Direct Thermal Print/Rewind Systems	27,720
Two Compaq 386 Microcomputers (80386 microprocessors, 50 MB hard drives, 1 MB RAM)	16,500
Nine MSI PDT-3 Portable Data Collection Terminals with Model LS700 II HeNe Laser Scanners	41,640
One Matthews MATTNET Network Manager (RS232/RS485 converter, 16K data buffer, 110–19,200 baud)	1,000
Fifteen Matthews Serial Interface Adaptors	14,000
Six Fixed Position Terminals with Scanning Equipment	28,000
Four Fixed Position 2424-2 Terminals (2-line, 24-column LCD displays)	3,000
One Software System Using Matthews MATTNET LAN	10,000
Installation and Training	16,000
DEC MODIS Distribution System Software (with modules for sales order processing, inventory management, load planning, and report generation)	49,600
Total System Cost at Each Distribution Center	$226,220
Total System Cost (all three centers)	$678,660

Having discussed the details of the proposed network system, the telecommunications team summarizes how it will benefit Mori Industries. (See Exhibit M4.3.) Again, the team uses a logical sequence, presenting arguments in the exact same order as they initially discussed problems: first, the corporate telecommunications and information system, then the distribution system, then the customer interface, and finally, the inventory control system.

This global telecommunications system will allow headquarters direct access to information on local sales, making revenue forecasts more accurate. This network will allow each distributor to place orders through a Fujitsu personal computer connected to the mainframe at headquarters. The order information is not only transmitted to the distribution center, but the shipment of the order, and, at the same time, the recording of the sale are prompted. Headquarters is informed almost immediately.

Feedback from distribution centers to headquarters on inventory levels will be transmitted instantly. This gives headquarters complete, accurate, and timely data.

This telecommunications network has global coverage (except the link between the sales offices and the distributors). It will reduce overall

Exhibit M4.3

Areas of Improvement through Telecommunications Strategy

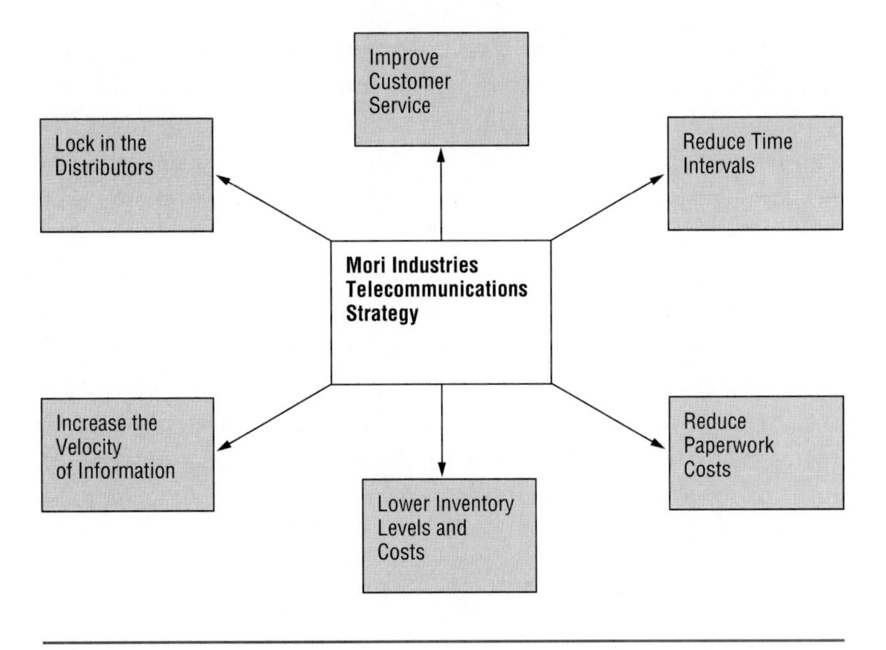

telecommunications costs within four years of installation. We assume that the telecommunications costs of the present system total approximately U.S.$14 million per year. The annual cost of the new global network system should be approximately U.S.$7 million. Assuming further that the initial investment for the entire system would be about U.S.$20 million, depreciated over 5 years (by the ACRS method), this project will break even in the fourth year.

The distribution system will be improved, giving us the ability to track each unit throughout our system, thus allowing us to serve our customers better. As soon as a unit is produced, it is tagged with an identification label. As it travels through our system, finally reaching customers, we will have a record of the part location. If a customer calls about an ordered part, we can locate it easily and expedite it, if necessary.

The customer interface will be greatly improved. Almost 100 percent of our customers will have immediate access to the spare parts they require. Mori Industries will no longer have to suffer opportunity costs from not having product available where and when it is needed. We expect a 10 percent increase in market share.

Sales office staff can easily access the inventory levels of every local sales office and warehouse through the internal network. In case of a shortage in stock, the local sales office can quickly transfer available parts from another regional center warehouse. Even though some sales offices will have to import parts, the maximum time for a customer to obtain those parts will be 5 days.

In addition, distributors will update their inventory levels and sales results through the firm's inventory control system constantly, and therefore, the firm can quite accurately forecast demand using information given by the distributors. This will drastically reduce the chance of stockouts. This, too, will help give customers quick access to parts. Due to the increased accuracy in the inventory control system, based on information sent directly from the distributors, the firm should not lose its opportunity to sell.

Distributor ordering will be smooth and sound. Distributors will no longer have to worry about their optimal inventory levels. Mori Industries will be able to forecast product demand more accurately. The distributors will place orders through PCs subsidized by Mori Industries. They will enjoy our smooth order process and the availability of periodic order summaries from their PCs. This alone should help keep distributors loyal to Mori. In addition, as the firm provides a sophisticated inventory control system and up-to-date demand forecasts, the distributors do not need to worry about either excess stock or stockouts. This also enhances the loyalty from the distributors.

An additional benefit will be available for sales offices and distributors. They will be connected with local banks through our value added network. All the accounts payable/receivable and other financial information will be exchanged through the VAN. As a result, they will not have to spend as much time confirming their balance of payments over the telephone. This further helps to lock in the distributors.

The system for inventory control will also benefit from greatly increased efficiency. Order processing time from invoice to shipment will be reduced to 2 to 3 hours. When an order is received via the new leased line, the MODIS distribution software package will process the order, prepare a pick list and a bill of lading, and transmit this information to the warehouse PC. Bar codes applied to all bills of lading will allow us to track inventory that is in-transit. The software package will also manage inventory by optimizing warehouse space and it will control stock rotation by tracking the locations of finished goods by production date. The operator will scan the product to be picked and download this information into a nearby terminal, which will transmit the data to the corporate mainframe. As the goods are loaded onto trucks, they will be scanned again. Real-time information will be sent to the on-line terminal, then to the control PC, and ultimately to the host mainframe, reducing order processing time while providing more accurate control information.

The bar code inventory control system will improve customer service by reducing the number of orders shipped late and/or incomplete. The system will streamline transaction processing and improve the accuracy of finished goods inventory, enabling tighter control and improved logistical planning. This on-line access to real-time sales order, inventory, and shipping information will result in improved customer service. In addition, the system will be able to provide information to customers regarding expected delivery dates and times, shipping origins and other pertinent product information.

The MODIS software package in conjunction with the bar code inventory control system will reduce costs internally by $1.8 million per year. Improved inventory control and logistical planning will result in lower inventory levels, providing cost savings of $600,000 per year across the three distribution centers. Product ticketing, storage, and retrieval will be automated, enabling the reassignment of 24 employees for annual savings of $840,000. In addition, we will save distribution costs.

The Critique

The telecommunications strategy of Mori Industries was to build a system to completely tie in each of its distributors to its centralized inventory control system. In so doing, the company faced the problem

of getting the customers to accept a personal computer to serve as a data terminal. This problem was solved by providing the personal computers at a very deep discount. This telecommunications plan has some important strengths:

- *It very carefully examines the relationship between the information technology required and the supporting telecommunications infrastructure.*
- *It reviews all of the possible sources of savings from the strategy, and estimates these savings (with the exception of distribution cost savings).*
- *The telecommunications group presents the ideas in a highly structured, logical format, and in sequence, so each of the key relationships is clearly explained.*
- *The plan contains detailed cost estimates and budgets, including miscellaneous items such as installation and training, for each of the major types of equipment that are needed.*
- *The telecommunications group has clearly worked very closely with the MIS group in formulating a joint strategy.*

Critical analysis shows that the telecommunications group might have improved its presentation in a few places:

- *They made no clear analysis of the costs and benefits of different types of equipment that could serve the same purpose. This leaves one to wonder if additional savings might have been available.*
- *They made no estimate of the data traffic and volumes in relation to the types of channels they want to use, leaving one to wonder if the type of carrier solution they chose—the value added network—was the best alternative.*
- *They included no discussion of how the network will be phased in, how long this will take, and any problems that might occur during this process.*
- *It is clear that the type of equipment being chosen will allow for voice and data integration, and yet they provide no discussion of the savings possible through this integration. They also give little discussion of the implications of a new PBX strategy for all locations.*

On the whole, however, this telecommunications group has come very close to presenting a comprehensive plan that would convince almost any management group that the potential savings are well worth the investment.

Minicase 5

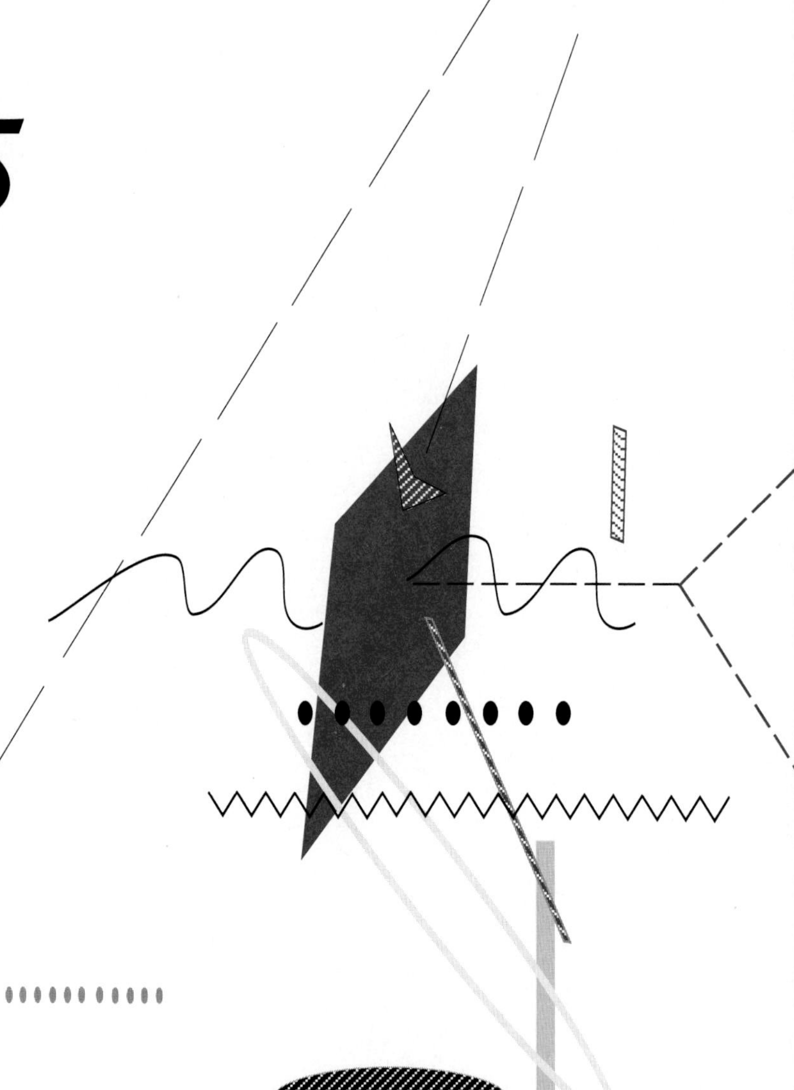

Super Charge Credit Card

Setting Up Credit Card Authorization in Brazil

The Challenge

You are on the telecommunications planning staff of a major credit card company. You have been told that the competitive cost pressures in your industry are great, and that the company is attempting aggressively to increase market share in Latin America. As you know, a critical ingredient of any credit card network is the credit authorization which must be coordinated through a computer network. Problems arise in Brazil and other countries of Latin America. The government authorities there make it difficult to interconnect telecommunications equipment. You have trouble getting circuits, they are expensive, and delays take too much time. You must develop a Brazilian strategy for credit card authorization.

Your manager has hinted that if you fail in this project you will be terminated from your job, all your credit cards will be taken away, and you will be transferred to the branch office in Beirut, Lebanon in charge of new customer marketing.

In its analysis, the telecommunications team goes to great lengths to discuss both the political and economic situation in Brazil and to discuss the background of packet switching technologies. Next, it reviews the basics of credit card authorization systems, and defines many of the key terms involved in the analysis. The telecommunications team then discusses the international linkages available between the United States and Brazil, then the domestic linkages available within Brazil. Finally, the team compares two different possible solutions, then ends with its recommendations.

The Response

Our assignment poses a number of problems. Many factors—technological, political, even cultural—are involved. (See Exhibit M5.1.) Some forces at work in the international sphere are not encountered in the domestic environment in the United States, where we have the bulk of our operations. The most obvious of these is that foreign nations play by their own rules. Outsiders, regardless of whether or not they consider those rules fair, must abide by them. A company that refuses to adapt to new, unfamiliar, and at times perplexing circumstances will not last long in a foreign market.

This difference in business climate is especially pronounced in less developed countries. A common denominator among all less developed nations seems to be their fascination with bureaucracy. Complicated rules that develop may appear absurd, but nevertheless, one must abide by them. These rules can range from having to obtain special permits to outright denial of access to a market, without any significant commitment to fairness. One must take this "bureaucracy factor" into account when considering an international move.

Our task is to develop a strategy for credit card authorization in Brazil, the largest and most powerful of the developing nations. The first step, in our view, is to take a brief look at Brazil itself. What kind of market is it? Where is it likely to go? This will help us to determine how extensive the network ought to be.

The next step is to examine how credit card authorizations are handled domestically. We will then explain packet switched networks. At that point, we will examine our Brazilian options, and suggest options we feel best meet company needs. Finally, we will examine the benefits of our solution.

The telecommunications team then briefly reviews the political and economic climate in Brazil, with reference to the practical problems faced by the plan to build a credit card authorization network.

Before considering a system in Brazil, one must first ask just what kind of place Brazil is, and where it is likely to go. Brazil is the fifth largest nation in the world in terms of land area; it is larger than all of Western Europe, and nearly as large as the United States. Its population of 135 million is expected to grow to 180 million by the year 2000. Its $280

Exhibit M5.1

Factors Affecting Telecommunications in Developing Countries

Telecommunications strategies in developing countries must be based on more than just technological considerations.

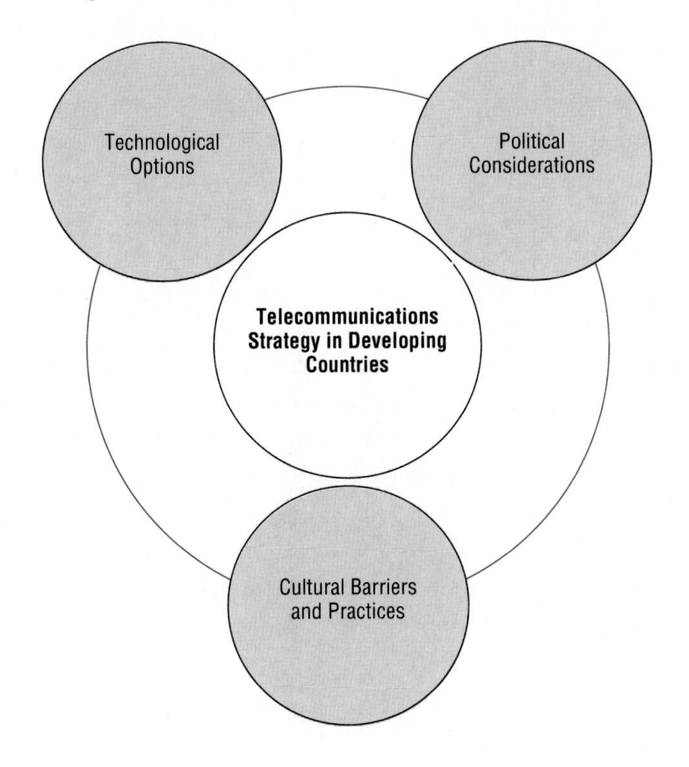

billion GDP makes Brazil's economy the eighth largest in the West. Economic growth averaged 7 percent per year from 1948 to 1961 and 8.9 percent between 1968 and 1982. After a recession in the early 1980s, the Brazilian economy bounced back and has averaged 7 percent per year since 1984. Brazil is presently in another recessionary phase, which, in our view, may turn around due to a change in political administration.

Brazil was controlled by its military from 1964 to 1985. In 1985, José Sarney came to power after being chosen indirectly by Brazil's electoral college. In November 1989, Brazil held its first democratic election for president since 1960. The election of the new president,

Fernando Collor, signaled that the economy would take a free-market approach, including expansion of some of the financial services sectors. The end result should be an increase in the demand for credit cards.

One of Brazil's biggest problems at this time is inflation. Inflation in June 1989, for example, reached 22 percent per month. Controlling inflation will be one of the major challenges of the new administration. If inflation can be brought down to a reasonable level, the overall outlook for the economy and, therefore, for credit card services, is good. For that reason we feel our network should be widespread throughout Brazil and poised to handle the large increase in volume of transactions that may well occur.

The telecommunications team then turns its discussion to the basics of credit card authorization systems.

Credit card authorization systems are a vital part of any credit card company. They enable the company to limit losses on lost or stolen cards. In the United States, packet switched networks are used to process card authorizations. Major companies, such as MasterCard, handle over 2 million authorizations from retailers on peak days. These companies have found that packet switched networks are economical, reliable, and responsive. For example, MasterCard uses a network known as "Banknet," developed by BBN Communications Corporation. Other examples are CATNET and CAFIS in Japan.

These systems work as follows. From a terminal placed at the point of sale (POS) inside the store, the packet of information regarding the individual card travels across data lines to a node, which routes it through the quickest and least congested pathway to a front-end processor and then to the mainframe computer database. The data are checked against a list of lost or stolen cards, and the answer is then immediately sent back to the source terminal through the same packet switched network. On the return trip, it may pass through different nodes, depending on the amount of usage in the system. In the United States, this whole process takes place very quickly; normally, an answer is received within 35 seconds of the request. The packets themselves arrive at their destinations in a fraction of a second.

Key Terms

Node	A major point in a network where lines from many sources meet and messages may be switched
Packet Assembler/Disassembler	A device used on a packet switched network to assemble information into packets and to convert received packets into continuous data streams
Protocol	The conventions used in a network for establishing communications compatibility between terminals, and for maintaining line discipline while connected to the network
Circuit	A transmission path between two points in a telecommunications system

The packet assembler/disassembler first segments each data message into short blocks of a specific maximum length, and provides each block with a header containing addressing and sequencing information. It then passes the data from node to node in just a fraction of a second. The nodes themselves do not archive, or store, the data. They forget a message as soon as the next node checks for errors and acknowledges receipt. The point where the information system connects to the data circuit-terminating equipment (DCE) is referred to as the "network gateway."

Some advantages of packet switching are the following:

- For data applications in which the amount of traffic between terminals cannot justify a dedicated circuit, packet switching can be more economical than transmission over private lines.
- Packet switching can also be more economical than dialed data for applications with data communication sessions shorter than a minimum chargeable time unit for a telephone call.
- Because the destination of the packet is inherent in the packet, large numbers of messages can be sent to many different destinations as quickly as the data terminal can turn them out.
- Because of the intelligence built into the network, each packet travels over the best available route for the packet at that time. This helps to minimize congestion and maximize efficiency.
- "Graceful degradation" is possible, that is, whenever a link or node fails, packets may be automatically rerouted around the defective portion of the network while repairs are carried out.

- Many telecommunications services are possible. Examples include error detection and correction, message delivery verification, group addressing, reverse billing, message sequence checking, and diagnostics.
- The standard for packet switching systems at this time is the networking architecture known as the X.25 standard.

After introducing a great deal of background on both Brazil and the nature of packet switched networks, the telecommunications team begins to consider the problem of linking Brazil to the United States.

To solve our problem, we must find a way to establish a credit card authorization system in Brazil. Since packet switching networks have become the standard choice of credit card companies in the United States, our task is to see whether such an arrangement can be made in Brazil. We must also arrange to establish a link between the United States and Brazil, connecting network to network.

Our first task is to make the international linkage. This can be done through two principal methods: satellite technology or undersea fiber optic cables. Our choice is to lease an existing circuit offered by MCI, AT&T, RCA, or another international record carrier. The alternative would be to attempt building our own system.

We have a further complication as well, in the need to deal with the Brazilian state-owned long distance carrier, EMBRATEL. This is where the third world bureaucracy factor comes into play. One must negotiate with EMBRATEL, and those negotiations can cause delays and frustrations.

Our next problem is what to do inside Brazil. Is a network available? Would it be possible to build our own network inside Brazil? Is such an alternative feasible?

It is important here to recognize some problems with Brazil's infrastructure. The domestic infrastructure is not sufficient to meet Brazil's local needs; demand surpasses capacity by approximately 9 million lines. One can come to own a line as a result of a favor, or even outright bribery. Private lines in Brazil are expensive, presently costing over $1,000. Even at this price, they are not 100 percent reliable. Even in a major city like Rio de Janeiro, a person must sometimes wait a minute or two just to get a dial tone due to busy circuits. In order to gain access to lines, we will have to negotiate with TELEBRAS, the authority in charge of local carriers.

Here, the telecommunications team introduces the very important political variable that will heavily influence any outcome of their strategy: the Brazilian information technology policy, one of the most radical and protective in the world.

Another problem will arise with the very nationalistic Brazilian information technology policy. A law passed in 1984 protects local industry and forbids the participation of foreign firms in the manufacture of computers and equipment that national companies are able to manufacture. Work is sometimes carried out slowly in order to ensure that nationally produced equipment is used. Brazilian-produced equipment can cost from two to four times the international market price. An example of this is found in front-end telecommunications processors. A company is obligated to use Brazilian equipment, such as that produced by ITAUTEK and CPM Informatics SA, even though IBM's 3705 or 3725 front-end processors would be much cheaper.

Having reviewed all of the various factors, the telecommunications team presents the solutions it has found. It recommends the international linkages, then discusses what type of a network is possible within Brazil.

The first solution has to do with the international link. According to our research, the best option would be to lease a point-to-point fiber optic line from MCI. The cost would be $3,800 per month on the U.S. end, and approximately $4,000 per month on the Brazilian end. The transmission rate would be 9,600 bps, which is high enough for credit card authorization, and the monthly charge would be locked in for a year. The system could be completely set up in less than 4 months.

This system compares with an AT&T satellite link, which runs $4,000 per month at the U.S. end and approximately $4,000 per month on the Brazilian end. The only other option with AT&T, copper cable, is far more expensive. We were unable to get a bid or comment on this project from GTE Sprint.

We do not feel that it would be wise at this point to invest in the construction of our own fiber optic line. The costs, in our view, are not justified considering our inability to gauge exactly how much business we will be doing in Brazil at this point. While we see long-term potential for tremendous growth, we also recognize that the economy is somewhat

Exhibit M5.2

Brazilian Credit Card Authorization System

International credit card authorization in Brazil requires the use of many different telecommunications authorities and services.

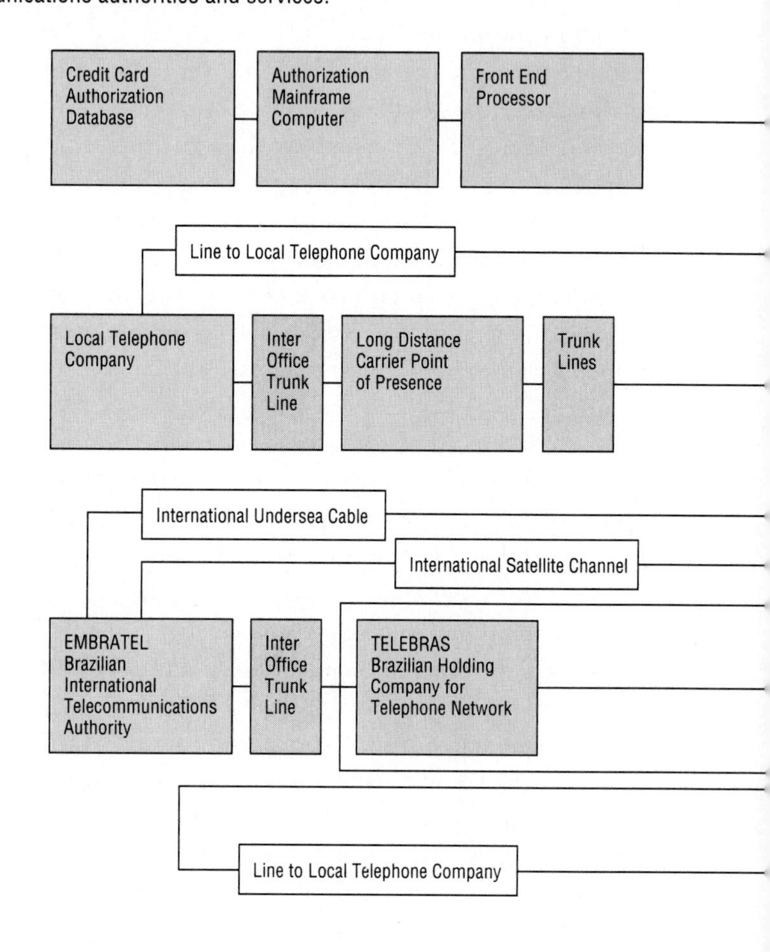

off course at present and that the political situation is somewhat undefined. While it appears likely that a new administration will stimulate economic growth, and therefore personal and business expenditures, we also wish to exercise caution and test the waters before making so heavy an investment as building our own international circuit. Our research shows, moreover, that at present it is not difficult

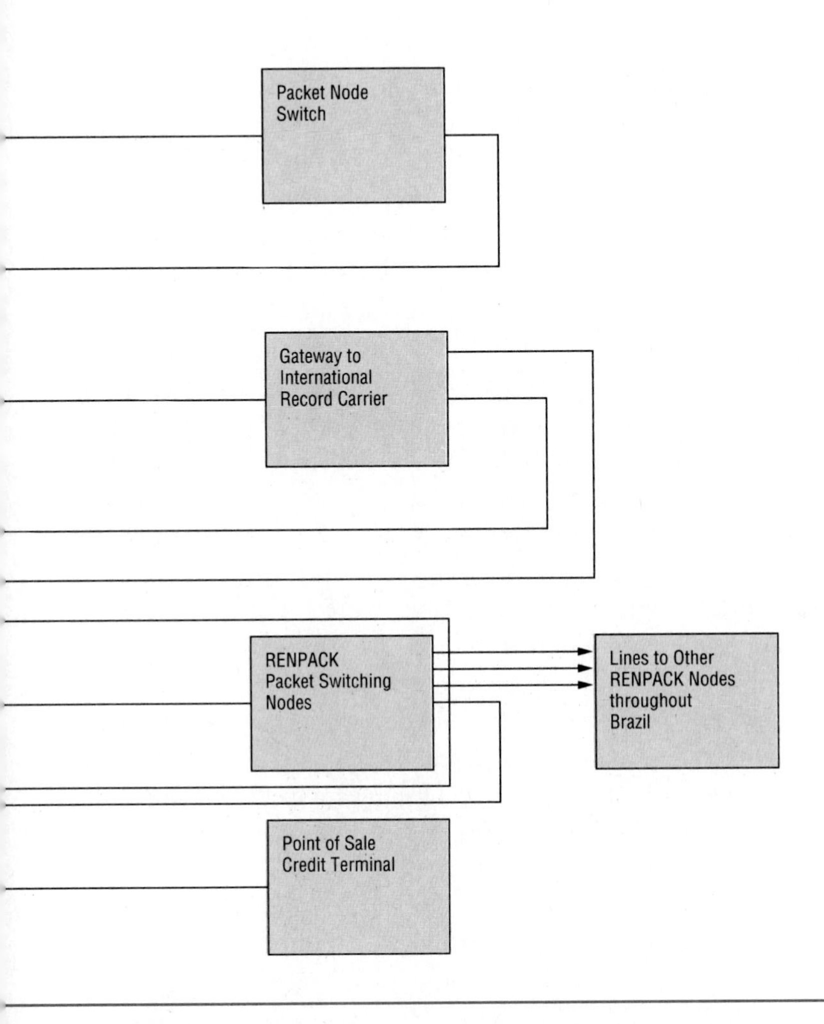

to obtain a leased line to Brazil. Because MCI offers the lowest rate and because an undersea cable is not subject to atmospheric interference, we choose this option.

The second solution has to do with the packet switched network in Brazil. Should we try to build our own network? Would we be able to lease an existing network? (See Exhibit M5.2.)

Because of Brazil's present information technology policy, building a network there would be complicated and very expensive. As mentioned previously, Brazil's policy requires that domestically manufactured equipment be used when such equipment is available. Therefore, equipment such as front-end processors would have to be of Brazilian manufacture and purchased at extremely inflated prices. In addition, obtaining permission to build such a network would likely require long bureaucratic delays.

There is another option. Brazil presently has a public packet switched network of the X.25 standard known as RENPACK. This network is linked up with INTERDATA. RENPACK has a total of six nodes: two in São Paulo, and one each in Rio de Janeiro, Brasília, Recife, and Pôrto Alegre. Brazil is also linked through Telenet to the United States and six other nations. Unfortunately, we were unsuccessful in our attempt to find out just how much it would cost to use RENPACK. Still, we feel that this would be a better option for our initial foray into the Brazilian credit card market than trying to build our own network, which would be extremely complicated due to Brazilian government regulations.

It is difficult to predict the economic course of a less developed nation as opposed to an industrial nation. We see tremendous market potential in Brazil, yet we are also aware that Brazil is presently in an awkward stage of development. It is still developing the infrastructure necessary for smooth, orderly growth. In spite of Brazil's present problems, however, we feel that in establishing our card and marketing it aggressively, we will open a profitable new market for our firm.

Leasing a fiber optic line from MCI will be cost effective and provide a reliable communications channel. By leasing for a one-year period we gain access to Brazil's market while keeping our options open in case our expansion plan does not work as foreseen. By using a local packet switched network, our credit cards can be authorized in much the same time frame as they are in the United States. Such networks in the United States have been shown to have a reliability factor of 99.95 percent. This would lead to the rapid, accurate transfer of information necessary in a business such as ours.

The Critique

This telecommunication plan has both strengths and weaknesses. On the strong side, it appears that the telecommunications team has basically solved the problem. They have determined that it is indeed possible to connect into an X.25 packet switched network in Brazil, and that such a network would make it possible to set up a credit card verification system with an adequate response time. Another strength of the approach is that it does not rely solely on the technical aspects of the solution. Political and, to a lesser extent, cultural factors are also taken into consideration. In summary, the strengths of the presentation were the following:

- *Political and cultural factors are taken into consideration in deciding the telecommunications strategy. This is necessary because Brazil is a developing country.*
- *A reasonable strategy has been worked out using international record carriers with linkages flowing into the Brazilian domestic telecommunications network.*

In a few areas, however—some of them very serious—this presentation could be improved:

- *There is virtually no substantial cost analysis. Particularly on the Brazilian side of the network, it is virtually impossible to determine what the costs would be.*
- *The report goes perhaps too deeply into the political and economic considerations, while not giving enough detail on the technical problems.*
- *It does not explain how the credit card verification terminals will be connected into the Brazilian packet switched network and what problems might arise, particularly in light of the difficult access to local phone lines.*

- ***A few technical details appear misinformed. Is there really an available undersea fiber optic circuit between the United States and Brazil?***

On the whole, it is hard to believe that this was a convincing presentation. The reader should be wary of ever submitting a report to management without a very careful cost analysis of the situation.

Minicase 6

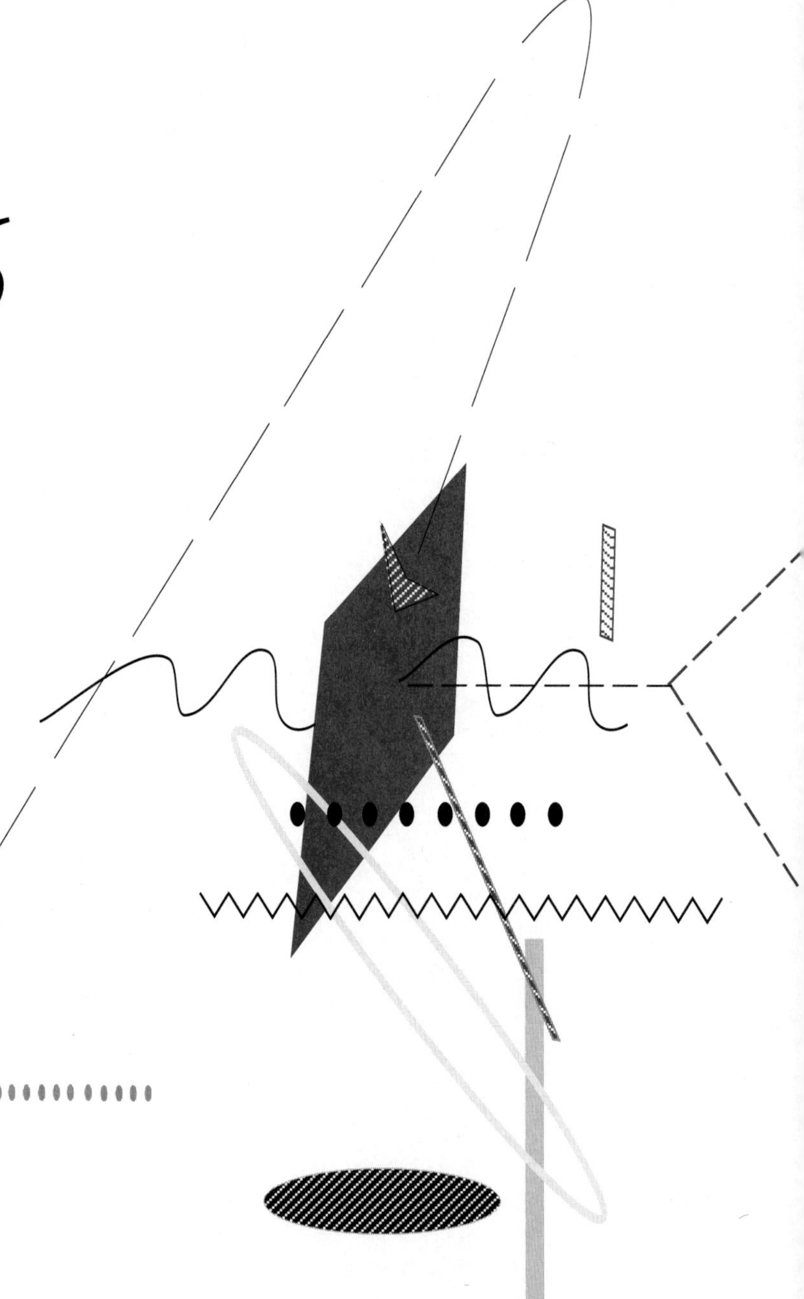

Trafalgar Investments Limited

International PBX Network

The Challenge

You are the telecommunications manager in a major securities trading organization that expanded its London and Tokyo operations during the recent boom. Linking your New York operation with London, Tokyo, and Hong Kong is a large network of IBM dedicated circuits, some using satellite technology, others using undersea cables.

You have been called in by top management and given the following challenge: Recently you have heard complaints that phone service from New York to London, Tokyo, and Hong Kong was not satisfactory to customers. When customers call New York, they might need information from overseas. Our representative has to hang up, then call overseas to get the information, then call the customer back. You have been asked if there is a way to link together the PBXs of the centers in New York, Tokyo, Hong Kong, and London, so that all offices appear as a single system, allowing the representative in New York to transfer the call to any of the other offices without having the customer hang up. The customer would see the call as being transferred to another office, and of course would enjoy making an international phone call without having to pay for it.

Unfortunately, you have noticed that the regulations and technical problems between the different countries make things difficult. You may have to use more than one telecommunications company to accomplish this task.

You must present a good plan within two weeks to the vice president of equities trading. She is known to have a terrible temper and frequently fires people on the spot when they don't produce answers.

After going through the analysis required to solve the problem for top management, the telecommunications team presented the following report, beginning with the background to the situation.

The Response

Trafalgar Investments Ltd. is a full-service international investment banking organization headquartered in New York City. The firm manages and participates in public securities offerings. It also makes markets in and trades securities and related instruments of all types. Trafalgar Investments advises clients on a variety of financial matters including asset-related financings, asset and liability management, stock repurchases, and secondary market trading. Net income for the year ended December 31, 1988 was $120 million. Primary earnings per share were $3.20.

Trafalgar's sophisticated sales and trading operation spans the world's time zones and currencies. Exhibit M6.1 illustrates the global office network. The performance in 1988 extended to Trafalgar's many geographic markets.

At this point, the telecommunications team begins to discuss some of the problems with the current system of telecommunications. The main problem is that, in spite of the expense of maintaining two separate telecommunications networks, one for voice, the other for data, Trafalgar still cannot offer the best customer service.

At the present time, our voice and data telecommunications are separated. The existing system links together the New York, London, Tokyo, and Hong Kong operations with a large network of IBM dedicated circuits. Some of them employ satellites, others use undersea cables. This system has improved Trafalgar's internal data access capabilities.

However, clients frequently complain about telephone service. Calls are placed via telecommunications carriers in the various regions of the world. These include the regional Bell operating companies in the domestic market and AT&T, Kokusai Denshin Denwa (KDD) in Japan, Cable & Wireless in Hong Kong, and British Telecom in the United Kingdom. Different technologies among these international carriers result in electromagnetic interference, and a great deal of subjectivity to weather conditions. Problems in interoffice telecommunications arise from these unacceptable noise levels.

Many customers are increasing their trading overseas and need up-to-date information. Presently a sales representative must put the

Exhibit M6.1

Trafalgar Investments' Office Network

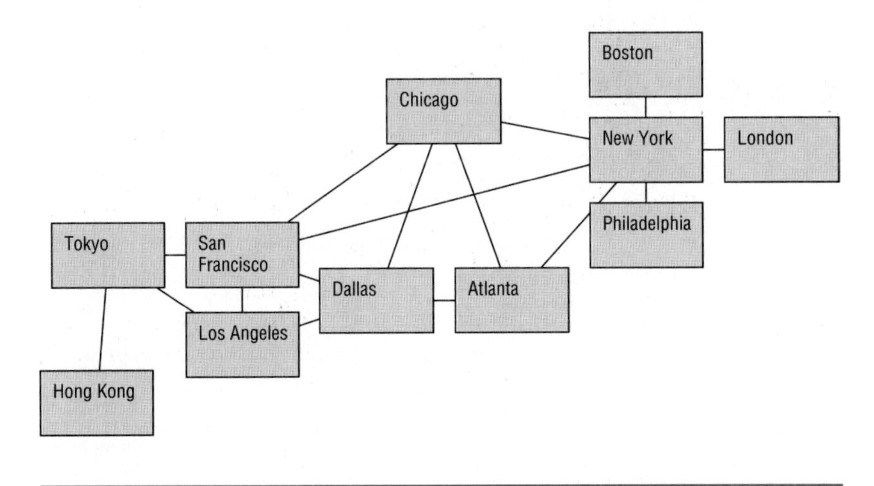

customer on hold, call the international office to obtain the necessary data, then relay this information back to the customer. If the task of getting the information is complex, the sales representative must hang up and then call the customer back after obtaining the information. The first priority is to improve telephone service connections between our domestic and international operations. In order to address this problem, we will introduce an integrated digital system that combines the telephone line and computer data on a software-defined network.

Due to time differences between our regional offices, illustrated in Exhibit M6.2, routing and conferencing will be investigated. Conferencing will keep the sales representative in close contact with clients rather than just transferring calls to field offices. Another possibility to improve customer service, especially between our domestic offices, Hong Kong, and Tokyo where office hours overlap very little, may be hourly batch runs of data pertaining to all events occurring in foreign markets. This will reduce the number of calls overseas for information.

Now that the basic objectives and problems in the voice telecommunications system have been set out, the telecommunications team turns to a discussion of the computer data systems.

Exhibit M6.2

Trading Hours around the World

Time zone differences result in trading almost 24 hours per day.

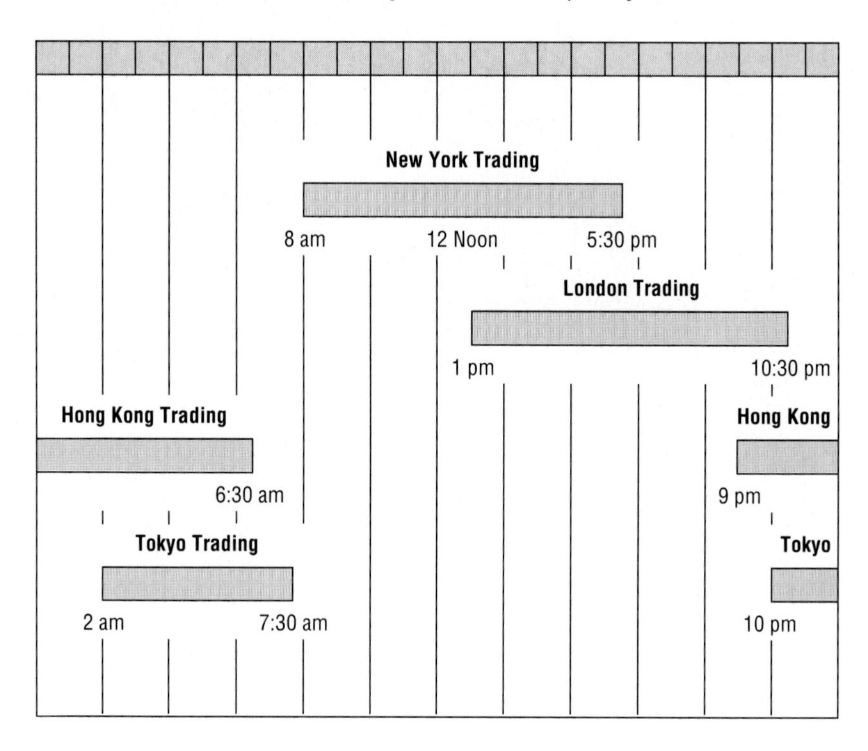

Times listed as Eastern Standard Time equivalents.

At present, we link our headquarters in New York to our three overseas branches through an IBM 3090 computer at each location. We connect the computers through network control processors (NCPs) with leased optical fiber undersea cables. To date, there is no undersea cable available to Hong Kong. Temporarily, we will link Hong Kong to our other regions by satellite. (See Exhibit M6.3.) We anticipate fiber optic connections to this area will be available sometime next year. Through this linkage we can transmit data (trading volumes, transaction confirmations, and market information) between offices rapidly.

The telecommunications team now reviews the limitations of the current telephone system.

Exhibit M6.3

Trafalgar Investments' Backbone Data Communications Network

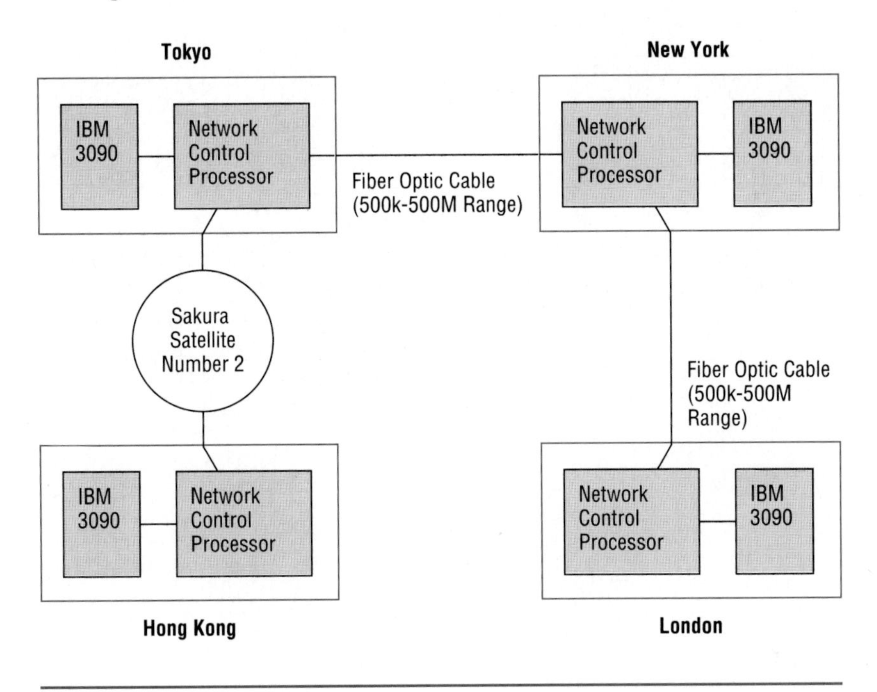

Whenever customers want overseas real-time market information our representatives must call overseas to get the information. This costs a lot of time and money. In financial markets, price fluctuations occur every second and time is critical. If customers cannot get the information they need to make their decisions immediately, they may incur either real losses or missed opportunities for profit. Since telephone calls are transferred by local telephone companies to designated long distance carriers, as Exhibit M6.4 shows, Trafalgar has little or no control over communication facilities, but is charged for access to and usage of their lines.

The telecommunications team now starts to discuss the solution it proposes and the economic assumptions that underlie this choice.

The new telecommunications system will incorporate data and voice transmissions on a software-defined network. The following

Exhibit M6.4

Trafalgar Investments' Telephone System

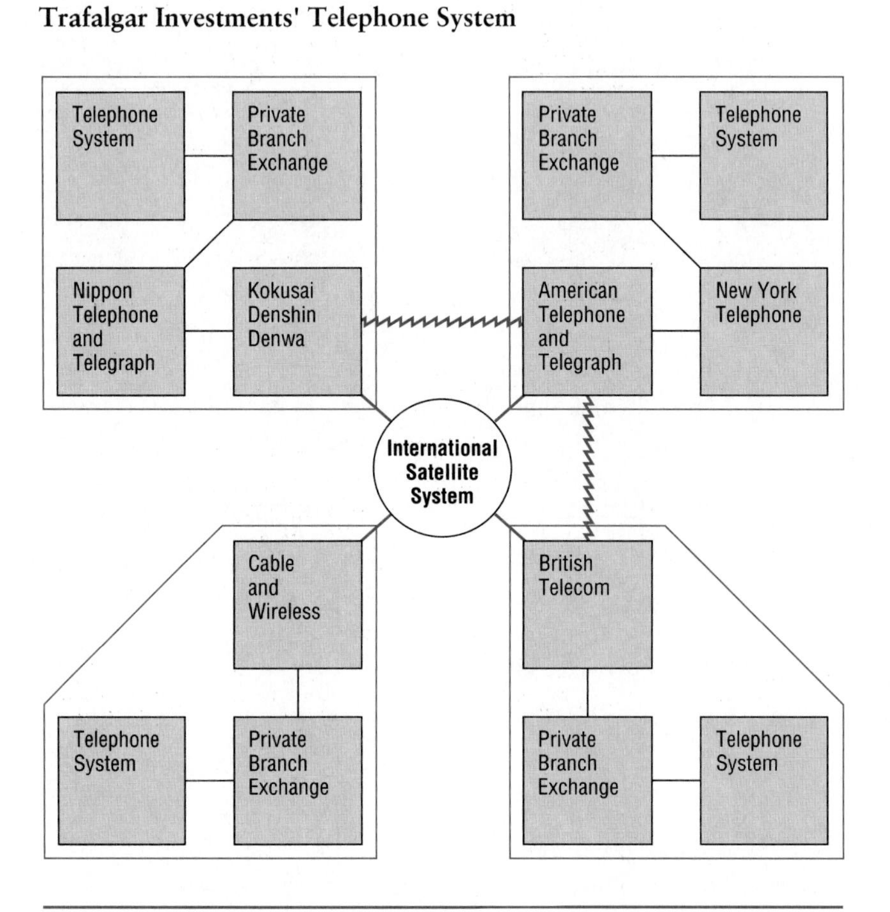

assumptions underlie this solution. We will change from using two separate communications lines for analog and digital transmissions to satellite and fiber optic linkages that will provide dedicated access lines through our own private branch network.

As a result of this virtually banded network we will enjoy a degree of insulation from fluctuating exchange rates:

Exchange Rates: June 20, 1989

British Pound	U.S.$1.54 = £1
Japanese Yen	U.S.$1 = ¥145
Hong Kong Dollar	U.S.$1 = HK7.7

We have assumed 10,000 calls originating in the United States, representing 75 percent of our monthly telephone usage. The new system will provide a potential annual savings on this traffic of close to $400,000. It is important to note that actual savings will depend on the monthly volume of calls placed from each region. The new system also provides value added features like more control over network facilities and savings on access charges for signal switching through regional Bell operating companies with direct lines to the long distance carriers, AT&T's points of presence in the domestic markets. Using the regional Bell offices to reach AT&T directly eliminates many local telephone charges. (See the accompanying table.) Also, by sending data and voice communications through a digital network, sound quality and customer relations will be improved. The only initial investment required is the purchase of four integrated digital network exchanges, which are capable through preprogrammed internal cards of routing calls throughout the system in addition to acting as front-end processors, bandwidth multiplexers, and protocol converters.

International Standard Voice Rates
(Based on MCI Network Service Rates—Global Value Plus)

United States to:	First Minute Charges	Additional Minute Charges	Number of Calls per Month	Total First Minute Charges	Total Additional Charges (10 min./call)	Total Charges—Average
Hong Kong	2.2592	0.1036	2,000	$4,518.40	$2,072.00	$6,590.40
Tokyo	2.2592	0.1036	3,000	6,777.60	3,108.00	9,885.60
London	0.8860	0.0780	5,000	4,430.00	3,900.00	8,330.00
Monthly Averages			10,000			$24,806.00

Note: Calls originating from the United States are estimated to represent 75 percent of the firm's monthly international telephone utilization. Calls originating from Hong Kong and Tokyo represent approximately 25 percent, estimated at $8,268.67, of the firm's monthly international telephone usage. The estimated total monthly international telephone costs are $33,074.67. The estimated total annual international telephone costs are $396,896.00.

The telecommunications team now introduces its technical solution to the problem. It has decided to use an Integrated Digital Network Exchange supplied by IBM.

The technical solution we propose is to install an integrated digital network exchange (IDNX) in front of the Network Control Processor (NCP) that connects to our private branch exchange (PBX) and mainframe. This will replace the use of international telephone companies such as AT&T, British Telecom, and KDD, by sending voice and data transmissions through the existing computer network, rather than over separate, costly circuits.

The integrated digital network exchange is made by IBM and has some very useful features:

- Wide range of transmissions
- Demand assigned bandwidths
- Dynamic, adaptive call routing
- Automatic circuit reconnection for service restoration
- PBX busy-back capability for both demand assigned and permanent voice calls
- Voice compression modes including PCM, ADPCM, and Digital Speech Interpolation

These sophisticated features are designed to reduce transmission fees, maximize call carrying capacity, ensure exceptionally high network ability, and provide a foundation for growth.

The Critique

The telecommunications team of Trafalgar has created a good analysis of the need for integration of the voice and data systems. This type of solution is very popular for large organizations that already have significant infrastructure and regular, nondiscretionary expenditures on telecommunications. The telecommunications team has made that analysis and concluded that something needed to be done.

Another strength of the team's analysis is that they have developed

their telecommunications strategy from the perspective of customer service. It is an extra plum when one can improve customer service and cut costs at the same time.

On the other hand, a few areas of improvement remain:

- *There is no reference in the report to the rates for data telecommunications. Without this information, it would be difficult to prepare a budget and understand what the ultimate telecommunications costs would be.*
- *It is unfortunate that the plan concentrates only on cutting the costs for international calls, and not for calls placed within the United States, where there is obviously a great deal of traffic and related expense.*
- *Although one of the critical factors in the solution is the ability to level the load during off-peak hours, there is no real analysis of the relationship between the time zone differences and how the utilization of the system would actually change, beyond simply asserting that such load levelling would take place.*
- *There is a fiber optic circuit linking Tokyo with Hong Kong, and the indication otherwise in the plan appears to demonstrate bad research. The cable is available through Kokusai Denshin Denwa.*
- *A critical factor missing from the analysis presented by the telecommunications team is a discussion of the additional circuit capacity that may be needed for the intensive transmission of voice. This is surely an additional cost, but it is not mentioned. Furthermore, there is no indication the telecommunications team has even a clue as to how much circuit capacity should be added.*
- *There is also no financial information about the cost of the additional equipment required to operate the new system, particularly the IDNX equipment. Without an estimate of the possible costs, it is impossible to do a cost/benefits analysis to show whether the decision is the right one. This the telecommunications team has ultimately failed to do.*
- *Finally, the telecommunications team does not know what the ultimate savings will be.*

These factors add up to a strong case against the presentation the telecommunications team is ready to give to top management. One can only wonder what will happen when difficult but basic questions like "How much will it cost?" come up in the presentation meeting.

Index